Lecture Notes in Control and Information Sciences

Volume 470

About this Series

This series aims to report new developments in the fields of control and information sciences—quickly, informally and at a high level. The type of material considered for publication includes:

1. Preliminary drafts of monographs and advanced textbooks
2. Lectures on a new field, or presenting a new angle on a classical field
3. Research reports
4. Reports of meetings, provided they are
   (a) of exceptional interest and
   (b) devoted to a specific topic. The timeliness of subject material is very important.

More information about this series at http://www.springer.com/series/642

Nathan van de Wouw · Erjen Lefeber
Ines Lopez Arteaga
Editors

Nonlinear Systems

Techniques for Dynamical Analysis and Control

Editors
Nathan van de Wouw
Department of Mechanical Engineering
Eindhoven University of Technology
Eindhoven, Noord-Brabant
The Netherlands

Ines Lopez Arteaga
Department of Mechanical Engineering
Eindhoven University of Technology
Eindhoven, Noord-Brabant
The Netherlands

Erjen Lefeber
Department of Mechanical Engineering
Eindhoven University of Technology
Eindhoven, Noord-Brabant
The Netherlands

ISSN 0170-8643 ISSN 1610-7411 (electronic)
Lecture Notes in Control and Information Sciences
ISBN 978-3-319-30356-7 ISBN 978-3-319-30357-4 (eBook)
DOI 10.1007/978-3-319-30357-4

Library of Congress Control Number: 2016938392

Printed on acid-free paper

This Springer imprint is published by Springer Nature
The registered company is Springer International Publishing AG Switzerland

Preface 1

This book consists of three parts. Each part focuses on a field of research that has been central to the scientific career of Henk Nijmeijer; namely (1) nonlinear control systems, (2) synchronization in networked systems, (3) control of nonlinear mechanical systems.

Part I on "Nonlinear Control Systems" commences with a contribution of Arjan van der Schaft entitled "Controlled Invariant Distributions and Differential Properties." Henk Nijmeijer and Arjan van der Schaft defended their Ph.D. theses on the same day and worked under the supervision of the same "promotor" (Jan C. Willems) on the topic of nonlinear geometric control theory, ultimately culminating in the well-known textbook entitled "Nonlinear Dynamical Control Systems," published by Springer in 1990. Chapter 1 of this book revisits the topic of controlled invariant distributions as also studied by Henk Nijmeijer and Arjan der Schaft in their earlier work and makes a direct link to the topics of convergent dynamics and contraction analysis that recently received wide attention in the systems and control community.

Chapter 2 is contributed by Tengfei Liu and Zhong-Ping Jiang and presents results on the distributed control of nonlinear systems. In particular, small-gain methods for the distributed control of nonlinear systems are proposed and an application to the distributed formation control problem of nonholonomic mobile robots is detailed. The tracking control of mobile robots has been a topic of fruitful collaboration between Henk Nijmeijer and Zhong-Ping Jiang for many years.

Chapter 3 is contributed by Alexey Pavlov and Nathan van de Wouw. This chapter reviews the class of nonlinear convergent systems and highlights many applications of the convergence property to nonlinear analysis and control problems, such as global output regulation, frequency domain analysis of nonlinear systems, model reduction, stable inversion, and extremum seeking control. Many of these recent results have culminated from earlier joint research of the authors with Henk Nijmeijer, when Alexey Pavlov performed his Ph.D. studies at the Eindhoven University of Technology in the early 2000s.

Part II of this book is devoted to one of Henk Nijmeijer's favorite topics of research: synchronization. It starts with a chapter written by Elena Panteley and Antonio Lora, which focusses on the synchronization and emergent behavior in networks of heterogeneous systems. The authors bring forth a new perspective on how to analyze synchronization for a network of systems with nonidentical dynamics.

Toshiki Oguchi authors Chap. 5, in which the topics of state predictors and anticipating synchronization for nonlinear delay systems are addressed. These topics have been a mutual and on-going research interest of the author and Henk Nijmeijer.

Chapter 6, contributed by Wim Michiels, focuses on delay systems. It centers on delays effects in dynamical systems and provides insightful analysis and control interpretations. Extensions of these results towards networks of interconnected nonlinear dynamical systems are discussed, with a focus on the synchronization problem.

Part II of the book closes with a contribution entitled "Emergence of Oscillations in Networks of Time-Delay Coupled Inert Systems" authored by Erik Steur and Sasha Pogromski. It presents results of the emergence of oscillations in networks of nonlinear single-input single-output systems that interact via linear, time-delayed coupling functions.

Part III of this book concerns the control of nonlinear mechanical systems. It opens with Chap. 8 authored by Dennis Belleter and Kristin Pettersen on "Leader-Follower Synchronisation for a Class of Underactuated Systems." It deals with the control of marine vehicles, such as underactuated autonomous surface vessels and autonomous underwater vehicles, a topic of joint interest to Henk Nijmeijer and Kristin Pettersen.

Chapter 9, entitled "Position Control via Force Feedback for a Class of Standard Mechanical Systems in the Port-Hamiltonian Framework," is authored by Mauricio Munoz-Arias, Jacquelien Scherpen, and Daniel Dirksz. It presents position control strategies for standard mechanical systems in the port-Hamiltonian framework via force feedback.

The book closes with a contribution of Krzysztof Tchoń "The Endogenous Configuration Space Approach: An Intersection of Robotics and Control Theory." The endogenous configuration approach is a control theory oriented methodology of robotics research, dedicated to mobile manipulators, which leads to motion planning algorithms.

Eindhoven
February 2016

Nathan van de Wouw
Erjen Lefeber
Ines Lopez Arteaga

Preface 2

This book has been written at the occasion of the 60th birthday of Prof. Henk Nijmeijer on March 16, 2015, and commemorates the role of Henk Nijmeijer in both the Dutch and the international (nonlinear) systems and control communities.

At the occasion of this birthday, the Dynamics and Control group of the Department of Mechanical Engineering of the Eindhoven University of Technology, the Netherlands, (chaired by Henk Nijmeijer) has taken the initiative to organize a one-day international workshop on "Nonlinear Systems" in honor of Henk Nijmeijer's sixtieth birthday, which was organized on January 21, 2016, in the Auditorium of the Eindhoven University of Technology, Eindhoven, the Netherlands.

During this workshop, Henk's colleagues, international collaborators, and former students have contributed by giving a seminar on a topic related to their joint research.

This book collects research contributions of many international scientists and former students of Henk, with whom he has had extensive research collaborations and/or who have been inspired and supported by him to take on challenging and

exciting research problems in the fields of nonlinear control systems, synchronization, coordinated and distributed control, model reduction and the dynamic analysis, and control of mechanical systems.

After his M.Sc. graduation (cum laude) in 1979 at the Rijksuniversiteit Groningen (RUG), the Netherlands, Henk Nijmeijer performed his Ph.D. studies (from 1980 to 1983) at the Center for Mathematics and Informatics (CWI) in Amsterdam, the Netherlands. His Ph.D. research focused on nonlinear geometric control theory and resulted, under the supervision of Jan van Schuppen and promotor Jan C. Willems (University of Groningen), in the Ph.D. thesis entitled "Nonlinear multivariable control: a differential geometric approach." At that time, Henk had intensive research collaborations with Arjan van der Schaft, who was also performing his Ph.D. research under the supervision of Jan C. Willems. This collaboration has ultimately led to the well-known textbook "Nonlinear Dynamical Control Systems", published by Springer in 1990, based on their lecture notes for a yearly taught DISC (Dutch Institute for Systems and Control) graduate course since 1987. Notably, a new edition of this book will appear in 2016, further evidencing the lasting impact of this book on the international nonlinear systems community.

In 1983, Henk took on a position as Assistant Professor, and later as Associate Professor, at the Department of Applied Mathematics of the University of Twente, Enschede, the Netherlands. His work remained focused on nonlinear control systems, but in the 1990s he also increasingly engaged in research related to the control of nonlinear mechanical systems, such as mobile robots and robotic manipulators.

The combination of his strong expertise on the mathematical foundations of nonlinear control theory and his interest in the dynamics and control of mechanical systems has undoubtedly played a role in Henk taking on a part-time Full Professor position in 1997 at the Department of Mechanical Engineering of the Eindhoven University of Technology, Eindhoven, the Netherlands. This part-time position ultimately led to a full-time professor position at this department in 2000, where he started the "Dynamics and Control" group, which he still chairs today. The name of the group reflects Henk's unique affinity with both the modeling and analysis of dynamical systems and control theory. Since 2000, Henk has continued to pursue fundamental research in nonlinear systems and control on topics such as output regulation, synchronization, hybrid systems, networked and delay systems, and model reduction, but combined this successfully with an increasing focus on engineering applications, such as vehicle dynamics, robotics, mechanical design, acoustics, mechatronics, and cooperative and automated driving. In doing so, he has been a mentor and an inspiring colleague for many of his co-workers in the group, which grew to be an internationally recognized center for research and teaching in dynamical systems and control, as evidenced by the excellent ratings from international review committees concerning the review periods of 2001–2006 and 2007–2012. Affiliated to Mechanical Engineering Department at the Eindhoven University of Technology, Henk has engaged in collaborations with both Dutch industries and industries abroad, such as for example ASML, Philips, Bosch Rexroth B.V., Océ, Kulicke and Soffa, Ford, DAF, FEI Company, VDL, CCM, Shell, Statoil, NXP, Prodrive, Vredestein, Honeywell, MAN, MTT, and many more. The Dynamics and Control group also generated the spin-off companies Sorama, focusing on sound imaging, and Rose B.V., focusing on robotics for care. As such, Henk effectively strives to combine fundamental research with its valorization in industry and society.

Henk has also strongly contributed to the education of new generations of engineers and scientists in the field of dynamics and control, for which there is a great need in the high-tech "Brainport" region of Eindhoven in the Netherlands. Under the supervision of Henk Nijmeijer, 35 Ph.D. students and over 250 M.Sc. students have graduated since the year 2000 at the Eindhoven University of Technology. Moreover, he has also played an important role in the Dutch Institute for Systems and Control (DISC), the Dutch national graduate school for systems and control. In fact, he was already playing an important role in the Dutch Network of Systems and Control, which has offered a national graduate program in systems and control since 1987 and is the predecessor of DISC. He has been involved in DISC since its founding in 1995 by lecturing a course on nonlinear control systems with Arjan van der Schaft for many years, and by organizing summer schools on the control of mechanical systems. Since 2000, he has been a DISC board member and, as of 2015, Henk became Scientific Director of DISC.

Henk has published a large number of journal and conference papers, and several books, including the classical "Nonlinear Dynamical Control Systems" (Springer Verlag, 1990, co-author Arjan van der Schaft), "Synchronization of Mechanical Systems" (2003, World Scientific, together with Alejandro Rodriguez-Angeles), "Dynamics and Bifurcations of Non-smooth Mechanical Systems" (Springer Verlag, 2004, together with Remco Leine), "Uniform Output Regulation of Nonlinear Systems: A Convergent Dynamics Approach" (Birkhäuser, 2005, together with Alexey Pavlov and Nathan van de Wouw), and many more.

Henk has also been strongly committed to the international systems and control community and has served our community in many ways. He is Council Member of the International Federation of Automatic Control (IFAC), and is/has been the organizer and/or IPC Chair of numerous international conferences and workshops. He is editor-in-chief of the Journal of Applied Mathematics, editor of the Communications on Nonlinear Systems and Numerical Simulation, corresponding editor of the SIAM Journal on Control and Optimization, and is/has been board member of the International Journal of Control, Automatica, the Journal of Dynamical Control Systems, the International Journal of Bifurcation and Chaos, the International Journal of Robust and Nonlinear Control, the Journal of Nonlinear Dynamics, and the Journal of Applied Mathematics and Computer Science.

Henk became a Fellow of the IEEE in 2000 and was awarded the IEE Heaviside Premium in 1987. Together with Kristin Pettersen, he received the 2006 IEEE Transactions on Control Systems Technology Outstanding Paper Award. Recently, he also received the 2015 IEEE Control Systems Technology Award (together with Marcel Heertjes, Alexey Pavlov and Nathan van de Wouw). These awards are exemplary of his remarkable achievements and the impact of his work in the field.

Dear Henk, this book is dedicated to you by all editors and authors. Thank you very much for inspiring us to take on exciting new research challenges in nonlinear dynamics and control, for your everlasting support and mentorship. It has been a great pleasure working with you and we hope to continue to do so for many years to come.

Eindhoven
February 2016

Nathan van de Wouw
Erjen Lefeber
Ines Lopez Arteaga

Acknowledgments

Many colleagues, friends and former students contributed to this book and the workshop. We thank them all, and especially Geertje Janssen-Dols, Tom van der Sande, and Joris Michielsen for all their invaluable work in the organization of the workshop and Arjan van der Schaft for the careful reading of the Preface 2.

Contents

Part III Control of Nonlinear Mechanical Systems

Part I
Nonlinear Control Systems

Chapter 1
Controlled Invariant Distributions and Differential Properties

Arjan van der Schaft

Abstract The theory of (controlled) invariant (co-)distributions is reviewed, emphasizing the theory of liftings of vector fields, one-forms and (co-)distributions to the tangent and cotangent bundle. In particular, it is shown how invariant distributions can be equivalently described as invariant submanifolds of the tangent and cotangent bundle. This naturally leads to the notion of an invariant Lagrangian subbundle of the Whitney sum of tangent and cotangent bundle, which amounts to a special case of the central equation of contraction analysis. The interconnection of the prolongation of a nonlinear control system (living on the tangent bundle of the state space manifold) with its Hamiltonian extension (defined on the cotangent bundle) is shown to result in a differential Hamiltonian system. The invariant submanifolds of this differential Hamiltonian system corresponding to Lagrangian subbundles are seen to result in general differential Riccati and differential Lyapunov equations. The established framework thus yields a geometric underpinning of recent advances in contraction analysis and convergent dynamics.

1.1 Introduction

In this chapter I will first recall the classical concepts of *invariant* and *controlled invariant* (co-)distributions, which are fundamental notions in nonlinear geometric control theory as developed starting from the seventies of the previous century. This will be done in the spirit of the paper [17] and of Chap. 13 in the book [18], by using the notions of liftings of vector fields, functions, and one-forms from a manifold to its tangent and cotangent bundle in a systematic way. Furthermore, I will provide two equivalent characterizations of invariance, and elucidate the necessary and sufficient condition for (local) controlled invariance as obtained in [17].

A.J. van der Schaft (✉)
Johann Bernoulli Institute for Mathematics and Computer Science,
Jan C. Willems Center for Systems and Control, University of Groningen,
Groningen, The Netherlands
e-mail: a.j.van.der.schaft@rug.nl

N. van de Wouw et al. (eds.), *Nonlinear Systems*, Lecture Notes
in Control and Information Sciences 470, DOI 10.1007/978-3-319-30357-4_1

Secondly, starting from [25], I will discuss how the notion of invariance of (co-) distributions can be generalized to invariance of *subbundles* of the Whitney sum of the tangent and cotangent bundle. By specializing to *Lagrangian* subbundles this is shown to lead to the basic equation of contraction theory, and thus under an additional zero-curvature condition to the classical Demidovich condition for *convergent dynamics*. Furthermore, I will argue that this geometric approach can be seen to be part of a larger story, where incremental analysis approaches are formulated geometrically using *differential properties*.

This chapter is dedicated to my colleague *from the very start* Henk Nijmeijer, at the occasion of the celebration of his 60th birthday. This seems to be fit, since the contents of this chapter start with the theory of our first joint publication [17] in 1982, and link, via our book *Nonlinear Dynamical Control Systems* [18], to later fundamental contributions of Henk within the area of convergent dynamics and synchronization. Furthermore, the chapter stresses the importance of a *geometric* language for nonlinear control.

The scientific work and careers of Henk and myself have been very close for a long time, with our Ph.D. defenses at the same day with the same "promotor" (Jan C. Willems) in the Aula of the University of Groningen, our joint Alma Mater. Furthermore, we were for a very long time close colleagues at the University of Twente, turning Twente into a center for nonlinear control theory. Moreover, despite the unavoidable turmoils in our careers, we have always remained good friends. Happy 60th birthday Henk!

1.2 Lifts to the Tangent and Cotangent Bundle

Throughout this chapter all objects (manifolds, functions, vector fields, one-forms, (co-)distributions, subbundles, etc.) will be assumed to be *smooth* (infinitely often differentiable).

Recall from [27] (see also [2, 4, 18]), how functions, vector fields and one-forms on a state space manifold $\mathscr{X}$ can be lifted to functions, vector fields and one-forms on the tangent and cotangent bundles $T\mathscr{X}$ and $T^*\mathscr{X}$. This will also lead to the definition of the lift of distributions and co-distributions to the tangent bundle.

First we introduce the notions of *complete* and *vertical* lifts of functions, vector fields and one-forms to the *tangent bundle*. Given a function h on $\mathscr{X}$, the *complete lift* of h to $T\mathscr{X}$, $h^{\mathrm{c}} : T\mathscr{X} \to \mathbb{R}$, is defined by $h^{\mathrm{c}}(x, \delta x) = \langle dh, \delta x \rangle (x)$, with $\langle \cdot, \cdot \rangle (x)$ denoting the duality pairing between elements of the cotangent space and the tangent space at $x \in \mathscr{X}$. In local coordinates $x = (x^1, \dots, x^n)$ for $\mathscr{X}$ and the induced local coordinates $(x, \delta x) = (x^1, \dots, x^n, \delta x^1, \dots, \delta x^n)$ for $T\mathscr{X}$ this reads

$$h^{\mathrm{c}}(x, \delta x) = \sum_{a=1}^{n} \frac{\partial h}{\partial x^a}(x)\, \delta x^a. \tag{1.1}$$

The *vertical lift* of a function h to a function on $T\mathscr{X}$, denoted by $h^{\mathrm{V}} : T\mathscr{X} \to \mathbb{R}$, is defined by $h^{\mathrm{V}}(x, \delta x) = h \circ \tau_{\mathscr{X}}$, where $\tau_{\mathscr{X}} : T\mathscr{X} \to \mathscr{X}$ denotes the tangent bundle projection $\tau_{\mathscr{X}}(x, \delta x) = x$. In local induced coordinates $h^{\mathrm{V}}(x, \delta x) = h(x)$.

Given a vector field f on $\mathscr{X}$, the *complete lift* f^{C} of f to $T\mathscr{X}$ is defined as the unique vector field satisying $L_{f^{\mathrm{C}}}h^{\mathrm{C}} = (L_f h)^{\mathrm{C}}$, for any function h on $\mathscr{X}$ (with $L_f h$ denoting the Lie-derivative of the function h along the vector field f, and similarly for $L_{f^{\mathrm{C}}}h^{\mathrm{C}}$). It can be proved [27] that this requirement uniquely determines f^{C}. Alternatively, if $\Phi_t : \mathscr{X} \to \mathscr{X}, t \in [0, \varepsilon)$, denotes the flow of f, then f^{C} is the vector field whose flow is given by $(\Phi_t)_* : T\mathscr{X} \to T\mathscr{X}$. In induced local coordinates $(x^1, \ldots, x^n, \delta x^1, \ldots, \delta x^n)$ for $T\mathscr{X}$,

$$f^{\mathrm{C}}(x, \delta x) = \sum_{a=1}^{n} f_a(x)\frac{\partial}{\partial x^a} + \sum_{a,b=1}^{n} \frac{\partial f_a}{\partial x^b}(x)\delta x^b \frac{\partial}{\partial(\delta x^a)}. \tag{1.2}$$

Finally, the *vertical lift* f^{V} of f to $T\mathscr{X}$ is the vector field on $T\mathscr{X}$ such that $L_{f^{\mathrm{V}}}h^{\mathrm{C}} = (L_f h)^{\mathrm{V}}$, for any function h. In induced local coordinates for $T\mathscr{X}$

$$f^{\mathrm{V}}(x, \delta x) = \sum_{a=1}^{n} f_a(x)\frac{\partial}{\partial(\delta x^a)}. \tag{1.3}$$

Thirdly, let α be a *differential* one-form on $\mathscr{X}$. The *complete lift* of α is the differential one-form on $T\mathscr{X}$ defined by setting

$$\alpha^{\mathrm{C}}(q)(Z) := Z(\hat{\alpha}), \quad Z \in T_z T\mathscr{X}, \quad z \in T\mathscr{X}, \tag{1.4}$$

where $\hat{\alpha} : T\mathscr{X} \to \mathbb{R}$ is the function given as

$$\hat{\alpha}(X_p) = \alpha(p)(X_p), \quad X_p \in T_p\mathscr{X}, \quad p \in \mathscr{X}. \tag{1.5}$$

If α is given in local coordinates as $\alpha = \sum_{i=1}^{n} \alpha_i(x)dx^i$, then it readily follows that

$$\alpha^{\mathrm{C}}(x, \delta x) = \sum_{i=1}^{n} \frac{\partial \alpha_i}{\partial x^j}(x)\delta x^j dx^i + \sum_{i=1}^{n} \alpha_i(x)d\delta x^i. \tag{1.6}$$

Finally the *vertical lift* of α is the differential one-form α^{V} on $T\mathscr{X}$ defined as

$$\alpha^{\mathrm{V}} := \pi_{\mathscr{X}}^* \alpha, \qquad \pi_{\mathscr{X}} : T\mathscr{X} \to \mathscr{X} \text{ projection}, \tag{1.7}$$

i.e., in local coordinates

$$\alpha^{\mathrm{V}}(x, \delta x) = \sum_{i=1}^{n} \alpha_i(x)dx^i. \tag{1.8}$$

Using the local coordinate expressions the following useful identities are easily verified.

Proposition 1.1 *For any function $h\colon \mathscr{X} \to \mathbb{R}$, any vector fields X, X_1 and X_2 on $\mathscr{X}$, and any differential one-form α on $\mathscr{X}$, we have*

$$X^{\mathrm{C}}(h^{\mathrm{V}}) = (X(h))^{\mathrm{V}} = X^{\mathrm{V}}(h^{\mathrm{C}}), \qquad X^{\mathrm{C}}(h^{\mathrm{C}}) = (X(h))^{\mathrm{C}}, \qquad X^{\mathrm{V}}(h^{\mathrm{V}}) = 0, \tag{1.9}$$

$$\alpha^{\mathrm{C}}(X^{\mathrm{V}}) = (\alpha(X))^{\mathrm{V}} = \alpha^{\mathrm{V}}(X^{\mathrm{C}}), \qquad \alpha^{\mathrm{C}}(X^{\mathrm{C}}) = (\alpha(X))^{\mathrm{C}}, \qquad \alpha^{\mathrm{V}}(X^{\mathrm{V}}) = 0, \tag{1.10}$$

$$[X_1^{\mathrm{C}}, X_2^{\mathrm{C}}] = [X_1, X_2]^{\mathrm{C}}, \qquad [X_1^{\mathrm{C}}, X_2^{\mathrm{V}}] = [X_1, X_2]^{\mathrm{V}}, \qquad [X_1^{\mathrm{V}}, X_2^{\mathrm{V}}] = 0, \tag{1.11}$$

$$dh^{\mathrm{C}} = (dh)^{\mathrm{C}}, \qquad dh^{\mathrm{V}} = (dh)^{\mathrm{V}}. \tag{1.12}$$

Liftings of *distributions* and *co-distributions* to the tangent bundle are now defined as follows. Recall [18] that a distribution D on $\mathscr{X}$ is a map that assigns to every $p \in \mathscr{X}$ a subspace $D(p) \subset T_p\mathscr{X}$ in such a way that around each $p \in \mathscr{X}$ there exist vector fields $X_i, i \in I$ (with I an arbitrary index set), such that for $q \in \mathscr{X}$ close to p the subspace $D(q)$ is given as $D(q) = \mathrm{span}\,\{X_i(q) \mid i \in I\}$. We will also call D a *subbundle* of the tangent bundle $T\mathscr{X}$.

Similarly [18], a co-distribution P on $\mathscr{X}$ (or, a subbundle of $T^*\mathscr{X}$) is locally given as $\mathrm{span}\,\{\alpha_i(q) \mid i \in I\}$, $q \in \mathscr{X}$, with α_i differential one-forms on $\mathscr{X}$.

Recall [18] that given a distribution D we can define the *annihilating* co-distribution $P = \mathrm{ann}\ D$. Dually, given a co-distribution P we can define the *kernel* distribution $D = \ker P$. For any D and P we have $D \subset \ker(\mathrm{ann}\ D)$ respectively $P \subset \mathrm{ann}(\ker D)$, while if D and P are *constant-dimensional* then equality in both expressions holds.

Definition 1.1 Let the distribution D on $\mathscr{X}$ be locally given as $D(q) = \mathrm{span}\,\{X_i(q) \mid i \in I\}$, $q \in \mathscr{X}$, with X_i vector fields on $\mathscr{X}$. Then the lift (also called prolongation) of D is the distribution $\dot{D}$ on $T\mathscr{X}$ defined as

$$\dot{D}(z) = \mathrm{span}\,\{X_i^{\mathrm{C}}(z), X_i^{\mathrm{V}}(z) \mid i \in I\}, \quad z \in T\mathscr{X}. \tag{1.13}$$

Analogously, let the co-distribution P on $\mathscr{X}$ be given as $P(q) = \mathrm{span}\,\{\alpha_i(q) \mid i \in I\}$, $q \in \mathscr{X}$, with α_i differential one-forms on $\mathscr{X}$, then the lift of P is the co-distribution $\dot{P}$ on $T\mathscr{X}$ defined as

$$\dot{P}(z) = \mathrm{span}\,\{\alpha_i^{\mathrm{C}}(z), \alpha_i^{\mathrm{V}}(z) \mid i \in I\}, \quad z \in T\mathscr{X}. \tag{1.14}$$

In case D and P are *involutive* and constant-dimensional we obtain the following simple local representations of $\dot{D}$ and $\dot{P}$. By Frobenius' Theorem we can find local coordinates $x = (x^1, \ldots, x^n)$ such that $D = \mathrm{span}\,\{\frac{\partial}{\partial x^1}, \ldots, \frac{\partial}{\partial x^k}\}$, $k = \dim D$. Then in the induced local coordinates $(x, \delta x)$ for $T\mathscr{X}$ we have

$$\dot{D} = \mathrm{span}\left\{\frac{\partial}{\partial x^1}, \ldots, \frac{\partial}{\partial x^k}, \frac{\partial}{\partial \delta x^1}, \ldots, \frac{\partial}{\partial \delta x^k}\right\}. \tag{1.15}$$

Similarly if $P = \text{span}\{dx^{l+1}, \dots, dx^n\}$, $n - l = \dim P$, then

$$\dot{P} = \text{span}\{dx^{l+1}, \dots, dx^n, d\delta x^{l+1}, \dots, d\delta x^n\}. \tag{1.16}$$

In general we obtain:

Proposition 1.2 *Let D be a distribution, and P be a co-distribution on $\mathscr{X}$. Then*

(a) *If $D = \ker P$, then $\dot{D} = \ker \dot{P}$.*
(b) *If $P = \text{ann}\ D$, then $\dot{P} = \text{ann}\ \dot{D}$.*
(c) *If D (resp. P) has constant dimension then $\dot{D}$ (resp. $\dot{P}$) has constant dimension.*
(d) *If D (resp. P) is involutive then $\dot{D}$ (resp. $\dot{P}$) is involutive.*
(e) *If $X^{\mathrm{C}} \in \dot{D}$ for a vector field X on $\mathscr{X}$, then $X \in D$.*

Next, the lifts of functions and vector fields to the *cotangent bundle* are defined; see again [27]. As before, for the tangent bundle case, the *vertical lift* $h^{\mathrm{V}} : T^*\mathscr{X} \to \mathbb{R}$ of a function $h : \mathscr{X} \to \mathbb{R}$ is defined by $h^{\mathrm{V}} = h \circ \pi_{\mathscr{X}}$, where $\pi_{\mathscr{X}} : T^*\mathscr{X} \to \mathscr{X}$ denotes the cotangent bundle projection $\pi_{\mathscr{X}}(x, p) = x$. In induced local coordinates $(x, p) := (x^1, \dots, x^n, p_1, \dots, p_n)$ for $T^*\mathscr{X}$ we have $h^{\mathrm{V}}(x, p) = h(x)$.

Since there is a natural symplectic form on the cotangent bundle $T^*\mathscr{X}$ we can define the Hamiltonian vector field on $T^*\mathscr{X}$ corresponding to h^{V}, denoted by $X_{h^{\mathrm{V}}}$, and called the *vertical Hamiltonian lift*. In induced local coordinates (x, p) for $T^*\mathscr{X}$

$$X_{h^{\mathrm{V}}} = -\sum_{a=1}^{n} \frac{\partial h}{\partial x^a}(x) \frac{\partial}{\partial (p_a)}. \tag{1.17}$$

Furthermore, for any vector field f on $\mathscr{X}$ define the *Hamiltonian function* $H^f : T^*\mathscr{X} \to \mathbb{R}$ as

$$H^f(x, p) = \langle p, f(x) \rangle. \tag{1.18}$$

In local induced coordinates (x, p) we have $H^f(x, p) = p^T f(x)$. The corresponding Hamiltonian vector field on $T^*\mathscr{X}$, denoted by X_{H^f}, is called the *complete Hamiltonian lift*. In induced local coordinates for $T^*\mathscr{X}$

$$X_{H^f} = \sum_{a=1}^{n} f_a(x) \frac{\partial}{\partial x^a} - \sum_{a,b=1}^{n} \frac{\partial f_b}{\partial x^a}(x) p_b \frac{\partial}{\partial (p_a)}. \tag{1.19}$$

For later use we mention that given a lift (complete or vertical) of a vector field f to the tangent bundle, as well as a lift (Hamiltonian or vertical) to the cotangent bundle, we can *combine* the two lifts into a vector field defined on the Whitney sum $T\mathscr{X} \oplus T^*\mathscr{X}$ (that is, the base manifold $\mathscr{X}$ together with the fiber space $T_x\mathscr{X} \times T_x^*\mathscr{X}$ at any point $x \in \mathscr{X}$). An example (to be used later on) is the combination of the complete lift f^{C} on $T\mathscr{X}$ with the Hamiltonian extension X_{H^f} on $T^*\mathscr{X}$, which defines a vector field on $T\mathscr{X} \oplus T^*\mathscr{X}$, which will be denoted as $f^{\mathrm{C}} \oplus X_{H^f}$. Furthermore, since the vertical lifts f^{V} (to $T\mathscr{X}$) and $X_{h^{\mathrm{V}}}$ (to $T^*\mathscr{X}$) do not have components on the base manifold $\mathscr{X}$ we may also define the combined vector field $f^{\mathrm{V}} \oplus X_{h^{\mathrm{V}}}$ on the Whitney sum $T\mathscr{X} \oplus T^*\mathscr{X}$ for any vector field f and function h.

1.3 Invariance of Distributions and Co-Distributions

First recall the standard definitions of invariance of a (co-)distribution with respect to a vector field from geometric control theory, see e.g., [18]. Consider an n-dimensional state space manifold $\mathscr{X}$ with tangent bundle $T\mathscr{X}$ and cotangent bundle $T^*\mathscr{X}$. Consider furthermore a vector field f on $\mathscr{X}$. A distribution D on $\mathscr{X}$ is called *invariant* with respect to f if $L_f X \in D$ for any vector field X in D, or equivalently, around p it holds that $L_f X_i \in D, i \in I$. Here L_f denotes the Lie-derivative with respect to f; i.e., $L_f X = [f, X]$. Invariance of D with respect to f will be abbreviated to $L_f D \subset D$.

Similarly, a co-distribution P on $\mathscr{X}$ (or, a subbundle of $T^*\mathscr{X}$) locally given as span $\{\alpha_i(q) \mid i \in I\}, q \in \mathscr{X}$, with α_i differential one-forms on $\mathscr{X}$, is called invariant if $L_f P \subset P$, that is, $L_f \alpha \in P$ for any one-form α in P, or equivalently around p, $L_f \alpha_i \in P, i \in I$. Invariance of P with respect to f will be abbreviated to $L_f P \subset P$.

Note that $L_f D \subset D$ implies $L_f(\text{ann } D) \subset \text{ann } D$ for any distribution D, and that $L_f P \subset P$ implies $L_f(\ker P) \subset \ker P$ for any co-distribution P.

Using the geometric notions of lifting discussed in the previous section, the invariance of (co-)distributions can be described in the following equivalent ways. Let us start with the case of distributions. Note that any distribution D on $\mathscr{X}$ can be also described by the following *submanifold* of $T\mathscr{X}$:

$$\mathscr{D} := \{(x, \delta x) \in T\mathscr{X} \mid \delta x \in D(x)\}. \tag{1.20}$$

Secondly, note that a vector field f on $\mathscr{X}$ is described by the map $F : \mathscr{X} \to T\mathscr{X}$ given as $F(x) = (x, f(x)) \in T\mathscr{X}$.

Proposition 1.3 *The distribution D on $\mathscr{X}$ is invariant with respect to the vector field f on $\mathscr{X}$ if and only if one of the following two equivalent conditions is satisfied:*

(i) $F_* D \subset \dot{D}$,
(ii) f^{C} is tangent to $\mathscr{D}$.

Proof The equivalence with (i) is already proved in [18]. For the equivalence with condition (ii) consider local coordinates x and induced local coordinates $x, \delta x$ for $T\mathscr{X}$. Then the vector field f^{C} at a point $(x, \delta x = X(x)) \in \mathscr{D} \subset T\mathscr{X}$ is given by

$$\begin{bmatrix} f(x) \\ \frac{\partial f}{\partial x}(x)X(x) \end{bmatrix} = \begin{bmatrix} f(x) \\ \frac{\partial X}{\partial x}(x)f(x) - L_f X(x) \end{bmatrix} = \begin{bmatrix} f(x) \\ \frac{\partial X}{\partial x}(x)f(x) \end{bmatrix} - \begin{bmatrix} 0 \\ L_f X(x) \end{bmatrix}, \tag{1.21}$$

where the first vector in the last term denotes a tangent vector to $\mathscr{D}$. Hence if $L_f D \subset D$ then the vector field f^{C} is tangent to $\mathscr{D}$. Conversely, if f^{C} is tangent to $\mathscr{D}$ then this implies that the second vector in the last term is tangent to $\mathscr{D}$ for all $X(x) \in D(x)$, which amounts to $L_f D \subset D$. □

Note that the equality expressed in (1.21) can be written in a coordinate-free way as

$$F_* X = X^{\mathrm{C}} - [f, X]^{\mathrm{V}} \tag{1.22}$$

for any vector field X.

Similarly, any co-distribution P on $\mathscr{X}$ can be described by the following *submanifold* of $T^*\mathscr{X}$:

$$\mathscr{P} := \{(x, p) \in T^*\mathscr{X} \mid p \in P(x)\}. \tag{1.23}$$

Then we obtain:

Proposition 1.4 *The distribution P on $\mathscr{X}$ is invariant with respect to the vector field f on $\mathscr{X}$ if and only if one of the following two equivalent conditions is satisfied:*

(i) $F^*\dot{P} = P$,
(ii) X_{Hf} *is tangent to* $\mathscr{P}$.

Proof Again, the equivalence with condition (i) is already shown in [18]. For the second condition, consider coordinates x and induced local coordinates (x, p) for $T^*\mathscr{X}$. Then the vector field X_{Hf} at a point $(x, p = \alpha(x)) \in \mathscr{P} \subset T^*\mathscr{X}$ is given by

$$\begin{bmatrix} f(x) \\ -(\frac{\partial f}{\partial x})^T(x)\alpha(x) \end{bmatrix} = \begin{bmatrix} f(x) \\ \frac{\partial \alpha}{\partial x}(x) f(x) - L_f\alpha(x) \end{bmatrix} = \begin{bmatrix} f(x) \\ \frac{\partial \alpha}{\partial x}(x) f(x) \end{bmatrix} - \begin{bmatrix} 0 \\ L_f\alpha(x) \end{bmatrix},$$

where the first vector in the last term denotes a tangent vector to $\mathscr{P}$. Thus if $L_f P \subset P$ holds then the vector field X_{Hf} is tangent to $\mathscr{P}$. Conversely, if X_{Hf} is tangent to $\mathscr{P}$ then this implies that the second vector in the last term is tangent to $\mathscr{P}$ for all $\alpha(x) \in P(x)$, which amounts to $L_f P \subset P$. □

Note the slight asymmetry with respect to the condition $F_* D \subset \dot{D}$ in the invariant distribution case. This is due to the definitions of $F(x) = (x, f(x))$ and $\dot{P}$ implying that for any P and f we have $P \subset F^*\dot{P}$.

Conditions (ii) in the above two propositions are especially appealing because they imply that invariance of *(co-)distributions* can be equivalently interpreted as invariance of *submanifolds* (of the tangent, respectively cotangent, bundle). In a certain sense this unifies the treatment of invariance in geometric control theory, where usually [18] the distinction is made between invariant (co-)distributions and invariant manifolds (of the state space manifold).

1.3.1 Controlled Invariance

The above geometric framework turned out to be especially useful for a generalization of the notion of invariance for distributions and co-distributions to *controlled invariance*, cf. [17] and Chap. 13 of [18]. Consider a nonlinear control system

$$\dot{x} = f(x, u), \quad x \in \mathscr{X}, u \in \mathscr{U}, \tag{1.24}$$

where $\mathscr{U}$ is an m-dimensional control manifold (e.g., $\mathscr{U} = \mathbb{R}^m$). A distribution D on the n-dimensional state space manifold $\mathscr{X}$ is called *invariant* with respect to the system (1.24) if D is invariant with respect to the vector fields $f(\cdot, u)$ for *every* $u \in \mathscr{U}$. Furthermore, the distribution D is called *locally controlled invariant* if there locally exists a feedback map $u = u(x, v)$, $v \in \mathscr{U}$, rank $\frac{\partial u}{\partial v} = m$, such that D is invariant with respect to the closed-loop dynamics $\dot{x} = f(x, u(x, v))$ with new inputs $v \in \mathscr{U}$.

The basic theorem is as follows. Define the system map $F\colon \mathscr{X} \times \mathscr{U} \to T\mathscr{X}$ by $F(x, u) = (x, f(x, u))$, and note the following consequence of Proposition 1.4.

Proposition 1.5 *Consider the nonlinear dynamics* (1.24) *given by the system map* $F\colon \mathscr{X} \times \mathscr{U} \to T\mathscr{X}$. *Let* D *be a distribution on* $\mathscr{X}$. *Define* D_e *as the unique distribution on* $\mathscr{X} \times \mathscr{U}$ *such that*

$$\pi_* D_e = D, \quad \bar{\pi}_* D_e = 0, \tag{1.25}$$

with π, *resp.* $\bar{\pi}$, *being the natural projection of* $\mathscr{X} \times \mathscr{U}$ *on* $\mathscr{X}$, *resp.* U. *Then* D *is invariant for* (1.24) *if and only if*

$$F_* D_e \subset \dot{D}. \tag{1.26}$$

(Local) controlled invariance is now characterized as follows.

Theorem 1.1 *Consider the nonlinear system* (1.24) *with system map* $F\colon \mathscr{X} \times \mathscr{U} \to T\mathscr{X}$. *Let* D *be an involutive distribution of constant dimension on* $\mathscr{X}$. *Define the vertical distribution* V *on* $\mathscr{X} \times \mathscr{U}$ *as the kernel of* π_*, *where* $\pi : \mathscr{X} \times \mathscr{U} \to \mathscr{X}$ *is the projection on* $\mathscr{X}$. *Assume that the distribution*

$$\tilde{V} := \{Z \in V \mid F_* Z \in \dot{D}\} \tag{1.27}$$

on $\mathscr{X} \times \mathscr{U}$ *has constant dimension. Then* D *is locally controlled invariant if and only if*

$$F_* D_e \subset \dot{D} + F_* V. \tag{1.28}$$

Remark 1.1 Notice that (1.28) may be equivalently replaced by the requirement

$$F_*(\pi_*^{-1}(D)) \subset \dot{D} + F_* V. \tag{1.28'}$$

From a geometric point of view the above theorem yields the following characterization of local controlled invariance. Let D satisfy the assumptions of Theorem 1.1, and denote $P = \text{ann } D$. Then $E := \ker F^* \dot{P}$ is an involutive distribution on $M \times U$. Moreover if (1.28) holds, then E is constant-dimensional and satisfies

$$\pi_* E = D. \tag{1.29}$$

Furthermore, by definition of E and by Proposition 1.2

$$F_* E = F_*(\ker F^* \dot{P}) \subset \ker \dot{P} = \dot{D}. \tag{1.30}$$

Then it is easy to see that, at least locally, there exists an involutive distribution D_e on $\mathscr{X} \times \mathscr{U}$ with $D_e \subset E$, $\pi_* D_e = D$ and with $\dim D_e = \dim D$. This distribution D_e defines a feedback map $u = u(x, v)$ that renders D invariant for the closed-loop dynamics by requiring that the leaves of D_e correspond to the submanifolds where v is constant.

1.4 Invariant Lagrangian Subbundles and Contraction Analysis

Remarkably, there is close geometric connection between the previous theory of (controlled) invariance of (co-)distributions with the theory of *convergent dynamics*, as originating in classical work in differential equations, see e.g., [10, 20] for treatments and references, and revisited in the theory of *contraction analysis*, see e.g., [12–16]. The key idea, see [25], is to interpret (co-)distributions as subbundles of $T\mathscr{X}$, respectively $T^*\mathscr{X}$, and to generalize the notion of invariance of (co-)distributions to invariance of subbundles of the Whitney sum $T\mathscr{X} \oplus T^*\mathscr{X}$.

Definition 1.2 A *subbundle* K of $T\mathscr{X} \oplus T^*\mathscr{X}$ is a vector bundle over $\mathscr{X}$ with fiber $K(x) \subset T_x\mathscr{X} \times T_x^*\mathscr{X}$ at any point $x \in \mathscr{X}$. The subbundle K is called *invariant* with respect to a vector field f on $\mathscr{X}$ if

$$(L_f X, L_f \alpha) \in K \ \text{ for any } (X, \alpha) \in K. \tag{1.31}$$

Remark 1.2 If K has only zero components in $T_x^*\mathscr{X}$ for any point $x \in \mathscr{X}$, then K can be identified with a *distribution* on $\mathscr{X}$. Alternatively, if K has only zero components in $T_x\mathscr{X}$ for any point $x \in \mathscr{X}$, then it can be identified with a *co-distribution* on $\mathscr{X}$. In these cases invariance of K amounts to invariance of the identified distribution, respectively co-distribution.

Remark 1.3 The above definition of invariance of K is formally identical to the definition of an *infinitesimal symmetry* of a *Dirac structure*; see [3, 6, 23] for details. (A Dirac structure is a subbundle of $T\mathscr{X} \oplus T^*\mathscr{X}$ which is maximally isotropic with respect to the duality product.)

Associated to the subbundle K define the *submanifold* $\mathcal{K}$ of $T\mathcal{X} \oplus T^*\mathcal{X}$ as follows

$$\mathcal{K} := \{(x, \delta x, p) \in T\mathcal{X} \oplus T^*\mathcal{X} \mid (\delta x, p) \in K(x)\}. \tag{1.32}$$

We have the following characterization of invariance of K. Recall from Sect. 1.2 the definition of the lift $f^{\mathrm{c}} \oplus X_{Hf}$ on the Whitney sum $T\mathcal{X} \oplus T^*\mathcal{X}$.

Proposition 1.6 *The subbundle K is invariant with respect to the vector field f on $\mathcal{X}$ if and only if the submanifold $\mathcal{K}$ is invariant for the vector field $f^{\mathrm{c}} \oplus X_{Hf}$ on $T\mathcal{X} \oplus T^*\mathcal{X}$.*

Proof In coordinates x and induced local coordinates for $T\mathcal{X}$ and $T^*\mathcal{X}$ the vector field $f^{\mathrm{c}} \oplus X_{Hf}$ at a point $(x, X(x), \alpha(x)) \in \mathcal{K} \subset T\mathcal{X} \oplus T^*\mathcal{X}$ is given by

$$\begin{aligned}
\begin{bmatrix} f(x) \\ \frac{\partial f}{\partial x}(x)X(x) \\ -(\frac{\partial f}{\partial x})^T(x)\alpha(x) \end{bmatrix} &= \begin{bmatrix} f(x) \\ \frac{\partial X}{\partial x}(x)f(x) - L_f X(x) \\ \frac{\partial \alpha}{\partial x}(x)f(x)) - L_f \alpha(x) \end{bmatrix} \\
&= \begin{bmatrix} f(x) \\ \frac{\partial X}{\partial x}(x)f(x) \\ \frac{\partial \alpha}{\partial x}(x)f(x) \end{bmatrix} - \begin{bmatrix} 0 \\ L_f X(x) \\ L_f \alpha(x) \end{bmatrix},
\end{aligned}$$

where the first vector in the last term denotes a tangent vector to $\mathcal{K}$. Thus if (1.31) holds then the vector field $f^{\mathrm{c}} \oplus X_{Hf}$ is tangent to $\mathcal{K}$. Conversely, if $f^{\mathrm{c}} \oplus X_{Hf}$ is tangent to $\mathcal{K}$ then this implies that the second vector in the last term is tangent to $\mathcal{K}$ for all $(X(x), \alpha(x)) \in K(x)$, which amounts to (1.31), i.e., invariance of K. □

In the rest of this section we will consider a special type of subbundle of the Whitney sum $T\mathcal{X} \oplus T^*\mathcal{X}$ defined as follows.

Definition 1.3 A subbundle K of $T\mathcal{X} \oplus T^*\mathcal{X}$ is called a *Lagrangian subbundle* if $K(x) \subset T_x\mathcal{X} \times T_x^*\mathcal{X}$ is a Lagrangian subspace (with respect to the canonical symplectic form on $T_x\mathcal{X} \times T_x^*\mathcal{X}$ [1]) for every $x \in \mathcal{X}$.

All subbundles

$$\begin{aligned}
K(x) = \{(\delta x, p) \mid V(x)p = U(x)\delta x, \\
V(x)U^T(x) = U(x)V^T(x), \operatorname{rank}\left[U(x)\ V(x)\right] = n\}
\end{aligned} \tag{1.33}$$

with $U(x)$, $V(x)$ $n \times n$ matrices depending on x (with n the dimension of the state space manifold $\mathcal{X}$), are Lagrangian, and conversely, all Lagrangian subbundles can be represented in this way.

Additionally assume that the projection of $K(x) \subset T_x\mathcal{X} \oplus T_x^*\mathcal{X}$ on $T_x\mathcal{X}$ is equal to the whole tangent space $T_x\mathcal{X}$ for all $x \in \mathcal{X}$. Then in any set of local coordinates $x^1, \ldots, x^n$ for $\mathscr{X}$ the Lagrangian subbundle K is spanned by pairs of vector fields and one-forms

$$\left(\frac{\partial}{\partial x^i}, \pi_i\right), \quad i = 1, \ldots, n,$$

where the one-forms $\pi_i(x) = \pi_{1i}(x)dx^I + \cdots \pi_{ni}(x)dx^n, i = 1, \ldots, n$ satisfy, because of the fact that K is Lagrangian, the symmetry property

$$\pi_{ji}(x) = \pi_{ij}(x), \quad i, j = 1, \ldots, n.$$

Conversely, all subbundles K with $K(x) = \{(\delta x, p) \mid p = \Pi(x)\delta x\}$, where $\Pi(x)$ is a *symmetric* matrix, are Lagrangian.

Now consider any such Lagrangian subbundle K with $K(x) = \{(\delta x, p) \mid p = \Pi(x)\delta x\}$, with $\Pi(x)$ symmetric. Invariance of such a Lagrangian subbundle K with respect to $f^c \oplus X_{H^f}$ can be seen to amount to [25]

$$\left(\frac{\partial f}{\partial x}\right)^T(x)\Pi(x) + \Pi(x)\frac{\partial f}{\partial x}(x) + \frac{\partial \Pi}{\partial x}(x)f(x) = 0. \tag{1.34}$$

This is a limiting case of the central equation of contraction analysis [10, 13, 21], which amounts to

$$\left(\frac{\partial f}{\partial x}\right)^T(x)\Pi(x) + \Pi(x)\frac{\partial f}{\partial x}(x) + \frac{\partial \Pi}{\partial x}(x)f(x) = -Q(x) \tag{1.35}$$

for some positive semi-definite matrix $Q(x)$. Geometrically, this means that the submanifold $\mathscr{K}$ corresponding to the Lagrangian subbundle K is invariant for the extended vector field $f^c \oplus X_{H^f} + q$, where q is the vector field in local coordinates x given by

$$q(x) = \sum_{i,j} Q_{ij}(x)\delta x^j \frac{\partial}{\partial p_i}. \tag{1.36}$$

We will call (1.35) a *differential Lyapunov equation*, cf. [25].

Alternatively, (1.34) can be interpreted from a Riemannian geometry point of view, by identifying it as the equation for a *Killing vector field* [21]; see also [2]. Indeed, the symmetric matrix $\Pi(x)$, in case it is positive-definite, defines a *Riemannian metric* on $\mathscr{X}$, in which case (1.34) amounts to the vector field f being a Killing vector field for this Riemannian metric, characterized by the property

$$L_f < X, Y >=< L_f X, Y > + < X, L_f Y >, \tag{1.37}$$

for any vector fields X, Y, with $< \cdot, \cdot >$ denoting the Riemannian metric defined by the positive-definite matrix $\Pi(x)$. Finally, if this Riemannian metric has curvature zero, then in appropriate coordinates x the matrix $\Pi(x)$ takes the form of a *constant* matrix P, and (1.35) reduces to the Demidovich condition [19, 20]

$$\left(\frac{\partial f}{\partial x}\right)^T(x)P + P\frac{\partial f}{\partial x}(x) = -Q(x). \tag{1.38}$$

As an open problem we mention the characterization of *locally controlled* invariant Lagrangian subbundles, in a similar geometric spirit as the characterization of locally controlled invariant distributions sketched before.

1.4.1 Prolongation of Nonlinear Control Systems to Tangent and Cotangent Bundle and Further Differential Properties

The previous geometric interpretation of the basic equation of contraction analysis is part of a larger story translating incremental analysis tools into *differential properties*. A basic notion in this larger story is the prolongation (or lift) of a *nonlinear control system* to the tangent and cotangent bundle, using the notions of the lifts of functions and vector fields to tangent and cotangent bundle as described in the previous subsection. This can be done as in [4], see also [2, 24].

Consider a nonlinear control system Σ with state space manifold $\mathscr{X}$, affine in the inputs u and with outputs y determined by the state x,

$$\Sigma : \begin{aligned} \dot{x} &= f(x) + \sum_{j=1}^{m} u_j g_j(x) \\ y_j &= h_j(x), \quad j = 1, \ldots, r, \end{aligned} \tag{1.39}$$

where $x \in \mathscr{X}$, and $u = (u_1, \ldots, u_m) \in \mathscr{U} \subset \mathbb{R}^m$. The set $\mathscr{U}$ is the input space, which is assumed to be an open subset of $\mathbb{R}^m$. Finally, $\mathscr{Y} = \mathbb{R}^r$ is the output space.

The prolongation of the nonlinear control system to the tangent bundle and the cotangent bundle is constructed as follows; cf. [4].

Given an initial state $x(0) = x_0$, take any coordinate neighborhood of $\mathscr{X}$ containing x_0. Let $t \in [0, T] \mapsto x(t)$ be the solution of (1.39) corresponding to the admissible input function $t \in [0, T] \mapsto u(t) = (u_1(t), \ldots, u_m(t))$ and the initial state $x(0) = x_0$, such that $x(t)$ remains within the selected coordinate neighborhood. Denote the resulting output by $t \in [0, T] \mapsto y(t) = (y_1(t), \ldots, y_r(t))$, with $y_j(t) = H_j(x(t))$. Then the *variational system* along the input-state-output trajectory $t \in [0, T] \mapsto (x(t), u(t), y(t))$ is given by the following time-varying system

$$\begin{aligned} \dot{\delta x}(t) = & \frac{\partial f}{\partial x}(x(t))\delta x(t) \\ & + \sum_{j=1}^{m} u_j(t)\frac{\partial g_j}{\partial x}(x(t))\delta x(t) \end{aligned}$$

$$+\sum_{j=1}^{m} \delta u_j(t) g_j(x(t)) \tag{1.40}$$

$$\delta y_j(t) = \frac{\partial h_j}{\partial x}(x(t))\delta x(t), \quad j = 1, \ldots, r,$$

with state $\delta x(t) \in T^*_{x(t)}\mathscr{X}$, where $\delta u = (\delta u_1, \ldots, \delta u_m)^T$, $\delta y = (\delta y_1, \ldots, \delta y_r)^T$ denote the input and output vectors of the variational system. (Note that $\frac{\partial h_j}{\partial x}(x)$ denotes a row vector.)

The reason behind the terminology "variational" comes from the following fact: let $(x(t, \varepsilon), u(t, \varepsilon), y(t, \varepsilon))$, $t \in [0, T]$, be a family of input-state-output trajectories of (1.39) parameterized by $\varepsilon \in (-c, c)$, $c > 0$, with $x(t, 0) = x(t)$, $u(t, 0) = u(t)$ and $y(t, 0) = y(t)$, $t \in [0, T]$. Then the infinitesimal variations

$$\delta x(t) = \frac{\partial x}{\partial \varepsilon}(t, 0), \quad \delta u(t) = \frac{\partial u}{\partial \varepsilon}(t, 0), \quad \delta y(t) = \frac{\partial y}{\partial \varepsilon}(t, 0), \tag{1.41}$$

satisfy (1.40).

Remark 1.4 For a *linear* system $\dot{x} = Ax + Bu$, $y = Cx$ the variational systems along any trajectory are simply given as $\dot{\delta x} = A\delta x + B\delta u$, $\delta y = C\delta x$.

The *prolongation* (or *prolonged system*) of (1.39) comprises the original system (1.39) *together* with its variational systems, that is the total system

$$\begin{aligned} \dot{x} &= f(x) + \textstyle\sum_{j=1}^{m} u_j g_j(x) \\ \dot{\delta x}(t) &= \tfrac{\partial f}{\partial x}(x(t))\delta x(t) + \\ &\quad \textstyle\sum_{j=1}^{m} u_j(t) \tfrac{\partial g_j}{\partial x}(x(t))\delta x(t) + \\ &\quad \textstyle\sum_{j=1}^{m} \delta u_j(t) g_j(x(t)) \\ y_j &= h_j(x), \quad j = 1, \ldots, r \\ \delta y_j(t) &= \tfrac{\partial h_j}{\partial x}(x(t))\, \delta x(t), \quad j = 1, \ldots, r, \end{aligned} \tag{1.42}$$

with inputs $u_j, \delta u_j$, $j = 1, \ldots, m$, outputs $y_j, \delta y_j$, $j = 1, \ldots, r$, and state vector x, δx.

Using the previous subsection, the prolonged system (1.42) on the tangent space $T\mathscr{X}$ can be intrinsically defined in the following coordinate-free way. Denote the elements of $T\mathscr{X}$ by $x_l = (x, \delta x)$, where $\tau_{\mathscr{X}}(x_l) = x \in \mathscr{X}$ with $\tau_{\mathscr{X}} : T\mathscr{X} \to \mathscr{X}$ again the tangent bundle projection.

Definition 1.4 ([4]) The prolonged system $\delta\Sigma$ of a nonlinear system Σ of the form (1.39) is defined as the system

$$\delta\Sigma : \begin{aligned} \dot{x}_l &= f^{\mathrm{c}}(x_l) + \textstyle\sum_{j=1}^{m} u_j g_j^{\mathrm{C}}(x_l) + \sum_{j=1}^{m} \delta u_j g_j^{\mathrm{V}}(x_l) \\ y_j &= h_j^{\mathrm{V}}(x_l), \quad j = 1, \ldots, r \\ \delta y_j &= h_j^{\mathrm{C}}(x_l), \quad j = 1, \ldots, r, \end{aligned} \tag{1.43}$$

with state $x_l = (x, \delta x) \in T\mathscr{X}$, inputs $u_j, \delta u_j$, $j = 1, \ldots, m$, and outputs $y_j, \delta y_j$, $j = 1, \ldots, r$.

Note that the prolonged system $\delta\Sigma$ has state space $T\mathscr{X}$, input space $T\mathscr{U}$ and output space $T\mathscr{Y}$. One can easily check that in any system of local coordinates x for $\mathscr{X}$, u for $\mathscr{U}$, and y for $\mathscr{Y}$, and the induced local coordinates $x, \delta x$ for $T\mathscr{X}$, $u, \delta u$ for $T\mathscr{U}$, $y, \delta y$ for $T\mathscr{Y}$, the local expression of the system (1.43) equals (1.42).

Remark 1.5 For a linear system $\dot{x} = Ax + Bu$, $y = Cx$ the prolonged system is simply the *product* of the system with the copy system $\dot{\delta x} = A\delta x + B\delta u$, $\delta y = C\delta x$.

The prolongation of the nonlinear control system Σ to the *cotangent bundle* is defined as follows. Associated to the variational system (1.40) there is the *adjoint variational system*, defined as

$$
\begin{aligned}
\dot{p}(t) &= -\left(\tfrac{\partial f}{\partial x}\right)^T (x(t))p(t) \\
&\quad -\textstyle\sum_{j=1}^m u_j(t)(\frac{\partial g_j}{\partial x})^T(x(t))p(t) \\
&\quad -\textstyle\sum_{j=1}^r du_j(t)\frac{\partial^T h_j}{\partial x}(x(t)) \\
dy_j(t) &= p^T g_j(x(t)), \quad j = 1, \ldots, m,
\end{aligned}
\tag{1.44}
$$

with state variables $p \in T^*_{x(t)}\mathscr{X}$, and adjoint variational inputs and outputs du_j, $j = 1, \ldots, r$, dy_j, $j = 1, \ldots, m$. Then, the original nonlinear system Σ together with all its adjoint variational systems defines the total system

$$
\begin{aligned}
\dot{x} &= f(x) + \textstyle\sum_{j=1}^m u_j g_j(x) \\
\dot{p}(t) &= -(\tfrac{\partial f}{\partial x})^T(x(t))p(t) \\
&\quad -\textstyle\sum_{j=1}^m u_j(t)(\frac{\partial g_j}{\partial x})^T(x(t))p(t) \\
&\quad -\textstyle\sum_{j=1}^r du_j(t)\frac{\partial^T h_j}{\partial x}(x(t)) \\
y_j &= h_j(x), \quad j = 1, \ldots, r \\
dy_j(t) &= p^T g_j(x(t)), \quad j = 1, \ldots, m,
\end{aligned}
\tag{1.45}
$$

with inputs u_j, du_j, outputs y_j, dy_j and state x, p. This total system is called the *Hamiltonian extension*. In a coordinate-free way the Hamiltonian extension is defined as follows.

Definition 1.5 ([4]) The Hamiltonian extension $d\Sigma$ of a nonlinear system Σ of the form (1.39) is defined as the system

$$
d\Sigma: \quad
\begin{aligned}
\dot{x}_e &= X_{H^f}(x_e) + \textstyle\sum_{j=1}^m u_j X_{H^{g_j}}(x_e) + \sum_{j=1}^r du_j X_{h_j^{\mathrm{v}}}(x_e) \\
y_j &= h_j^{\mathrm{v}}(x_e), \quad j = 1, \ldots, r \\
dy_j &= H^{g_j}(x_e), \quad j = 1, \ldots, m,
\end{aligned}
\tag{1.46}
$$

with state $x_e = (x, p) \in T^*\mathscr{X}$, inputs u_j, du_j and outputs y_j, dy_j.

Note that the Hamiltonian extension $d\Sigma$ has state space $T^*\mathscr{X}$, and combined input and output space $T^*\mathscr{U} \times T^*\mathscr{Y}$, where $(u, \delta y) \in T^*\mathscr{U}$ and $(y, \delta u) \in T^*\mathscr{Y}$.

Remark 1.6 For a linear system $\dot{x} = Ax + Bu$, $y = Cx$ the Hamiltonian extension is the product of the system with its adjoint system $\dot{p} = -A^T p - C^T du, dy = B^T p$.

The prolongation $\delta\Sigma$ of Σ to the tangent bundle can be *combined* with the Hamiltonian extension $d\Sigma$ of Σ to the cotangent bundle. This will define a system on the Whitney sum $T\mathscr{X} \oplus T^*\mathscr{X}$ with inputs $u, \delta u, du$, states $x, \delta x, p$ and outputs $y, \delta y, dy$. It can be immediately verified that

$$\frac{d}{dt}\langle p, \delta x\rangle = \langle dy, \delta u\rangle - \langle du, \delta y\rangle, \tag{1.47}$$

which equality is in fact underlying the definition of the adjoint variational system; see also [4].

1.4.2 Differential Lyapunov Equations

In [25] it was shown how *differential Riccati equations* can be obtained by interconnecting the prolonged system $\delta\Sigma$ with the Hamiltonian extension $d\Sigma$ through the interconnection laws

$$\delta u = -dy, \; du = \delta y, \tag{1.48}$$

and to consider invariant Lagrangian subbundles for the resulting *differential Hamiltonian system* living on the Whitney sum $T\mathscr{X} \oplus T^*\mathscr{X}$. This yields a geometric formulation of the approach developed e.g., in [11, 12, 14–16, 26].

Differential Lyapunov equations result from the case $g_j = 0$, $j = 1, \ldots, m$, for the original system Σ (i.e., no inputs). In this case the interconnection of the prolongation $\delta\Sigma$ and the Hamiltonian extension $d\Sigma$ reduces to $du = \delta y$, and can be seen to lead to the differential Hamiltonian system (see [25] for further details)

$$\begin{aligned} \dot{z} &= f^c \oplus X_{H^f}(z) - \textstyle\sum_{j=1}^r h_j^c X_{h_j^v}(z), \quad z \in T\mathscr{X} \\ y_j &= h_j(x), \quad j = 1, \ldots, r. \end{aligned} \tag{1.49}$$

The submanifold $\mathscr{K}$ corresponding to a Lagrangian subbundle K with $K(x) = \{(\delta x, p) \mid p = \Pi(x)\delta x\}$, where $\Pi(x)$ is a *symmetric* matrix, is an *invariant submanifold* for (1.49) if and only if $\Pi(x)$ satisfies the equation

$$\left(\frac{\partial f}{\partial x}\right)^T (x)\Pi(x) + \Pi(x)\frac{\partial f}{\partial x}(x) + \left(\frac{\partial h}{\partial x}\right)^T (x)\frac{\partial h}{\partial x}(x) + \frac{\partial \Pi}{\partial x}(x)f(x) = 0. \tag{1.50}$$

This is nothing else than the *differential Lyapunov equation* (1.35) as considered before, where the matrix $Q(x)$ in (1.35) is of a somewhat special form.

1.5 Conclusions and Outlook

In this chapter (controlled) invariance of (co-)distributions has been reviewed using the geometric notions of liftings to the (co-)tangent bundle, continuing upon the approach initiated in [17] and further developed in [18]. It has been shown how the geometric treatment of invariance of (co-)distributions can be extended to invariance of Lagrangian subbundles, making a direct link to the theory of *convergent dynamics* and *contraction analysis*.

Next to the classical topic of (controlled) invariant (co-)distributions, and the recent developments on differential Lyapunov and differential Riccati equations, the established geometric framework also opens other possibilities. An important avenue for further research is the extension of the classical passivity and small-gain framework, see e.g., [22], to the differential level. Indeed, in [7–9, 24] differential passivity was defined and explored in this spirit. Furthermore, in [15, 24] differential L_2-gain was initiated. Despite the beauty of this approach and the insight provided, the implications for control design of these recent developments are yet to be seen, and probably require the combination with dynamical analysis tools.

Acknowledgments I would like to thank the *Dutch Nonlinear Systems Group* for continuing inspiration and support.

References

1. Abraham, R.A., Marsden, J.E.: Foundations of mechanics, 2nd edn. Benjamin / Cummings, Reading, MA (1978)
2. Cortés, J., van der Schaft, A.J., Crouch, P.E.: Characterization of gradient control systems. SIAM J. Control Optim. **44**(4), 1192–1214 (2005)
3. Courant, T.J.: Dirac manifolds. Trans. Amer. Math. Soc. **319**, 631–661 (1990)
4. Crouch, P.E., van der Schaft, A.J.: Variational and Hamiltonian control systems. Lecture Notes in Control and Information Sciences, vol. 101. Springer, New York (1987)
5. Dalsmo, M., van der Schaft, A.J.: On representations and integrability of mathematical structures in energy-conserving physical systems. SIAM J. Control and Optimization **37**, 54–91 (1999)
6. Dorfman, I.: Dirac Structures and Integrability of Nonlinear Evolution Equations. Wiley, Chichester (1993)
7. Forni, F., Sepulchre, R.: A differential Lyapunov framework for contraction analysis. IEEE Trans. Autom. Control **59**(3), 614–628 (2014)
8. Forni, F., Sepulchre, R.: On differentially dissipative dynamical systems, Proceedings of 9th IFAC Symposium on Nonlinear Control Systems (NOLCOS2013), Toulouse, France, Sept. 4–6 (2013)
9. Forni, F., Sepulchre, R., van der Schaft, A.J.: On differential passivity of physical systems. In: Proceedings of 52nd IEEE Conference on Decision and Control (CDC), pp. 6580–6585. Florence, Italy (2013)
10. Jouffroy, J.: Some ancestors of contraction analysis. In: 44th IEEE Conference Decision and Control and European Control Conference (CDC-ECC '05), pp. 5450–5455 (2005)
11. Kawano, Y., Ohtsuka, T.: Observability analysis of nonlinear systems using pseudo-linear transformation. In: Proceedings of 9th IFAC Symposium on Nonlinear Control Systems (NOLCOS2013), Toulouse, France, Sept. 4–6 (2013)

12. Kawano, Y., Ohtsuka, T.: Nonlinear eigenvalue approach to analysis of generalised differential Riccati equations (submitted, 2014)
13. Lohmiller, W., Slotine, J.-J.: On contraction analysis for non-linear systems. Automatica **34**, 683–696 (1998)
14. Manchester, I.R., Slotine, J.J.: Control contraction metrics and universal stabilizability. 19th IFAC World Congress, Cape Town, South Africa (2014)
15. Manchester, I.R., Slotine, J.J.: Control contraction metrics: differential L^2 gain and observer duality. arXiv:1403.5364v1 (2014)
16. Manchester, I.R., Slotine, J.J.: Control contraction metrics: Convex and intrinsic criteria for nonlinear feedback design. arXiv:1503.03144v1 (2015)
17. Nijmeijer, H., van der Schaft, A.J.: Controlled invariance for nonlinear systems. IEEE Trans. Autom. Control **AC–27**(4), 904–914 (1982)
18. Nijmeijer, H., van der Schaft, A.J.: Nonlinear Dynamical Control Systems. Springer, New York (1990). corrected printing 2016
19. Pavlov, A., Pogromsky, A., van de Wouw, N., Nijmeijer, H.: Convergent dynamics: a tribute to Boris Pavlovich Demidovich. Syst. Control Lett. **52**, 257–261 (2004)
20. Pavlov, A., van de Wouw, N., Nijmeijer, H.: Uniform Output Regulation of Nonlinear Systems: A Convergent Dynamics Approach. Springer, Berlin (2005)
21. Simpson, J.W., Bullo, F.: Contraction theory on Riemannian manifolds. Syst. Control Lett. **65**, 74–80 (2014)
22. van der Schaft, A.J.: L_2-Gain and Passivity Techniques in Nonlinear Control, Lecture Notes in Control and Information Sciences, Vol. 218, 2nd edn. Springer, Berlin (1996), Springer, London (2000) (Springer Communications and Control Engineering series)
23. van der Schaft, A.J.: Implicit Hamiltonian systems with symmetry. Rep. Math. Phys. **41**, 203–221 (1998)
24. van der Schaft, A.J.: On differential passivity. In: Proceedings of 9th IFAC Symposium on Nonlinear Control Systems (NOLCOS2013), pp. 21–25. Toulouse, France, Sept. 4–6 (2013)
25. van der Schaft, A.J.: A geometric approach to differential Hamiltonian and differential Riccati equations. In: Proceedings of 54th IEEE Conference on Decision and Control, (CDC), pp. 7151–7156. Osaka, Japan (2015)
26. Tyner, D.R., Lewis, A.D.: Geometric Jacobian linearization and LQR theory. J. Geom. Mech. **2**(4), 397–440 (2010)
27. Yano, K., Ishihara, S.: Tangent and cotangent bundles. Marcel Dekker, New York (1973)

Chapter 2
Some Recent Results on Distributed Control of Nonlinear Systems

Tengfei Liu and Zhong-Ping Jiang

Abstract The spatially distributed structure of complex systems motivates the idea of distributed control. In a distributed control system, the subsystems are controlled by local controllers through information exchange with neighboring agents for coordination purposes. One of the major difficulties of distributed control is due to the complex characteristics such as nonlinearity, dimensionality, uncertainty, and information constraints. This chapter introduces small-gain methods for distributed control of nonlinear systems. In particular, a cyclic-small-gain result in digraphs is presented as an extension of the standard nonlinear small-gain theorem. It is shown that the new result is extremely useful for distributed control of nonlinear systems. Specifically, this chapter first gives a cyclic-small-gain design for distributed output-feedback control of nonlinear systems. Then, an application to formation control problem of nonholonomic mobile robots with a fixed information exchange topology is presented.

2.1 Introduction

Distributed control of multiagent systems under communication constraints has attracted tremendous attention from the control community over the past 10 years; see, for example, [45] using an adaptive gradient climbing strategy, [3, 5, 12, 21, 48] based on linear algebra and graph theory, [2, 53] using passivity and dissipativity theory, [4, 16, 19, 34, 42, 43, 50, 51] with Lyapunov methods, [17] using the nonlinear small-gain theorem, and [54, 57, 58] based on output regulation. The recent

T. Liu (✉)
State Key Laboratory of Synthetical Automation for Process Industries,
Northeastern University, Shenyang, China
e-mail: tfliu@mail.neu.edu.cn

Z.-P. Jiang
Department of Electrical and Computer Engineering,
Tandon School of Engineering, New York University, Brooklyn, NY 11201, USA
e-mail: zjiang@nyu.edu

N. van de Wouw et al. (eds.), *Nonlinear Systems*, Lecture Notes
in Control and Information Sciences 470, DOI 10.1007/978-3-319-30357-4_2

hot topics such as formation control, consensus, flocking, swarm, rendezvous and synchronization are all closely related to distributed coordinated control.

The distributed control problem for agents with second-order dynamics has been mainly studied from the perspective of second-order consensus and flocking. Considerable efforts have been devoted to solving the problems under switching information exchange topologies. Related results include [4, 12, 46, 47, 55]. Specifically, [12, 46] used potential functions to define Lyapunov functions and the topologies are allowed to be switching but undirected. Reference [47] presented a consensus result for double integrator systems based on a refined graph theoretical method. Reference [4] proposed a variable structure approach-based consensus design method for systems with switching but always connected information exchange topology. Several recent results on distributed control can also be found in [15, 32, 41, 49, 59]. It should be pointed out that most of the papers mentioned above do not consider systems under physical constraints (e.g., saturation of velocities), for which specific distributed nonlinear designs are expected.

As a practical application of distributed control, the formation control of autonomous mobile agents aims at forcing the agents to converge toward, and to maintain, specific relative positions, by using available information, e.g., relative position measurements. Recent formation control results can be found in [1, 7, 9, 10, 20, 30, 31, 44, 56], to name a few. The earlier results, e.g., [7, 56], assume a tree sensing structure to avoid the technical difficulties caused by the loop interconnections. In [9, 10, 31, 44], the assumption of tree sensing structures is relaxed at the price of using global position measurements. An exception is the wiggling controller developed by [33] to drive the robots to stationary points, which does not use global position measurements of the robots. In the results of coordinated path-following as presented in [1, 20, 30], the global position measurement issue can be easily addressed as each robot has access to its desired path. In our recent paper [38], thanks to the use of nonlinear small-gain techniques [25, 35], the requirement on global position measurements has been removed for formation control of unicycles with fixed sensing topologies.

The discussion in this chapter starts with an example of a multivehicle formation control system in which each vehicle is modeled by an integrator. In the case of leader-following with fixed topology, it is shown that the problem can be transformed into the stability problem of a specific dynamic network composed of ISS subsystems. This motivates a cyclic-small-gain result in digraphs, which is given in Sect. 2.2. It is shown that the new result is extremely useful for distributed control of nonlinear systems. Specifically, Sect. 2.3 presents a cyclic-small-gain design for distributed output-feedback control of nonlinear systems. In Sect. 2.4, we study the distributed formation control problem of nonholonomic mobile robots with a fixed information exchange topology.

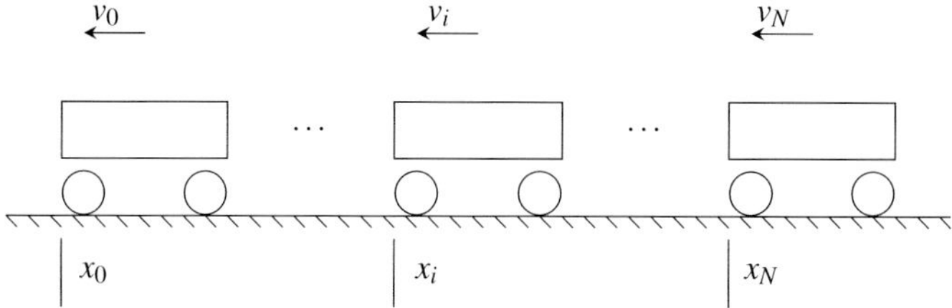

Fig. 2.1 A multivehicle system

2.2 A Cyclic-Small-Gain Result in Digraphs

Example 2.1 Consider a group of $N+1$ vehicles (multivehicle system) as shown in Fig. 2.1, with each vehicle modeled by an integrator:

$$\dot{x}_i = v_i, \quad i = 0, \ldots, N, \tag{2.1}$$

where $x_i \in \mathbb{R}$ is the position and $v_i \in \mathbb{R}$ is the velocity of the ith vehicle. The vehicle with index 0 is the leader while the other vehicles are the followers. The objective is to control the follower vehicles to specific positions relative to the leader by adjusting the velocities v_i for $i = 1, \ldots, N$. More specifically, it is required that

$$\lim_{t\to\infty} (x_i(t) - x_j(t)) = d_{ij}, \quad i, j = 0, \ldots, N, \tag{2.2}$$

where constants d_{ij} represent the desired relative positions. Clearly, to define the problem well, $d_{ij} = d_{ik} + d_{kj}$ for any $i, j, k = 0, \ldots, N$ and $d_{ij} = -d_{ji}$ for any $i, j = 0, \ldots, N$. Also, by default, $d_{ii} = 0$ for any $i = 0, \ldots, N$. In the literature of distributed control, the vehicles are usually considered as agents and the multivehicle system is studied as a multiagent system.

Compared with global positions, the relative positions between vehicles are often easily measurable in practice, and are used for feedback in this example. Considering the position information exchange, agent j is called a neighbor of agent i if $(x_i - x_j)$ is available to agent i, and $\mathcal{N}_i \subseteq \{0, \ldots, N\}$ is used to denote the set of agent i's neighbors. We consider the case where each vehicle only uses the position differences with the vehicles right before and after it, i.e., $\mathcal{N}_i = \{i-1, i+1\}$ for $i = 1, \ldots, N-1$ and $\mathcal{N}_N = \{N-1\}$.

Define $\tilde{x}_i = x_i - x_0 - d_{i0}$ and $\tilde{v}_i = v_i - v_0$. By taking the derivative of $\tilde{x}_i$, we have

$$\dot{\tilde{x}}_i = \tilde{v}_i, \quad i = 1, \ldots, N. \tag{2.3}$$

According to the definition of $\tilde{x}_i$, $\tilde{x}_i - \tilde{x}_j = x_i - x_j - d_{ij}$. Thus, the control objective is achieved if $\lim_{t\to\infty}(\tilde{x}_i - \tilde{x}_0) = 0$. Also, $(\tilde{x}_i - \tilde{x}_j)$ is available to the control of the $\tilde{x}_i$-subsystem if $(x_i - x_j)$ is available to agent i. This problem is normally known as the consensus problem. If the position information exchange topology has a spanning tree with agent 0 as the root, then the following distributed control law is effective:

$$\tilde{v}_i = k_i \sum_{j \in \mathcal{N}_i} (\tilde{x}_j - \tilde{x}_i), \tag{2.4}$$

where k_i is a positive constant. Moreover, if the velocities v_i are required to be bounded, one may modify (2.4) as

$$\tilde{v}_i = \varphi_i \left(\sum_{j \in \mathcal{N}_i} (\tilde{x}_j - \tilde{x}_i) \right), \tag{2.5}$$

where $\varphi_i : \mathbb{R} \to [\underline{v}_i, \overline{v}_i]$ with constants $\underline{v}_i < 0 < \overline{v}_i$ is a continuous, strictly increasing function satisfying $\varphi_i(0) = 0$. With control law (2.5), $v_i \in [v_0 + \underline{v}_i, v_0 + \overline{v}_i]$. The validity of the control laws defined by (2.4) and (2.5) can be directly verified using the state agreement result in [34].

With control law (2.5), each $\tilde{x}_i$-subsystem can be rewritten as

$$\dot{\tilde{x}}_i = \varphi_i \left(\sum_{j \in \mathcal{N}_i} \tilde{x}_j - N_i \tilde{x}_i \right) := f_i(\tilde{x}), \tag{2.6}$$

where N_i is the size of $\mathcal{N}_i$ and $\tilde{x} = [\tilde{x}_0, \ldots, \tilde{x}_N]^T$. Define $V_i(\tilde{x}_i) = |\tilde{x}_i|$ as an ISS-Lyapunov function candidate for the $\tilde{x}_i$-subsystem for $i = 1, \ldots, N$. It can be verified that for any $\delta > 0$, there exists a continuous, positive definite α such that

$$V_i(\tilde{x}_i) \geq \frac{1}{(1 - \delta_i) N_i} \sum_{j \in \mathcal{N}_i} V_j(\tilde{x}_j) \Rightarrow \nabla V_i(\tilde{x}_i) f_i(\tilde{x}) \leq -\alpha_i(V_i(\tilde{x}_i)) \ \text{ a.e.,} \tag{2.7}$$

where, for convenience of notation, $V_0(\tilde{x}_0) = 0$. This shows the ISS of each $\tilde{x}_i$-subsystem with $i = 1, \ldots, N$. If the network of ISS subsystems is asymptotically stable, then the control objective is achieved.

We employ a digraph $\mathcal{G}_f$ to represent the underlying interconnection structure of the dynamic network. The vertices of the digraph correspond to agents $1, \ldots, N$, and for $i, j = 1, \ldots, N$, directed edge (j, i) exists in the graph if and only if $\tilde{x}_j$ is an input of the x_i-subsystem. We use $\overline{\mathcal{N}}_i$ to represent the set of neighbors of agent i in $\mathcal{G}_f$. Then, it is directly verified that $\overline{\mathcal{N}}_i = \mathcal{N}_i \backslash \{0\}$. Recall that $V_0(\tilde{x}_0) = 0$. Then, the $\mathcal{N}_i$ in (2.7) can be directly replaced by $\overline{\mathcal{N}}_i$. Figure 2.2 shows the digraph $\mathcal{G}_f$ for the case in which each follower vehicle uses the position differences with the vehicles right before and after it.

Notice that for any positive constants $a_1, \ldots, a_n$ satisfying $\sum_{i=1}^n 1/a_i \leq n$, it holds that $\sum_{i=1}^n d_i = \sum_{i=1}^n (1/a_i) a_i d_i \leq n \max_{i=1,\ldots,n}\{a_i d_i\}$, for all $d_1, \ldots, d_n \geq 0$. Then, property (2.7) implies

$$1 \rightleftarrows 2 \rightleftarrows 3 \rightleftarrows \quad \cdots \quad \rightleftarrows N$$

Fig. 2.2 An example of information exchange digraph $\mathscr{G}_f$, for which each vehicle uses the position differences with the vehicles right before and after it. In this figure, $\overline{\mathscr{N}}_i = \{i-1, i+1\}$ for $i = 2, \ldots, N-1$, $\overline{\mathscr{N}}_1 = \{2\}$ and $\overline{\mathscr{N}}_N = \{N-1\}$

$$V_i(\tilde{x}_i) \geq \frac{\overline{N}_i}{(1-\delta_i)N_i} \max_{j \in \overline{\mathscr{N}}_i} \{a_{ij} V_j(\tilde{x}_j)\} \Rightarrow \nabla V_i(\tilde{x}_i) f_i(\tilde{x}) \leq -\alpha_i(V_i(\tilde{x}_i)), \tag{2.8}$$

where $\overline{N}_i$ is the size of $\overline{\mathscr{N}}_i$ and a_{ij} are positive constants satisfying $\sum_{j \in \overline{\mathscr{N}}_i} 1/a_{ij} \leq \overline{N}_i$. It can be observed that $N_i = \overline{N}_i + 1$ if $0 \in \mathscr{N}_i$ and $N_i = \overline{N}_i$ if $0 \notin \mathscr{N}_i$.

Given specific $a_{ij} > 0$, one can test the stability property of the closed-loop system by directly checking whether the cyclic-small-gain condition is satisfied. But, for a specific $\mathscr{G}_f$, can we find appropriate coefficients a_{ij} to satisfy the cyclic-small-gain condition, and how?

It should be noted that the effectiveness of control law (2.5) can be proved using the result in [34]. Here, our objective is to transform the problem into a stability problem of dynamic networks, and develop a result which is hopefully useful for more general distributed control problems.

The main result in this section answers the question in Example 2.1.

Consider a digraph $\mathscr{G}_f$ which has N vertices. For $i = 1, \ldots, N$, define $\overline{\mathscr{N}}_i$ such that if there is a directed edge (j, i) from the jth vertex to the i-th vertex, then $j \in \overline{\mathscr{N}}_i$. Each edge (j, i) is assigned a positive variable a_{ij}. For a simple cycle $\mathscr{O}$ of $\mathscr{G}_f$, denote $A_{\mathscr{O}}$ as the product of the positive values assigned to the edges of the cycle. For $i = 1, \ldots, N$, denote $\mathscr{C}(i)$ as the set of simple cycles of $\mathscr{G}_f$ through the i-th vertex.

Lemma 2.1 *If the digraph $\mathscr{G}_f$ has a spanning tree $\mathscr{T}_f$ with vertices $i_1^*, \ldots, i_q^*$ as the roots, then for any $\varepsilon > 0$, there exist $a_{ij} > 0$ for $i = 1, \ldots, N$, $j \in \overline{\mathscr{N}}_i$, such that*

$$\sum_{j \in \overline{\mathscr{N}}_i} \frac{1}{a_{ij}} \leq \overline{N}_i, \quad i = 1, \ldots, N \tag{2.9}$$

$$A_{\mathscr{O}} < 1 + \varepsilon, \quad \mathscr{O} \in \mathscr{C}(i_1^*) \cup \cdots \cup \mathscr{C}(i_q^*) \tag{2.10}$$

$$A_{\mathscr{O}} < 1, \quad \mathscr{O} \in \left(\bigcup_{i=1,\ldots,N} \mathscr{C}(i) \right) \Big\backslash \left(\mathscr{C}(i_1^*) \cup \cdots \cup \mathscr{C}(i_q^*) \right), \tag{2.11}$$

where $\overline{N}_i$ is the size of $\overline{\mathscr{N}}_i$.

Proof We only consider the case of $q = 1$. The case of $q \geq 2$ can be proved similarly. Denote i^* as the root of the tree.

Define $a^0_{ij} = 1$ for $1 \leq i \leq N$, $j \in \overline{\mathscr{N}}_i$. If $a_{ij} = a^0_{ij}$ for $1 \leq i \leq N$, $j \in \overline{\mathscr{N}}_i$, then

$$\sum_{j \in \overline{\mathscr{N}}_i} \frac{1}{a^0_{ij}} \leq \overline{N}_i, \quad i = 1, \ldots, N \tag{2.12}$$

$$A_{\mathscr{O}} = 1, \quad \mathscr{O} \in \bigcup_{i=1,\ldots,N} \mathscr{C}(i). \tag{2.13}$$

Consider one of the paths leading from root i^* in the spanning tree $\mathscr{T}_f$. Denote the path as $(p_1, \ldots, p_m)$ with $p_1 = i^*$.

One can find $a^1_{p_2p_1} = a^0_{p_2p_1} + \varepsilon^0_{p_2p_1} > 0$ with $\varepsilon^0_{p_2p_1} > 0$ and $a^1_{p_2j} = a^0_{p_2j} - \varepsilon_{p_2j} > 0$ with $\varepsilon_{p_2j} > 0$ for $j \in \overline{\mathscr{N}}_{p_2} \backslash \{p_1\}$ such that if $a_{ij} = a^1_{ij}$ for $i = p_2$ and $a_{ij} = a^0_{ij}$ for $i \neq p_2$, then (2.12) is satisfied, and also

$$A_{\mathscr{O}} < 1 + \varepsilon' \text{ for } \mathscr{O} \in \mathscr{C}(p_1), \tag{2.14}$$

$$A_{\mathscr{O}} < 1 \text{ for } \mathscr{O} \in \mathscr{C}(p_2) \backslash \mathscr{C}(p_1) \tag{2.15}$$

with $0 < \varepsilon' < \varepsilon$.

Then, one can find $a^1_{p_3p_2} = a^0_{p_3p_2} + \varepsilon^0_{p_3p_2} > 0$ with $\varepsilon^0_{p_3p_2} > 0$ and $a^1_{p_3j} = a^0_{p_3j} - \varepsilon^0_{p_3j} > 0$ with $\varepsilon^0_{p_3j} > 0$ for $j \in \overline{\mathscr{N}}_{p_3} \backslash \{p_2\}$ such that if $a_{ij} = a^1_{ij}$ for $i \in \{p_2, p_3\}$, and $a_{ij} = a^0_{ij}$ for $i \notin \{p_2, p_3\}$, then (2.12) is satisfied, and also

$$A_{\mathscr{O}} < 1 + \varepsilon'' \text{ for } \mathscr{O} \in \mathscr{C}(p_1), \tag{2.16}$$

$$A_{\mathscr{O}} < 1 \text{ for } \mathscr{O} \in (\mathscr{C}(p_2) \cup \mathscr{C}(p_3)) \backslash \mathscr{C}(p_1) \tag{2.17}$$

with $0 < \varepsilon' \leq \varepsilon'' < \varepsilon$.

By doing this for $i = p_2, \ldots, p_m$, we can find $a^1_{ij} > 0$ for $i \in \{p_2, \ldots, p_m\}$, $j \in \overline{\mathscr{N}}_i$, such that

$$A_{\mathscr{O}} < 1 + \varepsilon_1 \text{ for } \mathscr{O} \in \mathscr{C}(p_1), \tag{2.18}$$

$$A_{\mathscr{O}} < 1 \text{ for } \mathscr{O} \in (\mathscr{C}(p_2) \cup \cdots \cup \mathscr{C}(p_m)) \backslash \mathscr{C}(p_1) \tag{2.19}$$

with $0 < \varepsilon_0 < \varepsilon$.

By considering each path leading from the root i^* in the spanning tree one by one, we can find $a^1_{ij} > 0$ for $i \in \{1, \ldots, N\}$, $j \in \overline{\mathscr{N}}_i$, such that if $a_{ij} = a^1_{ij}$ for $i \in \{1, \ldots, N\}$, $j \in \overline{\mathscr{N}}_i$, then (2.12) and (2.11) are satisfied and

$$A_{\mathscr{O}} < 1 + \varepsilon^1 \text{ for } \mathscr{O} \in \mathscr{C}(i^*_1) \cup \cdots \cup \mathscr{C}(i^*_q), \tag{2.20}$$

where $0 < \varepsilon^1 < \varepsilon$.

Note that the left-hand sides of inequalities (2.9), (2.10), and (2.11) continuously depend on a_{ij} for $i \in \{1, \ldots, N\}$, $j \in \overline{\mathscr{N}}_i$. One can find $a^2_{ij} > 0$ for $i \in \{1, \ldots, N\}$,

$j \in \overline{\mathscr{N}}_i$, such that if $a_{ij} = a_{ij}^2$ for $i \in \{1, \ldots, N\}$, $j \in \overline{\mathscr{N}}_i$, then conditions (2.9), (2.10), and (2.11) are satisfied. □

Example 2.2 Continue Example 2.1. Define $\mathscr{L} = \{i \in \{1, \ldots, N\} : 0 \in \mathscr{N}_i\}$. Considering the relation between N_i and $\overline{N}_i$, and $\overline{N}_i \leq N$, the cyclic-small-gain condition can be satisfied by the network of ISS subsystems with property (2.8) if

$$A_{\mathscr{O}} < \frac{(1-\bar{\delta})^N (N+1)}{N}, \quad \mathscr{O} \in \bigcup_{i \in \mathscr{L}} \mathscr{C}(i), \tag{2.21}$$

$$A_{\mathscr{O}} < (1-\bar{\delta})^N, \quad \mathscr{O} \in \left(\bigcup_{i \in \{1,\ldots,N\}} \mathscr{C}(i) \right) \setminus \left(\bigcup_{i \in \mathscr{L}} \mathscr{C}(i) \right), \tag{2.22}$$

where $\bar{\delta} = \max_{i=1,\ldots,N}\{\delta_i\}$.

Using Lemma 2.1, if graph $\mathscr{G}_f$ has a spanning tree with the agents belonging to $\mathscr{L}$ as the roots, one can find a constant $\bar{\delta} > 0$ and constants $a_{ij} > 0$ satisfying $\sum_{j \in \overline{\mathscr{N}}_i} 1/a_{ij} \leq \overline{N}_i$ such that conditions (2.21) and (2.22) are satisfied. The graph shown in Fig. 2.2 satisfies this condition.

Lemma 2.1 proves very useful in constructing distributed controllers for nonlinear agents to achieve convergence of their outputs to an agreement value. It provides for a form of gain assignment in the network coupling.

2.3 Distributed Output-Feedback Control

In this section, the basic idea of cyclic-small-gain design for distributed control is generalized to high-order nonlinear systems. Consider a group of N nonlinear agents, of which each agent i ($1 \leq i \leq N$) is in the output-feedback form:

$$\dot{x}_{ij} = x_{i(j+1)} + \Delta_{ij}(y_i, w_i), \quad 1 \leq j \leq n_i - 1 \tag{2.23}$$

$$\dot{x}_{in_i} = u_i + \Delta_{in_i}(y_i, w_i) \tag{2.24}$$

$$y_i = x_{i1}, \tag{2.25}$$

where $[x_{i1}, \ldots, x_{in_i}]^T := x_i \in \mathbb{R}^{n_i}$ with $x_{ij} \in \mathbb{R}$ ($1 \leq j \leq n_i$) is the state, $u_i \in \mathbb{R}$ is the control input, $y_i \in \mathbb{R}$ is the output, $[x_{i2}, \ldots, x_{in_i}]^T$ is the unmeasured portion of the state, $w_i \in \mathbb{R}^{n_{w_i}}$ represents external disturbances, and Δ_{ij}'s ($1 \leq j \leq n_i$) are unknown locally Lipschitz functions.

The objective of this section is to develop a new class of distributed controllers for the multiagent system based on available information such that the outputs y_i for $1 \leq i \leq N$ converge to the same desired agreement value y_0. This problem is called the output agreement problem in this chapter.

Different from decentralized control, the major objective of distributed control is to control the agents in a coordinated way for some desired group behavior. For the output agreement problem, the objective is to control the agents so that the outputs converge to a desired common value. Information exchange between the agents is required for coordination purposes. In practice, the information exchange is subject to constraints. As considered in Example 2.1, the position x_0 of the leader vehicle is only available to some of the follower vehicles, and the formation control objective is achieved through information exchange between the neighboring vehicles.

For distributed control of the multiagent nonlinear system (2.23)–(2.25), we employ a digraph $\mathscr{G}^c$ to represent the information exchange topology between the agents. Digraph $\mathscr{G}^c$ contains N vertices corresponding to the N agents and M directed edges corresponding to the information exchange links. Specifically, if $y_i - y_k$ is available to the local controller design of agent i, then there is a directed link from agent k to agent i and agent k is called a neighbor of agent i; otherwise, there is no link from agent k to agent i. Set $\mathscr{N}_i \subseteq \{1, \ldots, N\}$ is used to represent agent i's neighbors. In this section, an agent is not considered as a neighbor of itself and thus $i \notin \mathscr{N}_i$ for $1 \le i \le N$. Agent i is called an informed agent if it has access to the knowledge of the agreement value y_0 for its local controller design. Let $\mathscr{L} \subseteq \{1, \ldots, N\}$ represent the set of all the informed agents.

The following assumption is made on the agreement value and system (2.23)–(2.25).

Assumption 2.1 There exists a nonempty set $\Omega \subseteq \mathbb{R}$ such that

1. $y_0 \in \Omega$;
2. for each $1 \le i \le N$, $1 \le j \le n_i$,

$$|\Delta_{ij}(y_i, w_i) - \Delta_{ij}(z_i, 0)| \le \psi_{\Delta_{ij}}(|[y_i - z_i, w_i^T]^T|) \tag{2.26}$$

for all $[y_i, w_i^T]^T \in \mathbb{R}^{1+n_{w_i}}$ and all $z_i \in \Omega$, where $\psi_{\Delta_{ij}} \in \mathscr{K}_\infty$ is Lipschitz on compact sets and known.

It should be noted that a priori information on the bounds of y_0 (and thus Ω) is usually known in practice. In this case, condition 2 in Assumption 2.1 can be guaranteed if for each z_i, there exists a $\psi_{\Delta_{ij}}^{z_i} \in \mathscr{K}_\infty$ that is Lipschitz on compact sets such that

$$\begin{aligned} |\Delta_{ij}(y_i, w_i) - \Delta_{ij}(z_i, 0)| &= |\Delta_{ij}((y_i - z_i) + z_i, w_i) - \Delta_{ij}(z_i, 0)| \\ &\le \psi_{\Delta_{ij}}^{z_i}(|[y_i - z_i, w_i^T]^T|). \end{aligned} \tag{2.27}$$

Then, $\psi_{\Delta_{ij}}$ can be defined as $\psi_{\Delta_{ij}}(s) = \sup_{z_i \in \Omega} \psi_{\Delta_{ij}}^{z_i}(s)$ for $s \in \mathbb{R}_+$. In fact, there always exists a $\psi_{\Delta_{ij}}^{z_i} \in \mathscr{K}_\infty$ that is Lipschitz on compact sets to fulfill condition (2.27) if Δ_{ij} is locally Lipschitz.

It is also assumed that the external disturbances are bounded.

Assumption 2.2 For each $i = 1, \ldots, N$, there exists a $\bar{w}_i \geq 0$ such that

$$|w_i(t)| \leq \bar{w}_i \tag{2.28}$$

for all $t \geq 0$.

The basic idea is to design observer-based local controllers for the agents such that each controlled agent i is IOS, and moreover, has the UO property. Then, the cyclic-small-gain theorem in digraphs can be used to guarantee the IOS of the closed-loop multiagent system and then the achievement of output agreement.

By introducing a dynamic compensator

$$\dot{u}_i = v_i \tag{2.29}$$

and defining $x'_{i1} = y_i - y_0$ and $x'_{i(j+1)} = x_{i(j+1)} + \Delta_{ij}(y_0, 0)$ for $1 \leq j \leq n_i$, we can transform each agent i defined by (2.23)–(2.25) into the form of

$$\dot{x}'_{ij} = x'_{i(j+1)} + \Delta_{ij}(y_i, w_i) - \Delta_{ij}(y_0, 0), \quad 1 \leq j \leq n_i + 1 \tag{2.30}$$

$$\dot{x}'_{in_i} = v_i + \Delta_{in_i}(y_i, w_i) - \Delta_{in_i}(y_0, 0) \tag{2.31}$$

$$y'_i = x'_{i1} \tag{2.32}$$

with the output tracking error $y'_i = y_i - y_0$ as the new output and v_i as the new control input.

Moreover, the dynamic compensator (2.29) guarantees that the origin is an equilibrium of the transformed agent system (2.30)–(2.32) if it is disturbance-free, and the distributed control objective can be achieved if the equilibrium at the origin of each transformed agent system is stabilized.

The local controller for each agent i is designed by directly using the available y_i^m, defined as follows:

$$y_i^m = \frac{1}{N_i + 1}\left(\sum_{k \in \mathcal{N}_i}(y_i - y_k) + (y_i - y_0)\right), \quad i \in \mathscr{L} \tag{2.33}$$

$$y_i^m = \frac{1}{N_i}\sum_{k \in \mathcal{N}_i}(y_i - y_k), \quad i \in \{1, \ldots, N\}\backslash\mathscr{L}, \tag{2.34}$$

where N_i is the size of $\mathcal{N}_i$. For the convenience of discussions, we represent y_i^m with the new outputs as

$$y_i^m = y'_i - \mu_i \tag{2.35}$$

with

$$\mu_i = \frac{1}{N_i + 1} \sum_{k \in \mathcal{N}_i} y'_k, \quad i \in \mathcal{L} \tag{2.36}$$

$$\mu_i = \frac{1}{N_i} \sum_{k \in \mathcal{N}_i} y'_k, \quad i \in \{1, \ldots, N\} \backslash \mathcal{L}. \tag{2.37}$$

2.3.1 Distributed Output-Feedback Controller

Owing to the output-feedback structure, we design a local observer for each transformed agent system (2.30)–(2.32):

$$\dot{\xi}_{i1} = \xi_{i2} + L_{i2}\xi_{i1} + \rho_{i1}(\xi_{i1} - y_i^m) \tag{2.38}$$

$$\dot{\xi}_{ij} = \xi_{i(j+1)} + L_{i(j+1)}\xi_{i1} - L_{ij}(\xi_{i2} + L_{i2}\xi_{i1}), \quad 2 \leq j \leq n_i \tag{2.39}$$

$$\dot{\xi}_{i(n_i+1)} = v_i - L_{i(n_i+1)}(\xi_{i2} + L_{i2}\xi_{i1}), \tag{2.40}$$

where $\rho_{i1} : \mathbb{R} \to \mathbb{R}$ is an odd and strictly decreasing function, and $L_{i2}, \ldots, L_{in_i}$ are positive constants. In the observer, ξ_{i1} is an estimate of y'_i, and ξ_{ij} is an estimate of $x'_{ij} - L_{ij}y'_i$ for $2 \leq j \leq n_i + 1$.

Here, (2.38) is constructed to estimate y'_i using y_i^m which is influenced by the outputs y'_k ($k \in \mathcal{N}_i$) of the neighbor agents (see (2.35)). The nonlinear function ρ_{i1} in (2.38) is used to assign an appropriate *nonlinear* gain to the observation error system. As shown later, it is the key to make each controlled agent IOS with specific gains satisfying the cyclic-small-gain condition.

With the estimates, a nonlinear local control law is designed as

$$e_{i1} = \xi_{i1}, \tag{2.41}$$

$$e_{ij} = \xi_{ij} - \kappa_{i(j-1)}(e_{i(j-1)}), \quad 2 \leq j \leq n_i + 1 \tag{2.42}$$

$$v_i = \kappa_{i(n_i+1)}(e_{i(n_i+1)}), \tag{2.43}$$

where $\kappa_{i1}, \ldots, \kappa_{i(n_i+1)}$ are continuously differentiable, odd, strictly decreasing, and radially unbounded functions.

Consider $Z_i = [x'_{i1}, \ldots, x'_{i(n_i+1)}, \xi_{i1}, \ldots, \xi_{i(n_i+1)}]^T$ as the internal state of each controlled agent composed of the transformed agent system (2.30)–(2.32) and the local observer-based controller (2.38)–(2.43). The block diagram of controlled agent i with μ_i as the input and y'_i as the output is shown in Fig. 2.3.

The following proposition presents the UO and IOS properties of each controlled agent i.

Proposition 2.1 *Each controlled agent i composed of* (2.30)–(2.32) *and* (2.38)–(2.43) *has the following UO and IOS properties with μ_i as the input and y'_i as the*

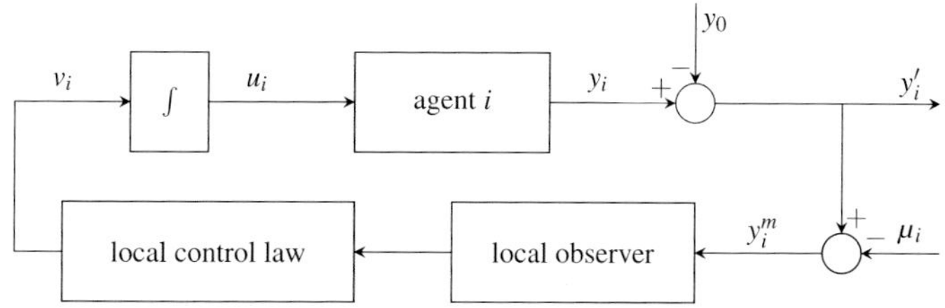

Fig. 2.3 The block diagram of each controlled agent i

output: for all $t \geq 0$,

$$|Z_i(t)| \leq \alpha_i^{UO}(|Z_{i0}| + \|\mu_i\|_{[0,t]}) \tag{2.44}$$

$$|y_i'(t)| \leq \max\left\{\beta_i(|Z_{i0}|, t), \chi_i(\|\mu_i\|_{[0,t]}), \gamma_i(\|w_i\|_{[0,t]})\right\}, \tag{2.45}$$

for any initial state $Z_i(0) = Z_{i0}$ *and any* μ_i, w_i, *where* $\beta_i \in \mathcal{KL}$ *and* $\chi_i, \gamma_i, \alpha_i \in \mathcal{K}_\infty$. *Moreover,* γ_i *can be designed to be arbitrarily small, and for any specified constant* $b_i > 1$, χ_i *can be designed such that* $\chi_i(s) \leq b_i s$, *for all* $s \geq 0$.

Due to space limitations, the proof of Proposition 2.1 is not provided here. The interested reader may consult [39] for reference.

2.3.2 Cyclic-Small-Gain Synthesis

With the proposed distributed output-feedback controller, the closed-loop multiagent system has been transformed into a network of IOS subsystems. This subsection presents the main result of output agreement and provides a proof based on the cyclic-small-gain result in digraphs.

Theorem 2.1 *Consider the multiagent system in the form of* (2.23)–(2.25) *satisfying Assumptions 2.1 and 2.2. If there is at least one informed agent, i.e.,* $\mathcal{L} \neq \emptyset$, *and the communication digraph* $\mathcal{G}^c$ *has a spanning tree with the informed agents as the roots, then we can design distributed observers* (2.38)–(2.40) *and distributed control laws* (2.29), (2.41)–(2.43) *such that all the signals in the closed-loop multiagent system are bounded, and the output* y_i *of each agent* i *can be steered to within an arbitrarily small neighborhood of the desired agreement value* y_0. *Moreover, if* $w_i = 0$ *for* $i = 1, \ldots, N$, *then each output* y_i *asymptotically converges to* y_0.

Proof Notice that for any constants $a_1, \ldots, a_n > 0$ satisfying $\sum_{i=1}^n (1/a_i) \leq n$, it holds that

$$\sum_{i=1}^n d_i = \sum_{i=1}^n \frac{1}{a_i} a_i d_i \leq n \max_{1 \leq i \leq n} \{a_i d_i\} \tag{2.46}$$

for all $d_1, \ldots, d_n \geq 0$.

Recall the definition of μ_i in (2.36) and (2.37). We have

$$|\mu_i| \leq \delta_i \max_{k \in \mathcal{N}_i} \{a_{ik} |y'_k|\}, \tag{2.47}$$

where $\delta_i = \frac{N_i}{N_i+1}$ if $i \in \mathcal{L}$, $\delta_i = 1$ if $i \notin \mathcal{L}$, and a_{ik} are positive constants satisfying

$$\sum_{k \in \mathcal{N}_i} \frac{1}{a_{ik}} \leq N_i. \tag{2.48}$$

Then, using the fact that the $\mathcal{N}_i$ in (2.47) are time invariant, property (2.45) implies

$$|y'_i(t)| \leq \max\left\{\beta_i(|Z_{i0}|, t), b_i \delta_i \max_{k \in \mathcal{N}_i} \{a_{ik} \|y'_k\|_{[0,t]}\}, \gamma_i(\|w_i\|_{[0,t]})\right\} \tag{2.49}$$

for any initial state Z_{i0} and any w_i, for all $t \geq 0$.

It can be observed that the interconnection topology of the controlled agents is in accordance with the information exchange topology, represented by digraph $\mathcal{G}^c$. For $i \in \mathcal{N}$, $k \in \mathcal{N}_i$, we assign the positive value a_{ik} to the edge (k, i) in $\mathcal{G}^c$. Denote $\mathcal{C}$ as the set of all simple cycles in $\mathcal{G}^c$ and $\mathcal{C}_{\mathcal{L}}$ as the set of all simple cycles through the vertices belonging to $\mathcal{L}$. Denote $A_{\mathcal{O}}$ as the product of the positive values assigned to the edges of the cycle $\mathcal{O} \in \mathcal{C}$.

Note that b_i can be designed to be arbitrarily close to one. By using the cyclic-small-gain theorem for networks of IOS systems, the closed-loop multiagent system is IOS if

$$A_{\mathcal{O}} \frac{N}{N+1} < 1, \quad \mathcal{O} \in \mathcal{C}_{\mathcal{L}} \tag{2.50}$$

$$A_{\mathcal{O}} < 1, \quad \mathcal{O} \in \mathcal{C} \backslash \mathcal{C}_{\mathcal{L}}. \tag{2.51}$$

If $\mathcal{G}^c$ has a spanning tree with vertices belonging to $\mathcal{L}$ as the roots, then according to Lemma 2.1, there exist positive constants a_{ik} satisfying (2.48), (2.50), and (2.51). Then, the closed-loop distributed system is UO and IOS with w_i as the inputs and y'_i as the outputs. With Assumption 2.2, the external disturbances w_i are bounded. The boundedness of the signals of the closed-loop distributed system can be directly verified under Assumption 2.2.

By designing the IOS gains γ_i arbitrarily small (this can be done according to Proposition 2.1), the influence of the external disturbances w_i is made arbitrarily small, and y'_i can be driven to within an arbitrarily small neighborhood of the origin. Equivalently, y_i can be driven to within an arbitrarily small neighborhood of y_0. In the case of $w_i = 0$ for $i = 1, \ldots, N$, each output y_i asymptotically converges to y_0. This ends the proof of Theorem 2.1. □

2.3.3 Robustness to Time Delays of Information Exchange

If there are communication delays, then y_i^m as defined in (2.33) and (2.34) should be modified as

$$y_i^m(t) = \frac{1}{N_i + 1}\left(\sum_{k\in\mathcal{N}_i}(y_i(t) - y_k(t - \tau_{ik}(t))) + (y_i(t) - y_0)\right), \quad i \in \mathcal{L} \tag{2.52}$$

$$y_i^m(t) = \frac{1}{N_i}\sum_{k\in\mathcal{N}_i}(y_i(t) - y_k(t - \tau_{ik}(t))), \quad i \in \{1, \ldots, N\}\backslash\mathcal{L}, \tag{2.53}$$

where $\tau_{ik} : \mathbb{R}_+ \to \mathbb{R}_+$ represents nonconstant time delays of exchanged information.

In this case, $y_i^m(t)$ can still be written in the form of $y_i^m(t) = y_i'(t) - \mu_i(t)$ with

$$\mu_i(t) = \frac{1}{N_i + 1}\sum_{k\in\mathcal{N}_i} y_k'(t - \tau_{ik}(t)), \quad i \in \mathcal{L} \tag{2.54}$$

$$\mu_i(t) = \frac{1}{N_i}\sum_{k\in\mathcal{N}_i} y_k'(t - \tau_{ik}(t)), \quad i \in \{1, \ldots, N\}\backslash\mathcal{L}. \tag{2.55}$$

We assume that there exists a $\bar{\tau} \geq 0$ such that, for $i = 1, \ldots, N$, $k \in \mathcal{N}_i$, $0 \leq \tau_{ik}(t) \leq \bar{\tau}$ holds for all $t \geq 0$. By considering μ_i and w_i as the external inputs, each controlled agent i composed of (2.30)–(2.32) and (2.38)–(2.43) is still UO and property (2.49) should be modified as

$$|y_i'(t)| \leq \max\left\{\beta_i(|Z_{i0}|, t), b_i\delta_i \max_{k\in\mathcal{N}_i}\{a_{ik}\|y_k'\|_{[-\bar{\tau},\infty)}\}, \gamma_i(\|w_i\|_{[0,\infty)})\right\} \tag{2.56}$$

for any initial state Z_{i0} and any w_i, for all $t \geq 0$.

Using the time-delay version of the cyclic-small-gain theorem, we can still guarantee the IOS of the closed-loop multiagent system with y_i' as the outputs and w_i as the inputs, following analysis similar to that for the proof of Theorem 2.1.

2.4 Formation Control of Nonholonomic Mobile Robots

Formation control of autonomous mobile agents is aimed at forcing agents to converge toward, and maintain, specific relative positions. Distributed formation control of multiagent systems based on available local information, e.g., relative position measurements, has attracted tremendous attention from the robotics and control communities.

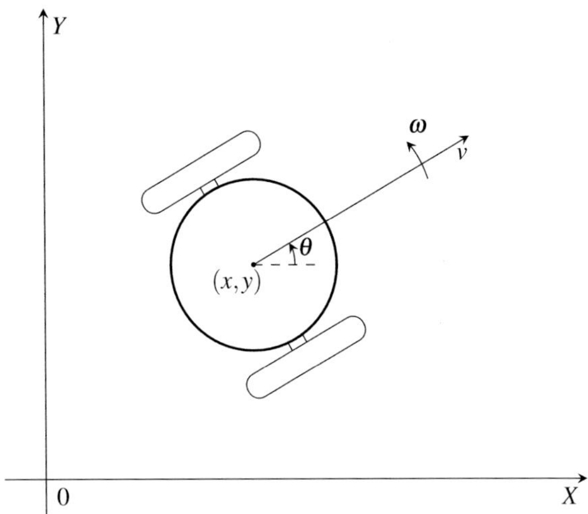

Fig. 2.4 Kinematics of the unicycle robot, where (x, y) represents the Cartesian coordinates of the center of mass of the robot, v is the linear velocity, θ is the heading angle, and ω is the angular velocity

Motivated by the cyclic-small-gain design for distributed output-feedback control of nonlinear systems in Sect. 2.3, this section proposes a class of distributed controllers for leader-following formation control of unicycle robots using the practically available relative position measurements. The kinematics of the unicycle robot are demonstrated by Fig. 2.4.

For this purpose, the formation control problem is first transformed into a state agreement problem of double integrators through dynamic feedback linearization. The nonholonomic constraint causes a singularity for the dynamic feedback linearization when the linear velocity of the robot is zero. This issue should be well taken into consideration for the validity of the transformed double integrator models. Then, distributed formation control laws are developed. To avoid the singularity problem caused by the nonholonomic constraint, saturation functions are introduced to the control design to restrict the linear velocities of the robots to be larger than zero. It should be noted that linear analysis methods may not be directly applicable due to the employment of the saturation functions. Then, the closed-loop system is transformed into a dynamic network of IOS systems. The cyclic-small-gain result in digraphs is used to guarantee the IOS of the dynamic network and thus the achievement of formation control.

With the effort mentioned above, the proposed design has three advantages:

1. The proposed distributed formation control law does not use global position measurements or assumes tree position sensing structures.
2. The formation control objective can be practically achieved in the presence of position measurement errors.
3. The linear velocities of the robots can be designed to be less than certain desired values, as practically required.

This section considers the formation control problem of a group of $N+1$ mobile robots. For $i = 0, 1, \ldots, N$, the kinematics of the ith robot are described by the unicycle model:

$$\dot{x}_i = v_i \cos\theta_i \tag{2.57}$$

$$\dot{y}_i = v_i \sin\theta_i \tag{2.58}$$

$$\dot{\theta}_i = \omega_i, \tag{2.59}$$

where $[x_i, y_i]^T \in \mathbb{R}^2$ represent the Cartesian coordinates of the center of mass of the ith robot, $v_i \in \mathbb{R}$ is the linear velocity, $\theta_i \in \mathbb{R}$ is the heading angle, and $\omega_i \in \mathbb{R}$ is the angular velocity.

The robot with index 0 is the leader robot, and the robots with indices $1, \ldots, N$ are follower robots. The linear velocity v_i and the angular velocity ω_i are considered as the control inputs of the ith robot for $i = 1, \ldots, N$. For this system, the formation control objective is to control each ith follower robot such that

$$\lim_{t\to\infty} (x_i(t) - x_j(t)) = d_{xij} \tag{2.60}$$

$$\lim_{t\to\infty} (y_i(t) - y_j(t)) = d_{yij} \tag{2.61}$$

with d_{xij}, d_{yij} being appropriate constants representing the desired relative positions, and

$$\lim_{t\to\infty} \big((\theta_i(t) - \theta_j(t)) \bmod 2\pi\big) = 0 \tag{2.62}$$

for any $i, j = 0, \ldots, N$, where mod represents the modulo operation. For convenience of notation, let $d_{xii} = d_{yii} = 0$ for any $i = 0, \ldots, N$. We assume that $d_{xij} = d_{xik} - d_{xkj}$ and $d_{yij} = d_{yik} - d_{ykj}$ for any $i, j, k = 0, \ldots, N$.

Assumption 2.3 on v_0 is made throughout this section.

Assumption 2.3 The linear velocity v_0 of the leader robot is differentiable with bounded derivative, i.e., $\dot{v}_0(t)$ exists and is bounded on $[0, \infty)$, and has upper and lower constant bounds $\overline{v}_0, \underline{v}_0 > 0$ such that $\underline{v}_0 \le v_0(t) \le \overline{v}_0$, for all $t \ge 0$.

One technical problem of controlling groups of mobile robots is that accurate global positions of the robots are usually not available for feedback, and relative position measurements should be used instead. A digraph can be employed to represent the relative position sensing structure between the robots. The position sensing digraph $\mathcal{G}$ has $N+1$ vertices with indices $0, 1, \ldots, N$ corresponding to the robots. If the relative position between robot i and robot j is available to robot j, then $\mathcal{G}$ has a directed edge from vertex i to vertex j; otherwise $\mathcal{G}$ does not have such an edge.

The goal of this section is to present a class of distributed formation controllers for mobile robots using local relative position measurements as well as the velocity and acceleration information of the leader. The basic idea of the design is to first transform the unicycle model into two double integrators through dynamic feedback linearization under constraints, and at the same time, to reformulate the formation control problem as a stabilization problem. Then, distributed control laws are designed to make each controlled mobile robot IOS. Finally, the cyclic-small-gain theorem is used to guarantee the achievement of the formation control objective.

2.4.1 Dynamic Feedback Linearization

In this subsection, the distributed formation control problem is reformulated with the dynamic feedback linearization technique. For details of dynamic feedback linearization and its applications to nonholonomic systems, please consult [6, 14].

For each $i = 0, \ldots, N$, introduce a new input $r_i \in \mathbb{R}$ such that

$$\dot{v}_i = r_i. \tag{2.63}$$

Define $v_{xi} = v_i \cos\theta_i$ and $v_{yi} = v_i \sin\theta_i$. Then, $\dot{x}_i = v_{xi}$ and $\dot{y}_i = v_{yi}$. Take the derivatives of v_{xi} and v_{yi}, respectively. Then,

$$\begin{pmatrix} \dot{v}_{xi} \\ \dot{v}_{yi} \end{pmatrix} = \begin{pmatrix} \cos\theta_i & -v_i \sin\theta_i \\ \sin\theta_i & v_i \cos\theta_i \end{pmatrix} \begin{pmatrix} r_i \\ \omega_i \end{pmatrix}. \tag{2.64}$$

In the case of $v_i \neq 0$, by designing

$$\begin{pmatrix} r_i \\ \omega_i \end{pmatrix} = \begin{pmatrix} \cos\theta_i & \sin\theta_i \\ -\frac{\sin\theta_i}{v_i} & \frac{\cos\theta_i}{v_i} \end{pmatrix} \begin{pmatrix} u_{xi} \\ u_{yi} \end{pmatrix}, \tag{2.65}$$

the unicycle model (2.57)–(2.59) can be transformed into two double integrators with new inputs u_{xi} and u_{yi}:

$$\dot{x}_i = v_{xi}, \quad \dot{v}_{xi} = u_{xi}, \tag{2.66}$$

$$\dot{y}_i = v_{yi}, \quad \dot{v}_{yi} = u_{yi}. \tag{2.67}$$

Define $\tilde{x}_i = x_i - x_0 - d_{xi}$, $\tilde{y}_i = y_i - y_0 - d_{yi}$, $\tilde{v}_{xi} = v_{xi} - v_{x0}$, $\tilde{v}_{yi} = v_{yi} - v_{y0}$, $\tilde{u}_{xi} = u_{xi} - u_{x0}$, and $\tilde{u}_{yi} = u_{yi} - u_{y0}$. Then,

$$\dot{\tilde{x}}_i = \tilde{v}_{xi}, \quad \dot{\tilde{v}}_{xi} = \tilde{u}_{xi}, \tag{2.68}$$

$$\dot{\tilde{y}}_i = \tilde{v}_{yi}, \quad \dot{\tilde{v}}_{yi} = \tilde{u}_{yi}. \tag{2.69}$$

The formation control problem is solvable, if we can design control laws for system (2.68)–(2.69) with $\tilde{u}_{xi}$ and $\tilde{u}_{yi}$ as the control inputs, so that $v_i \neq 0$ is guaranteed, and at the same time,

$$\lim_{t\to\infty} \tilde{x}_i(t) = 0, \tag{2.70}$$

$$\lim_{t\to\infty} \tilde{y}_i(t) = 0. \tag{2.71}$$

It should be noted that the validity of (2.66)–(2.67) (and thus (2.68)–(2.69)) for the unicycle model is under the condition that $v_i \neq 0$. Such requirement is basically caused by the nonholonomic constraint of the mobile robot. This leads to the major difference between this problem and the distributed control problem for double integrators.

To use (2.68)–(2.69) for control design, each follower robot should have access to u_{x0}, u_{y0}, which represent the acceleration of the leader robot. This requirement can be fulfilled if the leader robot can calculate u_{x0}, u_{y0} using $r_0, \omega_0, \theta_0, v_0$ according to (2.65) and transmit them to the follower robots. Note that ω_0, θ_0, v_0 are usually measurable, and r_0 is normally available as it is the control input of the leader robot.

2.4.2 A Class of IOS Control Laws

As an ingredient for the distributed control design, this subsection presents a class of nonlinear control laws for the following double integrator system with an external input, such that the closed-loop system is UO and IOS:

$$\dot{\eta} = \zeta \tag{2.72}$$

$$\dot{\zeta} = \mu \tag{2.73}$$

$$\hat{\eta} = \eta + w, \tag{2.74}$$

where $[\eta, \zeta]^T \in \mathbb{R}^2$ is the state, $\mu \in \mathbb{R}$ is the control input, $w \in \mathbb{R}$ represents an external input, $\hat{\eta}$ can be considered as a measurement of η subject to w, and only $(\hat{\eta}, \zeta)$ are used for feedback. As shown later, each controlled robot can be transformed into the form of (2.72)–(2.74) with w representing the interaction between the robots.

Lemma 2.2 *For system* (2.72)–(2.74), *consider a control law taking the form:*

$$\mu = -k_\mu(\zeta - \phi(\hat{\eta})). \tag{2.75}$$

For any constant $\overline{\phi} > 0$, one can find an odd, strictly decreasing, continuously differentiable function $\phi : \mathbb{R} \to [-\overline{\phi}, \overline{\phi}]$ and a positive constant k_μ satisfying

$$-\frac{k_\mu}{4} < \frac{d\phi(r)}{dr} < 0 \tag{2.76}$$

for all $r \in \mathbb{R}$, such that the closed-loop system (2.72)–(2.75) *is UO with zero offset, and is IOS with the identity function as the gain, i.e., the following properties hold:*

$$|\eta(t)| \leq \overline{\beta}(|[\eta(0), \zeta(0)]^T|, t) + \|w\|_t \tag{2.77}$$
$$|\zeta(t)| \leq |\zeta(0)| + \alpha_{UO}(\|\eta\|_t + \|w\|_t) \tag{2.78}$$

for some $\overline{\beta} \in \mathscr{KL}$, $\alpha_{UO} \in \mathscr{K}_\infty$, *and all* $t \geq 0$.

It is necessary to note that condition (2.76) is easily checkable for practical implementation of the control law (2.75).

2.4.3 Distributed Formation Controller Design and Small-Gain Analysis

As discussed in Sect. 2.4.1, for the validity of (2.68)–(2.69) of the formation control design, v_i should be guaranteed to be nonzero. For a specified λ_* satisfying $0 < \lambda_* < \underline{v}_0$, by designing a control law for the ith robot such that

$$\max\left\{|\tilde{v}_{xi}|, |\tilde{v}_{yi}|\right\} \leq \frac{\sqrt{2}}{2}(\underline{v}_0 - \lambda_*) \leq \frac{\sqrt{2}}{2}(v_0 - \lambda_*), \tag{2.79}$$

it can be guaranteed that $|v_i| = \sqrt{v_{xi}^2 + v_{yi}^2} = \sqrt{(v_{x0} + \tilde{v}_{xi})^2 + (v_{y0} + \tilde{v}_{yi})^2} \geq \lambda_* > 0$ and thus $v_i \neq 0$. In this way, singularity is avoided.

Practically, the linear velocity of each robot is usually required to be less than a desired value. For any given $\lambda^* > \overline{v}_0$, we can also guarantee $|v_i| \leq \lambda^*$ by designing a control law such that

$$\max\left\{|\tilde{v}_{xi}|, |\tilde{v}_{yi}|\right\} \leq \frac{\sqrt{2}}{2}(\lambda^* - \overline{v}_0). \tag{2.80}$$

For specified constants $\lambda_*, \lambda^*, \underline{v}_0, \overline{v}_0$ satisfying $0 < \lambda_* < \underline{v}_0 < \overline{v}_0 < \lambda^*$, we define

$$\lambda = \min\left\{\frac{\sqrt{2}}{2}(\underline{v}_0 - \lambda_*), \frac{\sqrt{2}}{2}(\lambda^* - \overline{v}_0)\right\}. \tag{2.81}$$

Then, conditions (2.79) and (2.80) can be satisfied if

$$\max\left\{|\tilde{v}_{xi}|, |\tilde{v}_{yi}|\right\} \leq \lambda. \tag{2.82}$$

The proposed distributed control law is composed of two stages: (a) initialization and (b) formation control. The initialization stage is employed because the formation control stage cannot solely guarantee $v_i \neq 0$ if (2.79) is not satisfied at the beginning of the control procedure. With the initialization stage, the linear velocity and the heading direction of each follower robot can be controlled to satisfy (2.82) within some finite time. Then, the formation control stage is triggered, and thereafter, the satisfaction of (2.82) is guaranteed, and at the same time, the formation control objective is achieved.

Initialization Stage

For this stage, we design the following control law:

$$\omega_i = \phi_{\theta i}(\theta_i - \theta_0) + \omega_0 \tag{2.83}$$

$$r_i = \phi_{vi}(v_i - v_0) + \dot{v}_0 \tag{2.84}$$

for each ith follower robot, where $\phi_{\theta i}, \phi_{vi} : \mathbb{R} \to \mathbb{R}$ are nonlinear functions.

Define $\tilde{\theta}_i = \theta_i - \theta_0$ and $\tilde{v}_i = v_i - v_0$. Taking the derivatives of $\tilde{\theta}_i$ and $\tilde{v}_i$, respectively, and using (2.83) and (2.84), we have

$$\dot{\tilde{\theta}}_i = \phi_{\theta i}(\tilde{\theta}_i), \tag{2.85}$$

$$\dot{\tilde{v}}_i = \phi_{vi}(\tilde{v}_i). \tag{2.86}$$

By designing $\phi_{\theta i}, \phi_{vi}$ such that $-\phi_{\theta i}(s), \phi_{\theta i}(-s), -\phi_{vi}(s), \phi_{vi}(-s)$ are positive definite for $s \in \mathbb{R}_+$, we can guarantee the asymptotic stability of systems (2.85) and (2.86). Moreover, there exist $\beta_{\tilde{\theta}}, \beta_{\tilde{v}} \in \mathcal{KL}$ such that $|\tilde{\theta}(t)| \le \beta_{\tilde{\theta}}(|\tilde{\theta}(0)|, t)$ and $|\tilde{v}(t)| \le \beta_{\tilde{v}}(|\tilde{v}(0)|, t)$.

By directly using the property of continuous functions, there exist $\overline{\delta}_{v0} > 0$ and $\overline{\delta}_{\theta 0} > 0$ such that, for all $v_0 \in [\underline{v}_0, \overline{v}_0]$, $\theta_0 \in \mathbb{R}$, $|\delta v_0| \le \overline{\delta}_{v0}$ and $|\delta_{\theta 0}| \le \overline{\delta}_{\theta 0}$,

$$|(v_0 + \delta_{v0})\cos(\theta_0 + \delta_{\theta 0}) - v_0 \cos\theta_0| \le \lambda, \tag{2.87}$$

$$|(v_0 + \delta_{v0})\sin(\theta_0 + \delta_{\theta 0}) - v_0 \sin\theta_0| \le \lambda. \tag{2.88}$$

Recall that for any $\beta \in \mathcal{KL}$, there exist functions $\alpha_1, \alpha_2 \in \mathcal{K}_\infty$ such that $\beta(s,t) \le \alpha_1(s)\alpha_2(e^{-t})$, for all $s, t \in \mathbb{R}_+$ according to [52, Lemma 8]. With control law (2.83)–(2.84), there exists a finite time T_{Oi} for the ith robot such that $|\theta_i(T_{Oi}) - \theta_0(T_{Oi})| \le \overline{\delta}_{\theta 0}$ and $|v_i(T_{Oi}) - v_0(T_{Oi})| \le \overline{\delta}_{v0}$, and thus condition (2.82) is satisfied at time T_{Oi}.

It should be noted that if $v_i(0) \le \lambda^*$, then control law (2.84) guarantees $v_i(t) \le \lambda^*$ for $t \in [0, T_{Oi}]$ because of $v_0(t) \le \overline{v}_0 < \lambda^*$.

Formation Control Stage

At time T_{Oi}, the distributed control law for the ith follower robot is switched to the formation control stage.

In this stage, we design

$$\tilde{u}_{xi} = -k_{xi}(\tilde{v}_{xi} - \phi_{xi}(z_{xi})) \tag{2.89}$$

$$\tilde{u}_{yi} = -k_{yi}(\tilde{v}_{yi} - \phi_{yi}(z_{yi})), \tag{2.90}$$

where $\phi_{xi}, \phi_{yi} : \mathbb{R} \to [-\lambda, \lambda]$ are odd, strictly decreasing, and continuously differentiable functions and k_{xi}, k_{yi} are positive constants satisfying

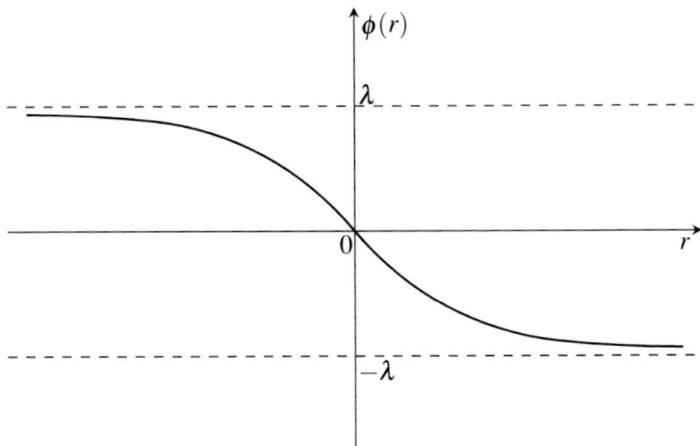

Fig. 2.5 An example for ϕ_{xi} and ϕ_{yi}

$$-k_{xi}/4 < d\phi_{xi}(r)/dr < 0 \tag{2.91}$$

$$-k_{yi}/4 < d\phi_{yi}(r)/dr < 0 \tag{2.92}$$

for all $r \in \mathbb{R}$. An example for ϕ_{xi} and ϕ_{yi} is shown in Fig. 2.5.

The variables z_{xi} and z_{yi} are defined as

$$z_{xi} = \frac{1}{N_i} \sum_{j \in \mathscr{N}_i} (x_i - x_j - (d_{xi} - d_{xj})) \tag{2.93}$$

$$z_{yi} = \frac{1}{N_i} \sum_{j \in \mathscr{N}_i} (y_i - y_j - (d_{yi} - d_{yj})), \tag{2.94}$$

where N_i is the size of $\mathscr{N}_i$ with $\mathscr{N}_i$ representing the position sensing structure. If $j \in \mathscr{N}_i$, then the position sensing digraph $\mathscr{G}$ has a directed edge (j, i) from vertex j to vertex i. Note that $d_{xi} - d_{xj}, d_{yi} - d_{yj}$ in (2.93) and (2.94) represent the desired relative position between the ith robot and the jth robot. By default, $d_{x0} = d_{y0} = 0$.

In the formation control stage, the control inputs r_i and ω_i are defined as (2.65) with $u_{xi} = \tilde{u}_{xi} + u_{x0}$ and $u_{yi} = \tilde{u}_{yi} + u_{y0}$.

Consider the $(\tilde{v}_{xi}, \tilde{v}_{yi})$-system defined in (2.68) and (2.69). With condition (2.82) satisfied at time T_{Oi}, the boundedness of ϕ_{xi} and ϕ_{yi} together with the control law (2.89) and (2.90) guarantees the satisfaction of (2.82) after T_{Oi}. For the proof of this statement, we can consider $\{(\tilde{v}_{xi}, \tilde{v}_{yi}) : \max\{|\tilde{v}_{xi}|, |\tilde{v}_{yi}|\} \le \lambda\}$ as an invariant set of the $(\tilde{v}_{xi}, \tilde{v}_{yi})$-system.

The main result of distributed formation control is summarized below.

Theorem 2.2 *Consider the multirobot model* (2.57)–(2.59) *and the distributed control laws defined by* (2.63), (2.65), (2.83), (2.84), (2.89), *and* (2.90) *with parameters k_{xi}, k_{yi} satisfying* (2.91) *and* (2.92). *Under Assumption 2.3, if the position sensing digraph $\mathscr{G}$ has a spanning tree with vertex* 0 *as the root, then for any constants $d_{xi}, d_{yi} \in \mathbb{R}$ with $i = 1, \ldots, N$, the coordinates $(x_i(t), y_i(t))$ and the angle $\theta_i(t)$ of each ith robot asymptotically converge to $(x_0(t) + d_{xi}, y_0(t) + d_{yi})$ and*

$\theta_0(t) + 2k\pi$ with $k \in \mathbb{Z}$, respectively. Moreover, given any $\lambda^* > \bar{v}_0$, if $v_i(0) \le \lambda^*$ for $i = 1, \ldots, N$, then $v_i(t) \le \lambda^*$, for all $t \ge 0$.

The two-stage distributed control law results in a switching incident of the control signal for each follower robot during the control procedure. The trajectories of such systems can be well defined in the spirit of Rademacher (see, e.g., [13]), and the performance of the system can be analyzed by considering the two stages one by one.

2.4.4 Small-Gain Analysis and Proof of Theorem 2.2

Recall the definition of λ in (2.81). With condition (2.82) satisfied after T_{Oi}, we have $v_i \neq 0$ and thus the validity of (2.65), for all $t \ge T_{Oi}$. Under the condition of $v_i(0) \le \lambda^*$, the boundedness of $v_i(t)$, i.e., $v_i(t) \le \lambda^*$, can also be directly proved based on the discussions in Sect. 2.4.3.

Denote $\tilde{x}_0 = 0$ and $\tilde{y}_0 = 0$. We equivalently represent z_{xi} and z_{yi} as

$$\begin{aligned} z_{xi} &= \frac{1}{N_i} \sum_{j \in \mathcal{N}_i} (x_i - d_{xi} - x_0 - (x_j - d_{xj} - x_0)) \\ &= \frac{1}{N_i} \sum_{j \in \mathcal{N}_i} (\tilde{x}_i - \tilde{x}_j) = \tilde{x}_i - \frac{1}{N_i} \sum_{j \in \mathcal{N}_i} \tilde{x}_j \end{aligned} \tag{2.95}$$

and similarly,

$$z_{yi} = \tilde{y}_i - \frac{1}{N_i} \sum_{j \in \mathcal{N}_i} \tilde{y}_j. \tag{2.96}$$

Denote

$$\omega_{xi} = \frac{1}{N_i} \sum_{j \in \mathcal{N}_i} \tilde{x}_j, \tag{2.97}$$

$$\omega_{yi} = \frac{1}{N_i} \sum_{j \in \mathcal{N}_i} \tilde{y}_j. \tag{2.98}$$

Then, control laws (2.89) and (2.90) are in the form of (2.75).

In the following proof, we only consider the $(\tilde{x}_i, \tilde{v}_{xi})$-system (2.68). The $(\tilde{y}_i, \tilde{v}_{yi})$-system (2.69) can be studied in the same way.

Define $T_O = \max_{i=1,\ldots,N}\{T_{Oi}\}$. Using Lemma 2.2, for each $i = 1, \ldots, N$, the closed-loop system composed of (2.68) and (2.89) has the following properties: for any $\tilde{x}_{i0}, \tilde{v}_{xi0} \in \mathbb{R}$, with $\tilde{x}_i(T_O) = \tilde{x}_{i0}$ and $\tilde{v}_{xi}(T_O) = \tilde{v}_{xi0}$,

$$|\tilde{x}_i(t)| \leq \beta_{xi}(|[\tilde{x}_{i0}, \tilde{v}_{xi0}]^T|, t - T_O) + \|\omega_{xi}\|_{[T_O,t]} \tag{2.99}$$

$$|\tilde{v}_{xi}(t)| \leq |\tilde{v}_{xi0}| + \alpha_{xi}(\|\tilde{x}_i\|_{[T_O,t]} + \|\omega_{xi}\|_{[T_O,t]}), \tag{2.100}$$

where $\beta_{xi} \in \mathcal{KL}$ and $\alpha_{xi} \in \mathcal{K}_\infty$.

Notice that for any constants $a_1, \ldots, a_n > 0$ satisfying $\sum_{i=1}^{n}(1/a_i) \leq n$, it holds that $\sum_{i=1}^{n} d_i = \sum_{i=1}^{n}(1/a_i)a_i d_i \leq n \max_{1\leq i\leq n}\{a_i d_i\}$, for all $d_1, \ldots, d_n \geq 0$. We have

$$|\omega_{xi}| \leq \delta_i \max_{j\in\overline{\mathcal{N}}_i}\{a_{ij}|\tilde{x}_j|\}, \tag{2.101}$$

where $\delta_i = (N_i - 1)/N_i$, $\overline{\mathcal{N}}_i = \mathcal{N}_i \backslash \{0\}$, and $\sum_{j\in\overline{\mathcal{N}}_i}(1/a_{ij}) \leq N_i - 1$ if $0 \in \mathcal{N}_i$; $\delta_i = 1$, $\overline{\mathcal{N}}_i = \mathcal{N}_i$, and $\sum_{j\in\overline{\mathcal{N}}_i}(1/a_{ij}) \leq N_i$ if $0 \notin \mathcal{N}_i$.

Then, properties (2.99) and (2.100) imply

$$|\tilde{x}_i(t)| \leq \beta_{xi}(|[\tilde{x}_{i0}, \tilde{v}_{xi0}]^T|, t - T_O) + \delta_i \max_{j\in\overline{\mathcal{N}}_i}\{a_{ij}\|\tilde{x}_j\|_{[T_O,t]}\}, \tag{2.102}$$

$$|\tilde{v}_{xi}(t)| \leq |\tilde{v}_{xi0}| + \alpha_{xi}(\|\tilde{x}_i\|_{[T_O,t]} + \delta_i \max_{j\in\overline{\mathcal{N}}_i}\{a_{ij}\|\tilde{x}_j\|_{[T_O,t]}\}). \tag{2.103}$$

Define the follower sensing digraph $\mathcal{G}_f$ as a subgraph of $\mathcal{G}$. Digraph $\mathcal{G}_f$ has N vertices with indices $1, \ldots, N$ corresponding to the vertices with indices $1, \ldots, N$ of $\mathcal{G}$ and representing the follower robots. From the definitions of $\overline{\mathcal{N}}_i$ and $\mathcal{G}_f$, it can be observed that, for $i = 1, \ldots, N$, if $j \in \overline{\mathcal{N}}_i$, then there is a directed edge (j, i) from the jth vertex to the ith vertex in $\mathcal{G}_f$. Clearly, $\mathcal{G}_f$ represents the interconnection topology of the network composed of the $(\tilde{x}_i, \tilde{v}_{xi})$-systems (2.68).

Define $\mathcal{F}_0 = \{i \in \{1, \ldots, N\} : 0 \in \mathcal{N}_i\}$. Denote $\mathcal{C}_f$ as the set of all simple cycles of $\mathcal{G}_f$, and denote $\mathcal{C}_0 \subseteq \mathcal{C}_f$ as the set of all simple cycles through the vertices with indices belonging to $\mathcal{F}_0$.

For $i = 1, \ldots, N$, $j \in \overline{\mathcal{N}}_i$, we assign the positive value a_{ij} to edge (j, i) in $\mathcal{G}_f$. For a simple cycle $\mathcal{O} \in \mathcal{C}_f$, denote $A_\mathcal{O}$ as the product of the positive values assigned to the edges of the cycle.

Consider $\tilde{x}_i$ with $i = 1, \ldots, N$ as the outputs of the network composed of the $(\tilde{x}_i, \tilde{v}_{xi})$-systems (2.68). Using the IOS small-gain theorem for general nonlinear systems in [27, 28], $\tilde{x}_i(t)$ with $i = 1, \ldots, N$ converge to the origin if

$$A_\mathcal{O}\frac{N-1}{N} < 1 \quad \text{for } \mathcal{O} \in \mathcal{C}_0, \tag{2.104}$$

$$A_\mathcal{O} < 1 \quad \text{for } \mathcal{O} \in \mathcal{C}_f \backslash \mathcal{C}_0. \tag{2.105}$$

Note that $A_\mathcal{O}(N-1)/N < 1$ is equivalent to $A_\mathcal{O} < N/(N-1) = 1 + 1/(N-1)$.

If $\mathcal{G}$ has a spanning tree with vertex 0 as the root, then $\mathcal{G}_f$ has a spanning tree with the indices of the root vertices belonging to $\mathcal{F}_0$. According to Lemma 2.1, there exist positive constants a_{ij} such that both conditions (2.104) and (2.105) are satisfied. For system (2.68), with the convergence of each $\tilde{x}_i$ to the origin and the boundedness of $\tilde{u}_{xi}$, we can guarantee the convergence of $\tilde{v}_{xi}$ to the origin using Barbălat's lemma [29]. Similarly, we can prove the convergence of $\tilde{v}_{yi}$ to the origin. Using the definitions of $\tilde{v}_{xi}$ and $\tilde{v}_{yi}$, the convergence of θ_i to $\theta_0 + 2k\pi$ with $k \in \mathbb{Z}$ can be concluded. This ends the proof of Theorem 2.2.

2.4.5 Robustness to Relative Position Measurement Errors

Measurement errors can decrease the performance of a nonlinear control system. In this section, we discuss the robustness of our distributed formation controller in the presence of relative position measurement errors.

It can be observed that the initialization stage of the distributed control law defined in (2.83) and (2.84) is not affected by the position measurement errors. Also, condition (2.82) still holds for $t \geq T_{Oi}$ for $i = 1, \dots, N$.

For the formation control stage, in the presence of relative position measurement errors, the z_{xi} and z_{yi} defined for the distributed control law (2.89) and (2.90) should be modified as

$$z_{xi} = \frac{1}{N_i} \sum_{j \in \mathcal{N}_i} \left(x_i - x_j - (d_{xi} - d_{xj}) + \omega_{ij}^x \right) \tag{2.106}$$

$$z_{yi} = \frac{1}{N_i} \sum_{j \in \mathcal{N}_i} \left(y_i - y_j - (d_{yi} - d_{yj}) + \omega_{ij}^y \right), \tag{2.107}$$

where N_i is the size of $\mathcal{N}_i$ and $\omega_{ij}^x, \omega_{ij}^y \in \mathbb{R}$ represent the relative position measurement errors corresponding to $(x_i - x_j)$ and $(y_i - y_j)$, respectively. Due to the boundedness of the designed ϕ_{xi} and ϕ_{yi} in (2.89) and (2.90), condition (2.82) is still satisfied in the presence of position measurement errors, which guarantees the validity of (2.65).

Here, we only consider each $\tilde{x}_i$-subsystem. The $\tilde{y}_i$-subsystems can be studied in the same way. By defining

$$\omega_{xi} = \frac{1}{N_i} \sum_{j \in \mathcal{N}_i} \left(\tilde{x}_j + \omega_{ij}^x \right), \tag{2.108}$$

we have $z_{xi} = \tilde{x}_i - \omega_{xi}$. With such definition, if the measurement errors ω_{ij}^x are piecewise continuous and bounded, then each $\tilde{x}_i$-subsystem still has the IOS and UO properties given by (2.99) and (2.100), respectively.

As in the discussion above (2.101), we have

$$
\begin{aligned}
|\omega_{xi}| &\leq \max\left\{\frac{\rho_i}{N_i}\sum_{j\in\mathcal{N}_i}(|\tilde{x}_j|), \frac{\rho_i'}{N_i}\sum_{j\in\mathcal{N}_i}(|\omega_{ij}^x|)\right\} \\
&:= \max\left\{\frac{\rho_i}{N_i}\sum_{j\in\mathcal{N}_i}(|\tilde{x}_j|), \omega_{xi}^e\right\} \\
&\leq \max_{j\in\overline{\mathcal{N}}_i}\left\{\delta_i a_{ij}|\tilde{x}_j|, \omega_{xi}^e\right\},
\end{aligned} \tag{2.109}
$$

where $\rho_i, \rho_i' > 0$ satisfying $1/\rho_i + 1/\rho_i' \leq 1$, and $\delta_i = \rho_i(N_i - 1)/N_i$, $\overline{\mathcal{N}}_i = \mathcal{N}_i \backslash \{0\}$ and $\sum_{j\in\overline{\mathcal{N}}_i}(1/a_{ij}) \leq N_i - 1$ if $0 \in \mathcal{N}_i$; $\delta_i = \rho_i$, $\overline{\mathcal{N}}_i = \mathcal{N}_i$ and $\sum_{j\in\overline{\mathcal{N}}_i}(1/a_{ij}) \leq N_i$ if $0 \notin \mathcal{N}_i$.

In the existence of the relative position measurement errors, we can still guarantee the IOS of the closed-loop distributed system by using the cyclic-small-gain theorem. In this case, the cyclic-small-gain condition is as follows:

$$
A_{\mathcal{O}}\frac{\rho(N-1)}{N} < 1 \quad \text{for } \mathcal{O} \in \mathcal{C}_0, \tag{2.110}
$$

$$
A_{\mathcal{O}}\rho < 1 \quad \text{for } \mathcal{O} \in \mathcal{C}_f \backslash \mathcal{C}_0, \tag{2.111}
$$

where $\rho := \max_{i\in\{1,\ldots,N\}}\{\rho_i\}$ is larger than one according to $\frac{1}{\rho_i} + \frac{1}{\rho_i'} \leq 1$, and can be chosen to be very close to one. Lemma 2.1 can guarantee (2.110) and (2.111) if $\mathcal{G}$ has a spanning tree with vertex 0 as the root. Thus, the proposed distributed control law is robust with respect to relative position measurement errors.

2.4.6 A Numerical Example

Consider a group of 6 robots with indices $0, 1, \ldots, 5$. Notice that the robot with index 0 is the leader. The neighbor sets of the robots are defined as follows: $\mathcal{N}_1 = \{0, 5\}$, $\mathcal{N}_2 = \{1, 3\}$, $\mathcal{N}_3 = \{2, 5\}$, $\mathcal{N}_4 = \{3\}$, $\mathcal{N}_5 = \{4\}$.

By default, the values of all the variables in this simulation are in SI units. For convenience, we omit the units. The desired relative position of the follower robots are defined by $d_{x1} = -\sqrt{3}d/2$, $d_{x2} = -\sqrt{3}d/2$, $d_{x3} = 0$, $d_{x4} = \sqrt{3}d/2$, $d_{x5} = \sqrt{3}d/2$, $d_{y1} = -d/2$, $d_{y2} = -3d/2$, $d_{y3} = -2d$, $d_{y4} = -3d/2$, $d_{y5} = -d/2$ with $d = 30$. Figure 2.6 shows the position sensing graph of the formation control system. Clearly, the position sensing graph has a spanning tree with vertex 0 as the root.

It should be noted that the control law for each follower robot also uses the velocity and acceleration information of the leader robot, the communication topology of which is not shown in Fig. 2.6.

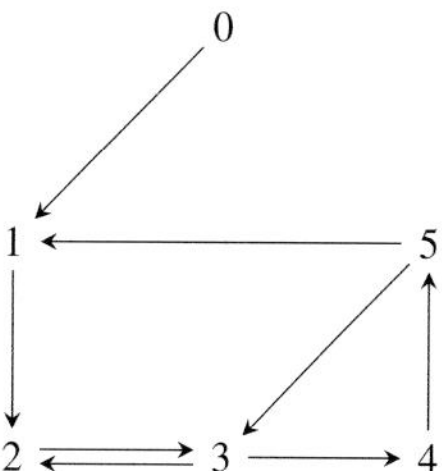

Fig. 2.6 The position sensing graph of the formation control system

The control inputs of the leader robot are $r_0(t) = 0.1\sin(0.4t)$ and $\omega_0(t) = 0.1\cos(0.2t)$. With such control inputs, the linear velocity v_0 with $v_0(0) = 3$ satisfies $\underline{v}_0 \leq v_0(t) \leq \overline{v}_0$ with $\underline{v}_0 = 3$ and $\overline{v}_0 = 3.5$.

Choose $\lambda_* = 0.45$ and $\lambda^* = 6.05$. The distributed control laws for the initialization stage are in the form of (2.83) and (2.84) with $\phi_{\theta i}(r) = \phi_{vi}(r) = -0.5(1 - \exp(-0.5r))/(1 + \exp(-0.5r))$ for $i = 1, \ldots, 5$. The distributed control laws for the formation control stage are in the form of (2.89) and (2.90) with $k_{xi} = k_{yi} = 2$ and $\phi_{xi}(r) = \phi_{yi}(r) = -1.8(1 - \exp(-0.5r))/(1 + \exp(-0.5r))$ for $i = 1, \ldots, 5$. With direct calculation, it can be verified that the designed $k_{xi}, k_{yi}, \phi_{xi}, \phi_{yi}$ satisfy (2.91) and (2.92). Also, $\phi_{xi}(r), \phi_{yi}(r) \in [-1.8, 1.8]$, for all $r \in \mathbb{R}$. With $\underline{v}_0 = 3$ and $\overline{v}_0 = 3.5$, the control laws can restrict the linear velocities of the follower robots to be in the range of $[3 - 1.8\sqrt{2}, 3.5 + 1.8\sqrt{2}] = [0.454, 6.046] \subset [\lambda_*, \lambda^*]$.

The initial states of the robots are chosen as

i	$(x_i(0), y_i(0))$	$v_i(0)$	$\theta_i(0)$
0	$(0, 0)$	3	$\pi/6$
1	$(-40, 10)$	4	π
2	$(-20, -40)$	3.5	$5\pi/6$
3	$(5, -40)$	2.5	0
4	$(50, -10)$	2	$-2\pi/3$
5	$(50, 10)$	3	0

The measurement errors are $\omega^x_{ij}(t) = 0.3(\cos(t + i\pi/6) + \cos(t/3 + i\pi/6) + \cos(t/5 + i\pi/6) + \cos(t/7 + i\pi/6))$ and $\omega^y_{ij}(t) = 0.3(\sin(t + i\pi/6) + \sin(t/3 + i\pi/6) + \sin(t/5 + i\pi/6) + \sin(t/7 + i\pi/6))$ for $i = 1, \ldots, N$, $j \in \mathcal{N}_i$.

The linear velocities and the angular velocities of the robots are shown in Fig. 2.7. The stage changes of the distributed controllers are shown in Fig. 2.8 with "0" representing initialization stage and "1" representing formation control stage. Figure 2.9 shows the trajectories of the robots. The simulation verifies the theoretical results.

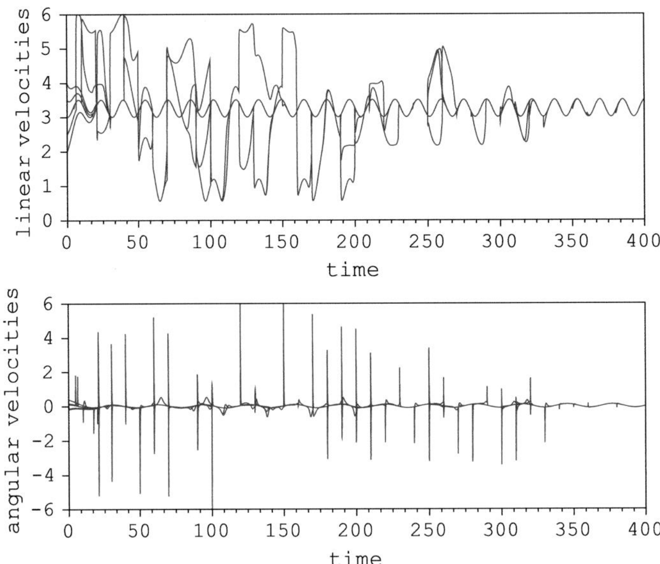

Fig. 2.7 The linear velocities and the angular velocities of the robots

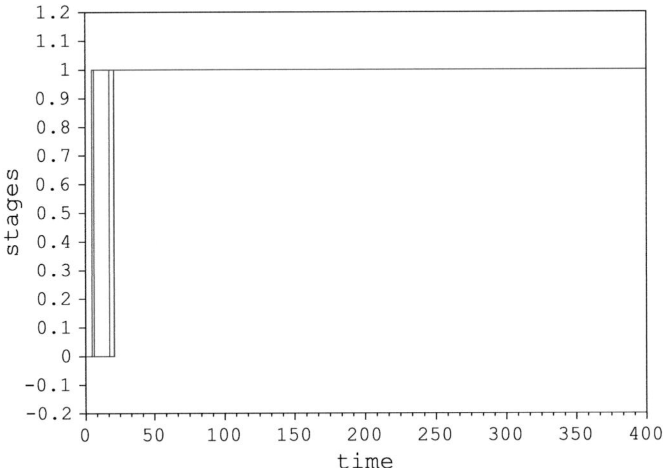

Fig. 2.8 The stages of the distributed controllers

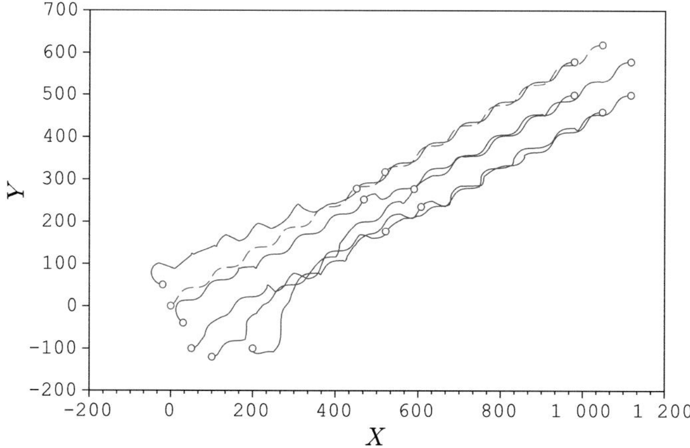

Fig. 2.9 The trajectories of the robots. The *dashed curve* represents the trajectory of the leader

2.5 Concluding Remarks

This chapter has presented cyclic-small-gain tools for distributed control of *nonlinear* multiagent systems.

In Sect. 2.3, with the proposed distributed observers and control laws, the outputs of the agents can be steered to within an arbitrarily small neighborhood of the desired agreement value under external disturbances. Asymptotic output agreement can be achieved if the system is disturbance-free. The robustness to bounded time delays of exchanged information can also be guaranteed. Section 2.3 only considers the case with time-invariant agreement value y_0. It is practically interesting to further study the distributed nonlinear control for agreement with a time-varying agreement value. Recent developments on the output-feedback tracking control of nonlinear systems (see, e.g., [22]) should be helpful for the research in this direction.

The distributed formation control law proposed in Sect. 2.4 uses relative position measurements without assuming a tree structure. For this purpose, the formation control problem is first transformed into a state agreement problem of double integrators with dynamic feedback linearization [6]. Then, a class of distributed nonlinear control laws is designed. With the proposed distributed nonlinear control law, the closed-loop system can be transformed into a network of IOS systems, and the achievement of the formation control objective can be guaranteed by using the cyclic-small-gain theorem. The special case in which there are only two robots and the desired relative positions are zero has been studied extensively in the past literature; see [8, 24] and the references therein.

By showing that a distributed control problem can be transformed into the stability/convergence problem of a dynamic network composed of IOS subsystems, Sects. 2.2–2.4 provide some partial answers to the question asked by Open Problem #5 in [28]: "Application of small-gain results for distributed feedback design of large-scale nonlinear systems." More discussions on the application of the cyclic-small-gain theorem to distributed control can be found in [36, 37, 39].

Acknowledgments This work has been supported partly by the National Science Foundations under grants ECCS-1230040 and ECCS-1501044, and partly by NSFC grants 61374042, 61522305, and 61533007.

References

1. Aguiar, A.P., Pascoal, A.M.: Coordinated path-following control for nonlinear systems with logic-based communication. In Proceedings of the 46th Conference on Decision and Control, pp. 1473–1479 (2007)
2. Arcak, M.: Passivity as a design tool for group coordination. IEEE Trans. Autom. Control **52**, 1380–1390 (2007)
3. Cao, M., Morse, A.S., Anderson, B.D.O.: Reaching a consensus in a dynamically changing environment: convergence rates, measurement delays, and asynchronous events. SIAM J. Control Optim. **47**, 601–623 (2008)
4. Cao, Y., Ren, W.: Distributed coordinated tracking with reduced interaction via a variable structure approach. IEEE Trans. Autom. Control **57**, 33–48 (2012)
5. Cortés, J., Martínez, S., Bullo, F.: Robust rendezvous for mobile autonomous agents via proximity graphs in arbitrary dimensions. IEEE Trans. Autom. Control **51**, 1289–1298 (2006)
6. d'Andréa-Novel, B., Bastin, G., Campion, G.:Dynamic feedback linearization of nonholonomic wheeled mobile robots. In Proceedings of the 1992 IEEE International Conference on Robotics and Automation, pp. 2527–2532 (1992)
7. Desai, J.P., Ostrowski, J.P., Kumar, V.: Modeling and control of formations of nonholonomic mobile robots. IEEE Trans. Robot. Autom. **17**, 905–908 (2001)
8. Dixon, W.E., Dawson, D.M., Zergeroglu, E., Behal, A.: Nonlinear Control of Wheeled Mobile Robots. Springer, London (2001)
9. Do, K.D.: Formation tracking control of unicycle-type mobile robots with limited sensing ranges. IEEE Trans. Control Syst. Technol. **16**, 527–538 (2008)
10. Dong, W., Farrell, J.A.: Decentralized cooperative control of multiple nonholonomic dynamic systems with uncertainty. Automatica **45**, 706–710 (2009)
11. Everett, H.R.: Sensors for Mobile Robots: Theory and Application. A. K, Peters, Natick (1995)
12. Fax, J.A., Murray, R.M.: Information flow and cooperative control of vehicle formation. IEEE Trans. Autom. Control **49**, 1465–1476 (2004)
13. Filippov, A.F.: Differential Equations with Discontinuous Righthand Sides. Kluwer Academic Publishers, Dordrecht (1988)
14. Fliess, M., Lévine, J.L., Martin, P., Rouchon, P.: Flatness and defect of non-linear systems: introductory theory and examples. Int. J. Control **61**, 1327–1361 (1995)
15. Fridman, E., Bar Am, N.: Sampled-data distributed h_∞ control of a class of parabolic systems. In Proceedings of the 51st IEEE Conference on Decision and Control, pp. 7529–7534 (2012)
16. Gazi, V., Passino, K.M.: Stability analysis of swarms. IEEE Trans. Autom. Control **48**, 692–697 (2003)
17. Ghabcheloo, R., Aguiar, A.P., Pascoal, A., Silvestre, C., Kaminer, I., Hespanha, J.: Coordinated path-following in the presence of communication losses and time delays. SIAM J. Control Optim. **48**, 234–265 (2009)
18. Hirschhorn, J.: Kinematics and Dynamics of Plane Mechanisms. McGraw-Hill Book Company, New York (1962)
19. Hong, Y., Gao, L., Cheng, D., Hu, J.: Lyapunov-based approach to multiagent systems with switching jointly connected interconnection. IEEE Trans. Autom. Control **52**, 943–948 (2007)

20. Ihle, I.-A.F., Arcak, M., Fossen, T.I.: Passivity-based designs for synchronized path-following. Automatica **43**, 1508–1518 (2007)
21. Jadbabaie, A., Lin, J., Morse, A.: Coordination of groups of mobile autonomous agents using nearest neighbor rules. IEEE Trans. Autom. Control **48**, 988–1001 (2003)
22. Jiang, Z.P.: Decentralized and adaptive nonlinear tracking of large-scale systems via output feedback. IEEE Trans. Autom. Control **45**, 2122–2128 (2000)
23. Jiang, Z.P., Mareels, I.M.Y.: A small-gain control method for nonlinear cascade systems with dynamic uncertainties. IEEE Trans. Autom. Control **42**, 292–308 (1997)
24. Jiang, Z.P., Nijmeijer, H.: Tracking control of mobile robots: a case study in backstepping. Automatica **33**, 1393–1399 (1997)
25. Jiang, Z.P., Teel, A.R., Praly, L.: Small-gain theorem for ISS systems and applications. Math. Control Signals Syst. **7**, 95–120 (1994)
26. Jiang, Z.P., Wang, Y.: A converse Lyapunov theorem for discrete-time systems with disturbances. Syst. Control Lett. **45**, 49–58 (2002)
27. Jiang, Z.P., Wang, Y.: A generalization of the nonlinear small-gain theorem for large-scale complex systems. In Proceedings of the 7th World Congress on Intelligent Control and Automation, pp. 1188–1193 (2008)
28. Karafyllis, I., Jiang, Z.P.: Stability and Stabilization of Nonlinear Systems. Springer, London (2011)
29. Khalil, H.K.: Nonlinear Systems (3rd Edition). Prentice-Hall, New Jersey (2002)
30. Lan, Y., Yan, G., Lin, Z.: Synthesis of distributed control of coordinated path following. IEEE Trans. Autom. Control **56**, 1170–1175 (2011)
31. Li, Q., Jiang, Z.P.: Flocking control of multi-agent systems with application to nonholonomic multi-robots. Kybernetika **45**, 84–100 (2009)
32. Li, Z., Liu, X., Ren, W., Xie, L.: Consensus control of linear multi-agent systems with distributed adaptive protocols. In Proceedings of the 2012 American Control Conference, pp. 1573–1578 (2012)
33. Lin, Z., Francis, B., Maggiore, M.: Necessary and sufficient graphical conditions for formation control of unicycles. IEEE Trans. Autom. Control **50**, 121–127 (2005)
34. Lin, Z., Francis, B., Maggiore, M.: State agreement for continuous-time coupled nonlinear systems. SIAM J. Control Optim. **46**, 288–307 (2007)
35. Liu, T., Hill, D.J., Jiang, Z.P.: Lyapunov formulation of ISS cyclic-small-gain in continuous-time dynamical networks. Automatica **47**, 2088–2093 (2011)
36. Liu, T., Jiang, Z.P.: Distributed nonlinear control of mobile autonomous multi-agents. Automatica **50**, 1075–1086 (2014)
37. Liu, T., Jiang, Z.P.: Distributed control of nonlinear uncertain systems: a cyclic-small-gain approach. *Acta Automatica Sinica*, 2016. In press
38. Liu, T., Jiang, Z.P.: Distributed formation control of nonholonomic mobile robots without global position measurements. Automatica **49**, 592–600 (2013)
39. Liu, T., Jiang, Z.P.: Distributed output-feedback control of nonlinear multi-agent systems. IEEE Trans. Autom. Control **58**, 2912–2917 (2013)
40. Liu, T., Jiang, Z.P., Hil, D.J.: Nonlinear Control of Dynamic Networks. CRC Press, Boca Raton (2014)
41. Martensson, K., Rantzer, A.: A scalable method for continuous-time distributed control synthesis. In Proceedings of the 2012 American Control Conference, pp. 6308–6313 (2012)
42. Moreau, L.: Stability of multiagent systems with time-dependent communication links. IEEE Trans. Autom. Control **50**, 169–182 (2005)
43. Nersesov, S.G., Ghorbanian, P., Aghdam, A.G.: Stabilization of sets with application to multi-vehicle coordinated motion. Automatica **46**, 1419–1427 (2010)
44. Ögren, P., Egerstedt, M., Hu, X.: A control Lyapunov function approach to multiagent coordination. IEEE Trans. Robot. Autom. **18**, 847–851 (2002)
45. Ögren, P., Fiorelli, E., Leonard, N.: Cooperative control of mobile sensor networks: adaptive gradient climbing in a distributed network. IEEE Trans. Autom. Control **49**, 1292–1302 (2004)

46. Olfati-Saber, R.: Flocking for multi-agent dynamic systems: algorithms and theory. IEEE Trans. Autom. Control **51**, 401–420 (2006)
47. Qin, J., Zheng, W.X., Gao, H.: Consensus of multiple second-order vehicles with a time-varying reference signal under directed topology. Automatica **47**, 1983–1991 (2011)
48. Qu, Z., Wang, J., Hull, R.A.: Cooperative control of dynamical systems with application to autonomous vehicles. IEEE Trans. Autom. Control **53**, 894–911 (2008)
49. Sadowska, A., Kostic, D., van de Wouw, N., Huijberts, H., Nijmeijer, H.: Distributed formation control of unicycle robots. In Proceedings of the 2012 IEEE International Conference on Robotics and Automation, pp. 1564–1569 (2012)
50. Scardovi, L., Sepulchre, R.: Synchronization in networks of identical linear systems. Automatica **45**, 2557–2562 (2009)
51. Shi, G., Hong, Y.: Global target aggregation and state agreement of nonlinear multi-agent systems with switching topologies. Automatica **45**, 1165–1175 (2009)
52. Sontag, E.D.: Comments on integral variants of ISS. Syst. Control Lett. **34**, 93–100 (1998)
53. Stan, G.-B., Sepulchre, R.: Dissipativity and global analysis of oscillators. IEEE Trans. Autom. Control **52**, 256–270 (2007)
54. Su, Y., Huang, J.: Output regulation of a class of switched linear multi-agent systems: a distributed observer approach. In Proceedings of the 18th IFAC World Congress, pp. 4495–4500 (2011)
55. Tanner, H., Jadbabaie, A., Pappas, G.: Flocking in fixed and switching networks. IEEE Trans. Autom. Control **52**, 863–868 (2007)
56. Tanner, H.G., Pappas, G.J., Kummar, V.: Leader-to-formation stability. IEEE Trans. Robot. Autom. **20**, 443–455 (2004)
57. Wang, X., Hong, Y., Huang, J., Jiang, Z.P.: A distributed control approach to a robust output regulation problem for multi-agent systems. IEEE Trans. Autom. Control **55**, 2891–2895 (2010)
58. Wieland, P., Sepulchre, R., Allgöwer, F.: An internal model principle is necessary and sufficient for linear output synchronization. Automatica **47**, 1068–1074 (2011)
59. Yu, H., Antsaklis, J.: Distributed formation control of networked passive systems with event-driven communication. In Proceedings of the 51st IEEE Conference on Decision and Control, pp. 3292–3297 (2012)

Chapter 3
Convergent Systems: Nonlinear Simplicity

Alexey Pavlov and Nathan van de Wouw

Abstract Convergent systems are systems that have a uniquely defined globally asymptotically stable steady-state solution. Asymptotically stable linear systems excited by a bounded time varying signal are convergent. Together with the superposition principle, the convergence property forms a foundation for a large number of analysis and (control) design tools for linear systems. Nonlinear convergent systems are in many ways similar to linear systems and are, therefore, in a certain sense simple, although the superposition principle does not hold. This simplicity allows one to solve a number of analysis and design problems for nonlinear systems and makes the convergence property highly instrumental for practical applications. In this chapter, we review the notion of convergent systems and its applications to various analyses and design problems within the field of systems and control.

3.1 Introduction

In many controller design problems, a controller is designed to ensure that some desired solution of the closed-loop system is asymptotically stable with a desired region of attraction. Traditionally, this is considered as a stabilization problem for the desired *solution*. However, if we take a step back and look at this design problem

A. Pavlov
Statoil ASA, P.b. 1004, 3905 Porsgrunn, Norway
e-mail: alepav@statoil.com

N. van de Wouw (✉)
Department of Mechanical Engineering, Eindhoven University of Technology, Eindhoven, Noord-Brabant, The Netherlands
e-mail: N.v.d.Wouw@tue.nl

N. van de Wouw
Department of Civil, Environmental and Geo-Engineering, University of Minnesota, Minneapolis, MN, USA

N. van de Wouw
Delft Center for Systems and Control, Delft University of Technology, Delft, The Netherlands

N. van de Wouw et al. (eds.), *Nonlinear Systems*, Lecture Notes in Control and Information Sciences 470, DOI 10.1007/978-3-319-30357-4_3

from a distance, we can see that the controller actually provides the closed-loop system with a *system* property: the closed-loop system has an asymptotically stable steady-state solution with the given region of attraction (e.g., local, global, or some given set of initial conditions). In addition to that, this steady-state solution has desired properties, e.g., desired output value.

This system point of view on the controller design problem allows one to define an important system property, which is common for linear asymptotically stable systems, but which is far from being straightforward for nonlinear systems: *the convergence property*. A system is called convergent if it has a unique, in a certain sense, globally asymptotically stable solution, called the steady-state solution. Originally, the term "convergence" was coined in the Russian literature in the 1950s. In [41], the notion was defined for nonstationary systems that are periodic in time. In that reference, such a system is called convergent if it has a unique globally asymptotically stable periodic solution. Later, in [10] (see also [34]) this definition was naturally extended to nonlinear systems with arbitrary (not necessarily periodic in time) right-hand sides. These references, together with [56] laid a foundation of basic results for establishing this property for nonlinear systems based on Lyapunov functions, matrix inequalities, and frequency domain methods. Almost 50 years later, notions similar to convergence received significant attention in the literature: contraction analysis, incremental stability and passivity, incremental input-to-state stability, etc. [1–3, 13, 15, 22, 27–29, 32, 44–46, 49, 50, 57]. A comparison establishing differences and similarities between incremental stability on the one hand and convergence on the other hand is provided in [46].

A brief historical overview on convergent systems and subsequent developments of this and similar notions can be found in [34]. Since that paper many new developments on convergent systems have appeared. In particular, sufficient conditions for convergence for different classes of systems have been pursued [8, 9, 25, 26, 35, 37, 39, 42, 43, 54]. Together with theoretical developments on convergent systems and related notions, the benefit of such system-level stability property has been demonstrated by its use to tackle fundamental system-theoretic problems. Further study of convergent systems indeed demonstrated that this notion is very instrumental for a number of design and analysis problems within nonlinear systems and control, such as synchronization, observer design, output tracking and disturbance rejection, the output regulation problem, model reduction, stable inversion of non-minimum phase systems, steady-state performance optimization of control systems, variable gain controller design and tuning and extremum seeking control. For linear systems many of these problems are solved in a relatively simple way. It turns out that this simplicity comes not only from the superposition principle, but also from the convergence property (for linear systems it is equivalent to asymptotic stability of the system with zero input). Unlike the superposition principle, which holds only for linear systems, the convergence property may still hold for a nonlinear system. It appears that convergent nonlinear systems enjoy, to a large extent, the simplicity inherent to linear asymptotically stable systems. This allows one to solve, based on the notion of convergence, a number of challenging nonlinear control and analysis problems.

In this chapter, we will revisit the notion of convergence and review its application to various problems within systems and control. We deliberately omit technical details and generality, keeping focus on the ideas. All technical details and general formulations can still be found in the corresponding references. The chapter is organized as follows. Definitions, sufficient conditions, and basic properties of convergent systems are given in Sect. 3.2. Applications of this notion are reviewed in Sects. 3.3–3.9. Conclusions are given in Sect. 3.10.

3.2 Convergent Systems

Consider the nonlinear system

$$\dot{x} = F(x,t), \quad x \in \mathbb{R}^n, \quad t \in \mathbb{R}, \tag{3.1}$$

where $F(x,t)$ is locally Lipschitz in x and piecewise continuous in t.[1]

Definition 3.1 ([10]) System (3.1) is called *convergent* if

(i) there is a unique solution $\bar{x}(t)$ that is defined and bounded for $t \in \mathbb{R}$,
(ii) $\bar{x}(t)$ is globally asymptotically stable.

If $\bar{x}(t)$ is uniformly (exponentially) asymptotically stable, then system (3.1) is called *uniformly (exponentially) convergent*.[2]

Since the time-varying component in a system is usually due to some input, we can define convergence for systems with inputs.

Definition 3.2 System

$$\dot{x} = F(x, w(t)), \quad w(t) \in \mathbb{R}^m, \quad t \in \mathbb{R}, \tag{3.2}$$

is (uniformly, exponentially) convergent for a class of inputs $\mathcal{I}$ if it is convergent for every input $w(t) \in \mathcal{I}$ from that class.

To emphasize the dependence of the steady-state solution on the input $w(t)$, it is denoted by $\bar{x}_w(t)$. Note that any solution of convergent system (3.2) forgets its initial conditions and converges to the steady-state solution, which is determined by the input $w(t)$. Relations between input $w(t)$ and steady-state solution $\bar{x}_w(t)$ can be further characterized by several additional properties.

---

[1]For simplicity, in this chapter, we consider only continuous-time systems with locally Lipschitz right-hand sides. Definitions and basic results for discrete-time systems and for continuous-time systems with discontinuous right-hand sides can be found in [18, 36, 38, 39, 56].

[2]A more general definition of convergent systems, where the steady-state solution has an arbitrary domain of attraction (not necessarily global as in this chapter) can be found in [35].

Definition 3.3 ([35]) The system (3.2) that is convergent for a class of piecewise continuous bounded inputs is said to have the Uniformly Bounded Steady-State (UBSS) property if for any $r > 0$ there exists $R > 0$ such that if a piecewise continuous input $w(t)$ satisfies $|w(t)| \leq r$ for all $t \in \mathbb{R}$, then the corresponding steady-state solution satisfies $|\bar{x}_w(t)| \leq R$ for all $t \in \mathbb{R}$.

Definition 3.4 ([35]) System (3.2) is called *input-to-state convergent* if it is uniformly convergent for the class of bounded piecewise continuous inputs and, for every such input $w(\cdot)$, system (3.2) is input-to-state stable[3] with respect to the steady-state solution $\bar{x}_w(t)$, i.e., there exist a $\mathscr{KL}$-function $\beta(r, s)$ and a $\mathscr{K}_\infty$-function $\gamma(r)$ such that any solution $x(t)$ of system (3.2) corresponding to some input $\hat{w}(t) := w(t) + \Delta w(t)$ satisfies

$$|x(t) - \bar{x}_w(t)| \leq \beta(|x(t_0) - \bar{x}_w(t_0)|, t - t_0) + \gamma\left(\sup_{t_0 \leq \tau \leq t} |\Delta w(\tau)|\right). \tag{3.3}$$

In general, the functions $\beta(r, s)$ and $\gamma(r)$ may depend on the particular input $w(\cdot)$.

3.2.1 Conditions for Convergence

Simple sufficient conditions for exponential convergence were given by B.P. Demidovich in [10, 34]. Here we present a slightly modified version of that result, which covers input-to-state convergence.

Theorem 3.1 ([10, 35]) *Consider system (3.2) with the function $F(x, w)$ being C^1 with respect to x and continuous with respect to w. Suppose there exist matrices $P = P^T > 0$ and $Q = Q^T > 0$ such that*

$$P\frac{\partial F}{\partial x}(x, w) + \frac{\partial F}{\partial x}^T(x, w)P \leq -Q, \quad \forall x \in \mathbb{R}^n, \quad w \in \mathbb{R}^m. \tag{3.4}$$

Then, system (3.2) is globally exponentially convergent with the UBSS property and input-to-state convergent for the class of bounded piecewise continuous inputs.

For systems of Lur'e-type form, sufficient conditions for exponential convergence were obtained by V.A. Yakubovich [56]. Consider the system

$$\begin{aligned} \dot{x} &= Ax + Bu + Hw(t) \\ y &= Cx + Dw(t) \\ u &= -\varphi(y, w(t)), \end{aligned} \tag{3.5}$$

where $x \in \mathbb{R}^n$ is the state, $y \in \mathbb{R}$ is the output, $w(t) \in \mathbb{R}^m$ is a piecewise continuous input and $\varphi(y, w)$ is a static nonlinearity.

[3] See [48] for a definition of the input-to-state stability property.

Theorem 3.2 ([35, 56]) *Consider system (3.5). Suppose the matrix A is Hurwitz, the nonlinearity $\varphi(y, w)$ satisfies*

$$|\varphi(y_2, w) - \varphi(y_1, w)| \leq K\, |y_2 - y_1|\,, \tag{3.6}$$

for all $y_1, y_2 \in \mathbb{R}$ and $w \in \mathbb{R}^m$, and the frequency response function $G_{yu}(j\omega) = C(j\omega I - A)^{-1}B$ from u to y satisfies

$$\sup_{\omega\in\mathbb{R}} |G_{yu}(j\omega)| =: \gamma_{yu} < \frac{1}{K}. \tag{3.7}$$

Then, system (3.5) is exponentially convergent with the UBSS property and input-to-state convergent for the class of piecewise continuous bounded inputs.[4]

Below follows an alternative result, not based on quadratic Lyapunov functions.

Theorem 3.3 ([35]) *Consider system (3.2). Suppose there exist C^1 functions $V_2(x)$ and $V_1(x_1, x_2)$, $\mathcal{K}$-functions $\alpha_2(s)$, $\alpha_3(s)$, $\alpha_5(s)$, $\gamma(s)$, and $\mathcal{K}_\infty$-functions $\alpha_1(s)$, $\alpha_4(s)$ satisfying the conditions*

$$\alpha_1(|x_1 - x_2|) \leq V_1(x_1, x_2) \leq \alpha_2(|x_1 - x_2|), \tag{3.8}$$

$$\frac{\partial V_1}{\partial x_1}(x_1, x_2)F(x_1, w) + \frac{\partial V_1}{\partial x_2}(x_1, x_2)F(x_2, w) \leq -\alpha_3(|x_1 - x_2|), \tag{3.9}$$

$$\alpha_4(|x|) \leq V_2(x) \leq \alpha_5(|x|), \tag{3.10}$$

$$\frac{\partial V_2}{\partial x}(x)F(x, w) \leq 0 \text{ for } |x| \geq \gamma(|w|) \tag{3.11}$$

for all $x_1, x_2, x \in \mathbb{R}^n$ and $w \in \mathbb{R}^m$. Then, system (3.2) is globally uniformly convergent and has the UBSS property for the class of bounded piecewise continuous inputs.

One can show that conditions of Theorems 3.1–3.3 imply incremental stability [2]. In fact, the proof of convergence in these results is based on two basic components:

(1) incremental stability: it guarantees global asymptotic stability of any solution,
(2) existence of a compact positively invariant set: it guarantees existence of a solution $\bar{x}(t)$ that is bounded on $\mathbb{R}$ [10, 56]. By virtue of (1), $\bar{x}(t)$ is globally asymptotically stable, which proves convergence.

Although here incremental stability is used in the sufficient conditions for convergence given above, in general these two properties are not equivalent. However, as

[4]This result is a particular case of a more general condition on $G_{yu}(j\omega)$ in the form of Circle criterion [56].

shown in [46], uniform convergence and incremental stability are equivalent on compact positively invariant sets. In the latter reference, also a necessary and sufficient condition for uniform convergence is formulated, which reads as follows.

Theorem 3.4 ([46]) *Assume that system (3.1) is globally uniformly convergent, with associated steady-state solution $\bar{x}(t)$. Assume that the function F is continuous in (x, t) and C^1 with respect to the x variable. Assume also that the Jacobian $\frac{\partial}{\partial x} f(x, t)$ is bounded, uniformly in t. Then there exist a C^1 function $V : \mathbb{R} \times \mathbb{R}^n \to \mathbb{R}_+$, functions α_1, α_2 and $\alpha_3 \in \mathcal{K}_\infty$, and a constant $c \geq 0$ such that*

$$\alpha_1(|x - \bar{x}(t)|) \leq V(t, x) \leq \alpha_2(|x - \bar{x}(t)|), \tag{3.12}$$

$$\frac{\partial V}{\partial t} + \frac{\partial V}{\partial x} F(x, t) \leq -\alpha_3(|x - \bar{x}(t)|) \tag{3.13}$$

and

$$V(t, 0) \leq c, \ t \in \mathbb{R}. \tag{3.14}$$

Conversely, if a differentiable function $V : \mathbb{R} \times \mathbb{R}^n \to \mathbb{R}_+$ and functions α_1, α_2 and $\alpha_3 \in \mathcal{K}_\infty$, and a constant $c \geq 0$ are given such that for some trajectory $\bar{x} : \mathbb{R} \to \mathbb{R}^n$ estimates (3.12)–(3.14) hold, then system (3.1) must be globally uniformly convergent and the solution $\bar{x}(t)$ is the unique bounded solution as in Definition 3.1.

For interconnections of convergent systems, one can obtain similar results as for interconnections of systems with a stable equilibrium. In particular, a series connection of input-to-state convergent (ISC) systems is again an ISC system [35]. Feedback interconnection of two ISC systems is again an ISC system under a small-gain condition on the gain functions $\gamma(r)$ for each subsystem, see details in [6]. Theorems 3.1–3.3 in combination with these interconnection properties provide a practical toolbox of sufficient conditions for the convergence property.

Sufficient conditions for convergence for other classes of systems, such as discrete-time nonlinear systems, piecewise affine systems, linear complementarity systems, switched systems, measure differential inclusions, delay differential equations, have been pursued in [8, 9, 25, 26, 35, 37, 39, 43, 54].

3.2.2 Basic Properties of Convergent Systems

The convergence property is an extension of stability properties of asymptotically stable linear systems with inputs:

$$\dot{x} = Ax + Bw, \tag{3.15}$$

where A is Hurwitz. For any piecewise continuous input $w(t)$ that is bounded on $\mathbb{R}$, this system has a unique solution $\bar{x}_w(t)$ which is defined and bounded on $\mathbb{R}$:

$$\bar{x}_w(t) := \int_{-\infty}^{t} \exp(A(t-s))Bw(s)ds. \tag{3.16}$$

This solution is globally exponentially stable since A is Hurwitz. Therefore, system (3.15) is exponentially convergent for the class of bounded piecewise continuous inputs. This example also illustrates the selection of the steady-state solution in the definition of convergent systems. The steady-state solution is not only a solution that attracts all other solutions in forward time–all solutions of system (3.15) have this property. It is key to realize that, among all these solutions, only one remains bounded both in forward and backward time. The selection of this bounded on $\mathbb{R}$ solution defines the steady state in a unique way for uniformly convergent systems [35]. The natural choice for the definition of the steady-state solution is further illustrated by the following property.

Property 3.1 ([10]) Suppose system (3.2) with a given input $w(t)$ is uniformly convergent. If the input $w(t)$ is constant, the corresponding steady-state solution $\bar{x}_w(t)$ is also constant; if $w(t)$ is periodic with period T, then the corresponding steady-state solution $\bar{x}_w(t)$ is also periodic with the same period T.

As it will be demonstrated in subsequent sections, the following two basic properties of convergent systems will be very instrumental in design and analysis problems within systems and control:

(i) a convergent system defines a steady-state operator $\mathscr{F}w(t) := \bar{x}_w(t)$ that maps bounded on $\mathbb{R}$ inputs to bounded on $\mathbb{R}$ steady-state solutions and periodic inputs to periodic steady-state solutions with the same period.
(ii) any solution of a uniformly convergent system starting in a compact positively invariant set $\mathscr{X}$ is uniformly asymptotically stable in $\mathscr{X}$.

Property (i) is highly instrumental in problems focused on steady-state dynamics, while property (ii) significantly simplifies stability proofs for particular solutions. The latter property follows from [46], where it is shown that for a compact positively invariant set uniform convergence and incremental stability are equivalent. In subsequent sections, we will demonstrate how these two basic properties can be used in various design and analysis problems.

3.3 Controller and Observer Design

In controller and observer design problems for nonstationary systems (e.g., systems with time-varying inputs), the common objective is to ensure, by means of controller design, that a certain solution with desired properties is asymptotically stable with a given domain of attraction (e.g., global). For example, in the observer design problem, the desired solution is the solution of the observed system. In the output tracking and disturbance rejection problem, the desired solution is the one that matches the desired output of the system regardless of the disturbance.

The conventional approach to prove whether a controller/observer solves these problems consists of the following steps:

(a) find a solution of the closed-loop system/observer $\bar{x}(t)$ with desired properties,
(b) translate that solution to the origin through the transformation $z(t) = x(t) - \bar{x}(t)$,
(c) prove asymptotic stability of $z(t) \equiv 0$ with a desired domain of attraction.

Although stability analysis of an equilibrium should be simpler, in many cases this simplicity is essentially reduced by the coordinate transformation: the right-hand side of the system in the new coordinates z typically depends on $\bar{x}(t)$. This makes the analysis challenging and in some cases even prohibitively complex as, for example, for piecewise affine systems [53]. On the other hand, the same design problems can be approached using the property of convergence:

(1) design a feedback controller that ensures uniform convergence of the closed-loop system: as a result, any solution starting in a compact positively invariant set is uniformly asymptotically stable in this set.
(2) design a feedforward controller that ensures that the system with the feedback and feedforward controllers has a solution $\bar{x}(t)$ with the desired properties.

Thus for any compact positively invariant set of initial conditions, the solution $\bar{x}(t)$ will be uniformly asymptotically stable in this set. This approach allows one to avoid the coordinate transformation $z = x - \bar{x}(t)$ and subsequent cumbersome stability analysis of the transformed system.[5]

Let us illustrate the benefit of the above convergence-based approach in the scope of the observer design problem. Consider the system

$$\begin{cases} \dot{x} = F(x, w), \\ y = h(x, w) \end{cases} \tag{3.17}$$

with input w and output y. The objective is to design an observer that asymptotically reconstructs from the measured input and output the state $x(t)$ starting at an unknown initial condition $x(t_0) = x_0$. A conventional way to design an observer is to construct it as a copy of the system dynamics with an output injection term:

$$\begin{cases} \dot{\hat{x}} = F(\hat{x}, w) + L(y, \hat{y}, w), \\ \hat{y} = h(\hat{x}, w), \end{cases} \tag{3.18}$$

where the injection function $L(y, \hat{y}, w)$ satisfies $L(y, y, w) \equiv 0$. The latter condition guarantees that the observer, if initiated in the same initial condition as system (3.17), $\hat{x}(t_0) = x_0$, has a solution $\hat{x}(t) \equiv x(t)$, i.e., condition (2) above is satisfied. If the injection term $L(y, \hat{y}, w)$ is designed in such a way that the observer (3.18) is uniformly convergent, then the desired solution $\hat{x}(t) \equiv x(t)$ is uniformly asymptotically stable in any compact positively invariant set of initial conditions. Hence, one

[5]This benefit has recently also been explicitly recognized in [13].

can think of the observer design as aiming to ensure that the observer is a convergent system rather than aiming at rendering the observer error dynamics asymptotically stable.

The problem of controlled synchronization (e.g., master–slave synchronization) has a lot in common with the observer design problem (see, e.g., [31]). Therefore, it can be treated in the same way, as the observer design problem. The same holds for the output tracking and disturbance rejection problems.

For a class of piecewise affine (PWA) systems, this convergence-based approach allows one to solve the output tracking, synchronization, and observer design problems in a relatively simple manner [53]. For PWA systems these nonstationary problems become very hard to solve by conventional methods if the number of cells with affine dynamics is larger than two.

For the tracking and disturbance rejection problems, in the approach mentioned above one needs to answer the following questions:

- How to find a feedback that makes the closed-loop system uniformly convergent?
- How to find a bounded feedforward input that shapes the output of the steady-state solution to a desired value, and whether such a feedforward exists at all?

For an answer to the first question the reader is referred, for example, to [35], where controller design tools based on quadratic stability, backstepping, separation principle and passivity were developed. The second question will be addressed in the next section.

3.4 Stable Inversion of Nonlinear Systems

The problem of finding a bounded input that ensures the existence of a bounded solution with a desired output is called the stable inversion problem. Conventionally, it is studied after transforming the system into a normal form:

$$\dot{\xi} = p(\xi, \bar{y}, u), \tag{3.19}$$

$$y^{(r)} = q(\bar{y}, \xi) + s(\bar{y}, \xi)u, \tag{3.20}$$

where $y \in \mathbb{R}$ is the output, $u \in \mathbb{R}$ is the input; $\xi \in \mathbb{R}^n$ and $\bar{y} := (y, \dot{y}, \ldots, y^{(r-1)})^T$ constitute the state of the system. The functions p, q and s are locally Lipschitz and $s(\bar{y}, \xi)$ is nonzero for all $\bar{y}$ and ξ. For simplicity of the presentation, we assume that the normal form (3.19), (3.20) is defined globally. The reference output trajectory is given by $y_d(t)$, which is bounded together with its r derivatives. From (3.20), we compute an input u corresponding to the reference output trajectory $y_d(t)$:

$$u = s(\bar{y}_d, \xi)^{-1}(y_d^{(r)} - q(\bar{y}_d, \xi)) =: U(\xi, \bar{y}_d, y_d^{(r)}), \tag{3.21}$$

where $\bar{y}_d := (y_d, \dot{y}_d, \ldots, y_d^{(r-1)})^T$. Substituting this control into (3.19), we obtain the tracking dynamics

$$\dot{\xi} = p(\xi, \bar{y}_d(t), U(\xi, \bar{y}_d(t), y_d^{(r)}(t))) =: \bar{p}(\xi, t). \tag{3.22}$$

If we can find a bounded solution $\bar{\xi}(t)$ of (3.22), then the corresponding bounded input $u_d(t)$ can be computed from (3.21) by substituting this $\bar{\xi}(t)$ for ξ. The desired bounded solution of (3.19), (3.20) equals $(\bar{\xi}^T(t), \bar{y}_d^T(t))^T$.

For minimum phase systems, bounded $\bar{y}_d(t)$, $y_d^{(r)}(t)$ ensure boundedness of any solution of the tracking dynamics in (3.22) in forward time. For non-minimum phase systems, this is not the case, since the tracking dynamics are unstable. However, there may still exist a bounded solution, as it has been shown in [11, 12] for the local case.

To extend that result to the nonlocal case, we can, first, observe that the unstable tracking dynamics can be convergent in backward time. In this case, all solutions except for one diverge to infinity as $t \to +\infty$. The only bounded on $\mathbb{R}$ solution is the steady-state solution from the definition of convergence.

We can apply similar reasoning to the case if (3.22) can be decomposed (after, possibly, a coordinate transformation) into a series connection of two systems:

$$\dot{\eta} = F(\eta, t), \quad \eta \in \mathbb{R}^{n_s}, \tag{3.23}$$

$$\dot{\zeta} = G(\zeta, \eta, t), \quad \zeta \in \mathbb{R}^{n_u}. \tag{3.24}$$

If system (3.23) is convergent and (3.24) with η as input is convergent in backward time for the class of bounded continuous inputs, one can easily verify that the bounded on $\mathbb{R}$ solution of (3.23), (3.24) is unique and it equals $(\bar{\eta}^T(t), \bar{\zeta}_{\bar{\eta}}^T(t))^T$, where $\bar{\eta}(t)$ is the steady-state solution of (3.23) and $\bar{\zeta}_{\bar{\eta}}(t)$ is the steady-state solution of (3.24) corresponding to $\bar{\eta}(t)$.

If the tracking dynamics can be represented as a feedback interconnection of a convergent system in forward time and a convergent system in backward time,

$$\dot{\eta} = F(\eta, \zeta, t), \quad \eta \in \mathbb{R}^{n_s}, \tag{3.25}$$

$$\dot{\zeta} = G(\zeta, \eta, t), \quad \zeta \in \mathbb{R}^{n_u}, \tag{3.26}$$

then one can still ensure the existence of a unique bounded on $\mathbb{R}$ solution if a certain small-gain condition is satisfied, as formalized in the result below.

Theorem 3.5 *Consider system (3.25). Suppose that*

1. *system (3.25) with ζ as input is convergent for the class of continuous bounded inputs with the corresponding steady-state operator $\mathscr{F}$ being Lipschitz continuous with a Lipschitz constant γ_F, i.e., $\|\mathscr{F}\zeta_1 - \mathscr{F}\zeta_1\|_\infty \le \gamma_F\|\zeta_1 - \zeta_2\|_\infty$,*
2. *system (3.26) with η as input is convergent in backward time for the class of continuous bounded inputs with the corresponding steady-state operator $\mathscr{G}$ being Lipschitz continuous with a Lipschitz constant γ_G, i.e., $\|\mathscr{G}\eta_1 - \mathscr{G}\eta_1\|_\infty \le \gamma_G\|\eta_1 - \eta_2\|_\infty$.*

If the small-gain condition

$$\gamma_F \gamma_G < 1, \tag{3.27}$$

is satisfied, then system (3.25) has a unique bounded on $\mathbb{R}$ *solution.*

Finding the Lipschitz constant for the steady-state operator as well as a numerical method for the calculation of the bounded solution are described in [33].

The simple convergence-based considerations presented above extend the local results from [11, 12] on stable inversion of non-minimum phase systems to the nonlocal nonlinear case.

3.5 The Output Regulation Problem

In [35], the notion of convergent systems was successfully applied to solve the output regulation problem for nonlinear systems in a nonlocal setting. Before that, the output regulation problem was solved for linear systems, see, e.g., [14], resulting in the well-known internal model principle, and for nonlinear systems in a local setting [7, 20]. The application of the convergent systems property allowed us to extend these results to nonlocal problem settings for nonlinear systems. In particular, necessary and sufficient conditions for the solvability of the global nonlinear output regulation problem were found [35].[6] These conditions included, as their particular case, the solvability conditions for the linear and the local nonlinear cases.

The output regulation problem can be treated as a special case of the tracking and disturbance rejection problem, where the reference signal for the output and/or disturbance are generated by an external autonomous system. Consider systems modeled by equations of the form

$$\dot{x} = f(x, u, w), \tag{3.28}$$

$$e = h_r(x, w), \tag{3.29}$$

$$y = h_m(x, w), \tag{3.30}$$

with state $x \in \mathbb{R}^n$, input $u \in \mathbb{R}^k$, regulated output $e \in \mathbb{R}^{l_r}$, and measured output $y \in \mathbb{R}^{l_m}$. The exogenous signal $w(t) \in \mathbb{R}^m$, which can be viewed as a disturbance in (3.28) or as a reference signal in (3.29), is generated by an external autonomous system

$$\dot{w} = s(w), \tag{3.31}$$

starting in a compact positively invariant set of initial conditions $\mathscr{W}_+ \subset \mathbb{R}^m$. System (3.31) is called an exosystem.

[6]These results were obtained in parallel with [21], where an alternative approach to nonlocal nonlinear output regulation problem was pursued.

The global uniform output regulation problem is to find, if possible, a controller of the form

$$\dot{\xi} = \eta(\xi, y), \quad \xi \in \mathbb{R}^q, \tag{3.32}$$

$$u = \theta(\xi, y), \tag{3.33}$$

for some $q \geq 0$ such that the closed-loop system

$$\dot{x} = f(x, \theta(\xi, h_m(x, w)), w), \tag{3.34}$$

$$\dot{\xi} = \eta(\xi, h_m(x, w)) \tag{3.35}$$

satisfies three conditions:

- *regularity:* the right-hand side of the closed-loop system is locally Lipschitz with respect to (x, ξ) and continuous with respect to w;
- *uniform convergence:* the closed-loop system is uniformly convergent with the UBSS property for the class of bounded continuous inputs;
- *asymptotic output zeroing:* for all solutions of the closed-loop system and the exo-system starting in $(x(0), \xi(0)) \in \mathbb{R}^{n+q}$ and $w(0) \in \mathscr{W}_+$ it holds that $e(t) = h_r(x(t), w(t)) \to 0$ as $t \to +\infty$.[7]

In conventional formulations of the output regulation problem, some other stability requirement on the closed-loop system is used instead of the requirement of uniform convergence, e.g., (global) asymptotic stability of the origin for zero input or boundedness of solutions. For linear and local nonlinear cases, it can be shown that these requirements are equivalent to the requirement of the uniform convergence. For nonlocal nonlinear problem settings, boundedness of solutions of the uniformly convergent system follows from the definition of uniform convergence and boundedness of $w(t)$. Thus the choice of uniform convergence as a "stability requirement" is natural in this problem. It leads to a necessary and sufficient solvability condition for the output regulation problem that includes, as its particular case, the solvability conditions for the linear and local nonlinear output regulation problems, see e.g., [20]. This fact indicates that the right problem formulation for the nonlocal output regulation problem is captured in this way.

From the controller design point of view, this problem can be addressed using the same approach as described in Sect. 3.3: design a controller such that the closed-loop system is (a) uniformly convergent with the UBSS property and (b) has a solution along which the asymptotic output zeroing condition holds, see Sect. 3.3. The questions to be addressed in this approach are, first, whether the structure of the controlled system and the exo-system allows for a solution of the problem; second, how to design a controller that makes the closed-loop system uniformly convergent; and, third, how to check that this controller ensures existence of a solution with the asymptotic output zeroing property.

---

[7] Other variants of the uniform output regulation problem can be found in [35].

While the second question is addressed in a number of papers, see, e.g., [35], the first and the third questions are answered by the following result.

Theorem 3.6 ([35]) *Consider system (3.28)–(3.30) and exo-system (3.31) with a compact positively invariant set of initial conditions* $\mathscr{W}_+ \subset \mathbb{R}^m$.

(i) The global uniform output regulation problem is solvable only if there exist continuous mappings $\pi(w)$ *and* $c(w)$ *defined in some neighborhood of* $\Omega(\mathscr{W}_+)$*– the* ω*-limit set for solutions of exo-system (3.31) starting in* $\mathscr{W}_+$*– and satisfying the regulator equations*

$$\frac{d}{dt}\pi(w(t)) = f(\pi(w(t)), c(w(t)), w(t)), \tag{3.36}$$

$$0 = h_r(\pi(w(t)), w(t)), \tag{3.37}$$

for all solutions of exo-system (3.31) satisfying $w(t) \in \Omega(\mathscr{W}_+)$ *for* $t \in \mathbb{R}$.

(ii) If controller (3.32), (3.33) makes the closed-loop system uniformly convergent with UBSS property for the class of bounded continuous inputs, then it solves the global uniform output regulation problem if and only if for any $w(t) \in \Omega(\mathscr{W}_+)$ *the controller has a bounded solution with input* $\bar{y}_w(t) := h_m(\pi(w(t)), w(t))$ *and output* $\bar{u}_w(t) = c(w(t))$.

Notice that solvability of the regulator equations implies that for any $w(t)$ from the omega-limit set $\Omega(\mathscr{W}_+)$, system (3.28) has a stable inversion $\bar{u}_w(t)$ for the desired output $e(t) \equiv 0$. The second condition implies that the controller, being driven by the output $\bar{y}_w(t)$, can generate the control signal $\bar{u}_w(t)$. Since all solutions of the exo-system starting in $\mathscr{W}_+$ converge to the omega-limit set $\Omega(\mathscr{W}_+)$, it is enough to verify conditions (i) and (ii) only on $\Omega(\mathscr{W}_+)$.

Here we see that the notion of uniform convergence allows us to extend the solvability conditions and controller design methods for the output regulation problem from the linear and local nonlinear cases to nonlocal nonlinear case.

3.6 Frequency Response Functions for Nonlinear Convergent Systems

Frequency response functions (FRF) for linear time invariant systems form a foundation for a large number of analysis and design tools. One can define FRF for linear systems through the Laplace transform. For nonlinear systems, however, the Laplace transform is not defined. If we notice that, for linear systems, FRF can also be viewed as a function that fully characterizes steady-state responses to harmonic excitations, we can extend the notion of FRF to nonlinear convergent systems of the form

$$\dot{z} = F(x, w), \tag{3.38}$$

$$y = h(x) \tag{3.39}$$

with $x \in \mathbb{R}^n$, $y \in \mathbb{R}$ and scalar input w. Recall that convergent systems have a uniquely defined periodic response to a periodic excitation (with the same period time). We can define the FRF as a mapping that maps input $w(t) = a \sin \omega t$ to the corresponding periodic steady-state solution $\bar{x}_{a,\omega}(t)$. As follows from the next result, this mapping has quite a simple structure.

Theorem 3.7 ([38]) *Suppose system (3.38) is uniformly convergent with UBSS property for the class of continuous bounded inputs $w(t)$. Then, there exists a continuous function $\alpha : \mathbb{R}^3 \to \mathbb{R}^n$ such that for any harmonic excitation of the form $w(t) = a \sin \omega t$, the corresponding (asymptotically stable) steady-state solution equals*

$$\bar{x}_{a\omega}(t) := \alpha(a \sin(\omega t), a \cos(\omega t), \omega). \tag{3.40}$$

As follows from Theorem 3.7, the function $\alpha(v_1, v_2, \omega)$ contains *all* information on the steady-state solutions of system (3.38) corresponding to harmonic excitations. For this reason, we give the following definition.

Definition 3.5 The function $\alpha(v_1, v_2, \omega)$ defined in Theorem 3.7 is called the *state frequency response function*. The function $h(\alpha(v_1, v_2, \omega))$ is called the *output frequency response function*.

In general, it is not easy to find such frequency response functions analytically. In some cases they can be found based on the following lemma.

Lemma 3.1 ([38]) *Under the conditions of Theorem 3.7, if there exists a continuous function $\alpha(v_1, v_2, \omega)$ differentiable in $v = [v_1, v_2]^T$ and satisfying*

$$\frac{\partial \alpha}{\partial v}(v, \omega) S(\omega) v = F(\alpha(v, \omega), v_1), \quad \forall\, v, \omega \in \mathbb{R}^2 \times \mathbb{R}, \tag{3.41}$$

then this $\alpha(v_1, v_2, \omega)$ is the state frequency response function. Conversely, if the state frequency response function $\alpha(v_1, v_2, \omega)$ is differentiable in v, then it is a unique solution of (3.41).

With this definition of the frequency response function, we can further define Bode magnitude plots for convergent systems that would map frequency ω and amplitude a of the harmonic input to a measure of the steady-state output (e.g., L_2 norm) normalized with the input amplitude a. This extension of the Bode plot enables graphical representation of convergent system steady-state responses at various frequencies and amplitudes of the excitation (due to nonlinearity it will depend on both). In this sense, such frequency response functions are instrumental in supporting frequency domain analysis of nonlinear convergent systems, similar to that employed for linear systems.

3.7 Steady-State Performance Optimization

For linear systems, Bode plots are commonly used to evaluate steady-state sensitivities of the closed-loop system to measurement noise, disturbances, and reference signals. If the performance of the closed-loop system, evaluated through the (frequency domain) sensitivity functions, is not satisfactory, controller parameters can be tuned to achieve desired or optimal steady-state performance. For nonlinear systems, such performance-based controller tuning is much more challenging. Even in the simple case of a convergent closed-loop system with a linear plant being controlled by a linear controller with a variable gain element [19, 55], this problem is far from straightforward. First, one needs to evaluate/calculate steady-state responses to the noise, disturbance and/or reference signals. In practice, this may be challenging already for excitations with only one harmonic (see previous section). In reality, the excitations will consist of multiple harmonics, and calculation of the steady-state solution to these excitations can be a challenge in itself. Second, after the steady-state solution is evaluated, one needs to find how to tune controller parameters to improve/optimize certain performance characteristics of the steady state responses.

For a subclass of nonlinear convergent systems, both of these problems can be solved numerically in a computationally efficient way. Let us consider Lur'e-type systems of the form

$$\dot{x} = Ax + Bu + Hw(t) \tag{3.42}$$

$$y = Cx + Dw(t) \tag{3.43}$$

$$u = -\varphi(y, w(t), \theta) \tag{3.44}$$

$$e = C_e x + D_e w(t), \tag{3.45}$$

where $x \in \mathbb{R}^n$ is the state, $y \in \mathbb{R}$ is the output, $w(t) \in \mathbb{R}^m$ is a piecewise continuous input, and $e \in \mathbb{R}$ is a performance output. We assume that the nonlinearity $\varphi : \mathbb{R} \times \mathbb{R}^m \times \Theta \to \mathbb{R}$ is memoryless and may depend on n_θ parameters collected in the vector $\theta = [\theta_1, \dots, \theta_{n_\theta}]^T \in \Theta \subset \mathbb{R}^{n_\theta}$. We also assume that $\varphi(0, w, \theta) = 0 \, \forall w \in \mathbb{R}^m$ and $\theta \in \Theta$. For simplicity, we only consider the case in which the parameters θ appear in the nonlinearity φ and none of the system matrices. An extension to the latter situation is relatively straightforward. The functions $G_{yu}(s)$, $G_{yw}(s)$, $G_{eu}(s)$ and $G_{ew}(s)$ are the corresponding transfer functions from inputs u and w to outputs y and e of the linear part of system (3.42)–(3.45).

In this section, we consider the case of periodic disturbances $w(t)$. Recall that if system (3.42)–(3.45) satisfies the conditions of Theorem 3.2 for all $\theta \in \Theta$, then it is exponentially convergent and thus for each periodic $w(t)$ it has a unique periodic steady-state solution $\bar{x}_w(t, \theta)$.

Once the steady-state solution is uniquely defined, we can define a performance measure to quantify the steady-state performance of the system for a particular T-periodic input $w(t)$ and particular parameter θ. For example, it can be defined as

$$J(\theta) = \frac{1}{T}\int_0^T \bar{e}_w(t,\theta)^2 dt, \tag{3.46}$$

where $\bar{e}_w(t,\theta)$ is the performance output response corresponding to the steady-state solution. If we are interested in quantifying simultaneously the steady-state performance corresponding to a family of disturbances, $w_1(t), w_2(t), \dots, w_N(t)$, with periods $T_1, \dots, T_N$, we, e.g., can choose a weighted sum of the functionals of the form (3.46). The choice of the performance objective strongly depends on the needs of the particular application.

System (3.42)–(3.45) may represent a closed-loop nonlinear control system with θ being a vector of controller parameters. Ultimately, we aim to optimize the steady-state performance of this system by tuning $\theta \in \Theta$. To this end, we propose to use gradient-like optimization algorithms, which provide a direction for decrease of $J(\theta)$ based on the gradient of $\partial J/\partial\theta(\theta)$. This approach requires computation of the gradient of $J(\theta)$. For the performance objective as in (3.46), the gradient equals

$$\frac{\partial J}{\partial\theta}(\theta) = \frac{2}{T}\int_0^T \bar{e}_w(t,\theta)\frac{\partial\bar{e}_w}{\partial\theta}(t,\theta)dt, \tag{3.47}$$

under the condition that $\bar{e}_w(t,\theta)$ is C^1 with respect to θ. Here we see that in order to compute the gradient of $J(\theta)$ we need to know both $\bar{e}_w(t,\theta)$ and $\partial\bar{e}_w/\partial\theta(t,\theta)$. The following theorem provides, firstly, conditions under which $\bar{x}_w(t,\theta)$ (and therefore $\bar{e}_w(t,\theta)$) is C^1 with respect to θ, and, secondly, gives us an equation for the computation of $\partial\bar{e}_w/\partial\theta(t,\theta)$.

Theorem 3.8 ([40]) *If system (3.42)–(3.45) satisfies the conditions of Theorem 3.2 for all $\theta \in \Theta$, and the nonlinearity $\varphi(y,w,\theta)$ is C^1 for all $y \in \mathbb{R}$, $w \in \mathbb{R}^m$ and θ in the interior of Θ, then the steady-state solution $\bar{x}_w(t,\theta)$ is C^1 in θ. The corresponding partial derivatives $\partial\bar{x}_w/\partial\theta_i(t,\theta)$ and $\partial\bar{e}_w/\partial\theta_i(t,\theta)$ are, respectively, the unique T-periodic solution $\bar{\Psi}(t)$ and the corresponding periodic output $\bar{\mu}(t)$ of the system*

$$\dot{\Psi} = A\Psi + BU + BW_i(t) \tag{3.48}$$

$$\lambda = C\Psi \tag{3.49}$$

$$U = -\frac{\partial\varphi}{\partial y}(\bar{y}(t,\theta), w(t), \theta)\lambda \tag{3.50}$$

$$\mu = C_e\Psi, \tag{3.51}$$

where $W_i(t) = -\partial\varphi/\partial\theta_i(\bar{y}_w(t,\theta), w(t), \theta)$.

To calculate the steady-state output $\bar{e}(t,\theta)$, notice that it is a solution of the following equation:

$$\bar{y} = \mathcal{G}_{yu} \circ \mathcal{F}\bar{y} + \mathcal{G}_{yw}w, \tag{3.52}$$

$$\bar{e} = \mathcal{G}_{eu} \circ \mathcal{F}\bar{y} + \mathcal{G}_{ew}w, \tag{3.53}$$

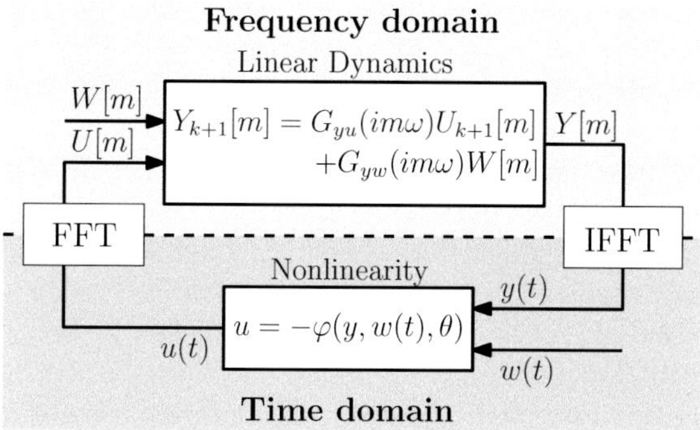

Fig. 3.1 Mixed-Time-Frequency algorithm to compute the steady-state solutions

where $\mathscr{G}_{yu}$, $\mathscr{G}_{yw}$, $\mathscr{G}_{eu}$ and $\mathscr{G}_{ew}$ are the linear operators mapping periodic inputs $u(t)$ and $w(t)$ to periodic steady-state outputs $y(t)$ and $e(t)$ of the linear part of system (3.42)–(3.45); and $\mathscr{F}$ is the operator defined by $\mathscr{F}y(t) := -\varphi(y(t), w(t), \theta)$. The conditions of Theorem 3.2 imply that the superposition operator $\mathscr{G}_{yu} \circ \mathscr{F}$ is a contraction operator acting from $L_2(T)$ to $L_2(T)$. Therefore, $\bar{y}$ (and then $\bar{e}$) can be calculated from the iterative process

$$u_{k+1} = \mathscr{F}y_k \tag{3.54}$$

$$y_{k+1} = \mathscr{G}_{yu}u_{k+1} + \mathscr{G}_{yw}w, \tag{3.55}$$

starting from an arbitrary initial guess y_0. To speed up the calculation, this iterative process can be implemented both in frequency domain (to compute $\mathscr{G}_{yu}u_{k+1}$ and $\mathscr{G}_{yw}w$) and in time domain (to compute the output of the nonlinearity $\mathscr{F}y_k$). This is schematically shown in Fig. 3.1, where Y, W and U denote the vectors of the Fourier coefficients (indexed by m) of the signals $y(t)$, $w(t)$ and $u(t)$, respectively. and (I)FFT denotes the (inverse) Fast Fourier Transform.

If at every iteration we truncate the Fourier coefficients for $u_k(t)$ and $w(t)$ to keep only the N first harmonics (which is inevitable in any numerical implementation of the algorithm), the algorithm will still converge from an arbitrary initial guess $y_0(t)$ to a unique solution $\bar{y}^N$. The error caused by the truncation can be estimated by [40]:

$$\|\bar{y} - \bar{y}^N\|_{L_2} \le \Big\{\sup_{|m|>N}|G_{yu}(im\omega)|\gamma_{yw}\tfrac{K\|w\|_{L_2}}{1-\gamma_{yu}K} \\ + \gamma_{yw}\|w - w^N\|_{L_2}\Big\}\tfrac{1}{1-\gamma_{yu}K}, \tag{3.56}$$

where $\gamma_{yw} := \sup_{m\in\mathbb{Z}}|G_{yw}(im\omega)|$ and $\|w - w^N\|_{L_2}$ is the error of truncation of harmonics in $w(t)$ higher than N. From this estimate, we can conclude that by choosing N high enough, one can reach any desired accuracy of approximation $\bar{y}(t)$ by the solution $\bar{y}^N$ of the algorithm with the truncation. Notice that this algorithm with truncation can be considered as a multiharmonic variant of the describing function method, described in [30] for *autonomous* systems.

After computing $\bar{y}_w$ and $\bar{e}_w$, one can then compute in the same way the partial derivatives $\partial\bar{e}_w/\partial\theta(t,\theta)$, since system (3.48)–(3.51) satisfies the same conditions as the original system (3.42)–(3.45) in Theorem 3.2. Then one can compute $\partial J/\partial\theta$ and proceed to gradient-like optimization of θ.

Details on implementation of these numerical algorithms can be found in [40], where they, in combination with a gradient optimization method, were applied for tuning parameters of a variable gain controller for wafer stage control. The results presented in [40] demonstrated very fast convergence of the algorithms for calculation of the steady-state solution and its gradients, as well as efficient performance of the gradient based tuning algorithm. This turns the algorithm into a powerful numerical method for optimizing steady-state performance of nonlinear closed-loop systems of Lur'e-type form.

In this section, we have shown how the convergence property can be instrumental in supporting the model-based performance optimization of nonlinear control systems. In the next section, we also consider the problem of performance optimization of nonlinear systems, where again performance is characterized by periodically time-varying steady-state solutions. However, now it is assumed that no model or disturbance information is available to support performance optimization and therefore a model-free optimization approach called extremum seeking is adopted.

3.8 Extremum Seeking Control

Extremum seeking control is a model-free, online approach for performance optimization for dynamical systems. The large majority of the works in extremum seeking is considering the case in which the performance of the system is quantified in terms of a (unknown) performance objective function depending on the *equilibrium state* of the system [23, 51, 52]. However, in many cases the performance of a system is characterized by *time-varying behaviors*; as an example, one can think of tracking control problems for high-tech positioning systems, such as industrial robots, wafer scanners or pick-and-place machines in which the machine's functioning relies on the accurate realization of time-varying (or periodic for repetitive tasks) reference trajectories.

In this section, we will show how the concept of convergence can be a key underlying property of the dynamical system subjected to an extremum seeker when the performance objective depends on periodic steady-state trajectories of the plant. For more details on extremum seeking for nonlinear plants with periodic steady-state solutions, we refer to [17].

Let us consider a nonlinear dynamical system of the form

$$\dot{x} = f(x, u, \theta, w(t)), \tag{3.57}$$

$$y = h(x, w(t)), \tag{3.58}$$

where $x \in \mathbb{R}^n$, $u \in \mathbb{R}^m$ are the state and the input, $w \in \mathbb{R}^l$ is an input disturbance and $\theta \in \mathbb{R}$ is a system parameter. The disturbances $w(t)$ are generated by an exo-system of the following form:

$$\dot{w} = \varphi(w). \tag{3.59}$$

We assume that for any initial condition $w(0) \in \mathbb{R}^l$, the solution of the exo-system (3.59) is uniformly bounded (in backward and forward time) and periodic with a known constant period $T_w > 0$.

Consider a state-feedback controller of the following form[8]

$$u = \alpha(x, \theta). \tag{3.60}$$

Now we assume that the closed-loop plant (3.57), (3.60) is uniformly convergent for any fixed $\theta \in \mathbb{R}$. The convergence property implies that for any fixed $\theta \in \mathbb{R}$, there exists a unique, bounded for all $t \in \mathbb{R}$, uniformly globally asymptotically stable steady-state solution $\bar{x}_{\theta,w}(t)$ of the closed-loop plant (3.57), (3.60).[9] As explained in Sect. 3.2, the convergence property implies that the steady-state response $\bar{x}_{\theta,w}(t)$ is periodic with period T_w, given the nature of the exo-system, which produces periodic disturbance inputs with period T_w.

We aim to find the fixed value of $\theta \in \mathbb{R}$ that optimizes the steady-state performance of the closed-loop plant (3.57), (3.60). To this end, we design a cost function that defines performance in terms of the system output y. As a stepping stone, we introduce various signal-norm-based performance measures of the following form:

$$L_p(y_d(t)) := \left(\frac{1}{T_w} \int_{t-T_w}^{t} |y(\tau)|^p d\tau \right)^{\frac{1}{p}}, \tag{3.61}$$

$$L_\infty(y_d(t)) := \max_{\tau \in [t-T_w, t]} |y(\tau)| \tag{3.62}$$

with $p \in [1, \infty)$. The argument $y_d(t)$ of the performance measures in (3.61), (3.62) represents a (past) function segment of the output, characterizing the performance, and is defined by $y_d(t) := y(t + \tau)$ for all $\tau \in [-t_d, 0]$, for some $t_d > T_w$, see [17] for details. We use one of the performance measures in (3.61), (3.62) in the design of the cost function, which is given by

$$q = Q_i(y_d(t)) := g \circ L_i(y_d(t)), \; i \in [1, \infty], \tag{3.63}$$

where the function $g(\cdot)$ further characterizes the performance cost.

Note that, by the grace of convergence, the cost function Q_i is constant in steady state. Finally, it is assumed that the steady-state performance map, i.e., the map from constant θ to q in steady state, exhibits a unique maximum. It is this maximum that we

[8] Sufficient smoothness of the functions f, h and α is assumed.

[9] In fact, a particular Lyapunov-based stability certificate is required for the solution $\bar{x}_{\theta,w}(t)$ in the scope of this section, see [17].

aim to find using an extremum seeking controller, without employing knowledge on the plant dynamics, the performance map or the amplitude or phase of the disturbance.

Next, we introduce the extremum seeker that will optimize (maximize) the steady-state performance output q. The total extremum seeking scheme is depicted schematically in Fig. 3.2 and consists of a gradient estimator (estimating the gradient of the cost function with respect to θ) and an optimizer (devised to steer the parameter θ to the optimum). The optimizer is given by

$$\dot{\hat{\theta}} = Ke, \tag{3.64}$$

where K is the optimizer gain and e is the gradient estimate provided by the gradient estimator. The gradient estimator employed here is based on a moving average filter called the mean-over-perturbation-period (MOPP) filter:

$$e = \frac{\omega}{a\pi} \int_{t-\frac{2\pi}{\omega}}^{t} q(\tau) \sin(\omega(t-\phi))d\tau, \tag{3.65}$$

where ω and a are the frequency and amplitude of the dither signal used to perturb the parameter input to the plant (therewith facilitating gradient estimation), see Fig. 3.2, and ϕ is a nonnegative constant. We note that both the performance measure in (3.61)–(3.63) and the MOPP estimation filter in (3.65) introduce delay in the closed-loop dynamics therewith challenging the analysis of stability properties of the resulting closed-loop system.

Still, it can be shown [17] that, under the assumptions posed above (in particular the convergence property), the total closed-loop system (3.57), (3.60), (3.63) including extremum seeking controller (3.64), (3.65) is semi-globally practically asymptotically stable in the sense that the parameter θ converges arbitrarily closely to its optimal value and the state solution of the plant converges arbitrarily closely to the optimal steady-state plant behavior, for arbitrarily large sets of initial conditions.

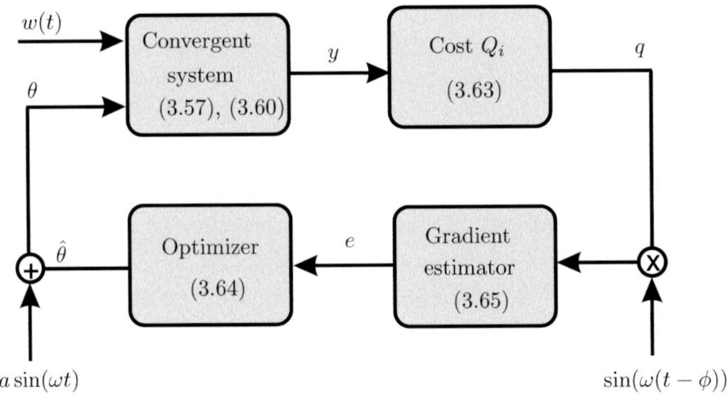

Fig. 3.2 Schematic representation of the extremum seeking scheme

The latter can be achieved by making the parameters a, ω, and K of the extremum seeker small enough.

Summarizing, the convergence property is instrumental in guaranteeing a unique and asymptotically stable periodic output response, which allows for a unique steady-state performance definition and, in turn, facilitates employing an extremum seeker to optimize the performance characterized by periodic steady-state solutions.

3.9 Model Reduction

In this section, we show how the convergence property can be instrumental in the scope of model reduction for a class of nonlinear systems.[10] The class of systems under consideration involves a feedback interconnection $\Sigma = (\Sigma_{lin}, \Sigma_{nl})$ between a linear system

$$\Sigma_{lin}: \begin{cases} \dot{x} &= Ax + B_u u + B_v v \\ y &= C_y x \\ w &= C_w x, \end{cases} \tag{3.66}$$

where $x \in \mathbb{R}^n$, $y \in \mathbb{R}^p$, $v \in \mathbb{R}^s$ and $w \in \mathbb{R}^q$, and a nonlinear system

$$\Sigma_{nl}: \begin{cases} \dot{z} &= g(z, w) \\ v &= h(z), \end{cases} \tag{3.67}$$

where, $z \in \mathbb{R}^r$, see the left part of Fig. 3.3.

As will be made more precise below, we assume that the plant Σ is (input-to-state) convergent and we will preserve such property after the model reduction. At a conceptual level, the system being convergent (before and after reduction) helps to reason about the quality of a reduced-order model. To understand this, suppose that the plant and its reduction are not convergent. Then, these systems may have complex nonlinear dynamics characterized by multiple (stable and/or unstable)

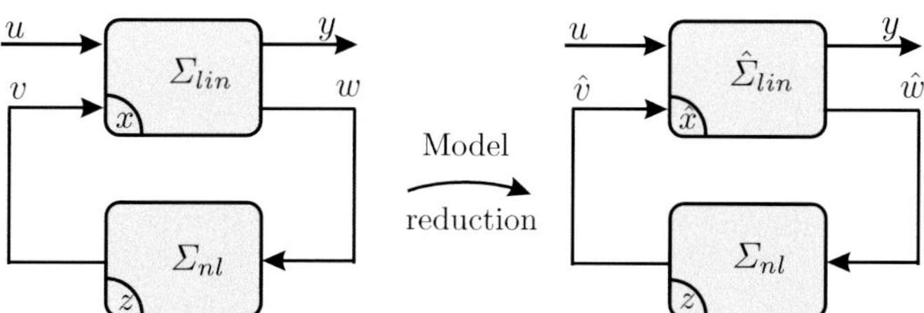

Fig. 3.3 Feedback interconnection plant dynamics and model reduction strategy

[10]The model reduction approach described here is based on [6].

attractors, such as e.g., equilibria and limit cycles, each associated with a potentially complex region of attraction. When reasoning about the quality of the reduction, one typically desires to compare solutions of the reduced-order and original system, especially when aiming to quantify a reduction error. It is hard to envision how such a comparison could be made if both systems have multiple attractors, even with regions of attraction defined on state spaces of different dimension (due to the reduction). The assumption of convergence facilitates the unique comparison of the output solutions of the reduced-order and original system, for a given identical input, as both systems now have a unique attractor (characterized by the unique steady-state solution); hence, the convergence property significantly simplifies establishing a clear definition of reduction error, as will be further explained below.

Figure 3.3 also expresses the fact that we pursue model reduction of the total system Σ by reducing the linear part Σ_{lin} of the dynamics and reconnecting the reduced-order linear dynamics $\hat{\Sigma}_{lin}$ to the nonlinear dynamics Σ_{nl}. This approach is inspired by practical applications in which, firstly, the high-dimensional nature of the dynamics is due to the linear dynamics and, second, the nonlinearities only act locally. Examples of such systems, e.g., can be found in mechanical systems in which the structural dynamics leads to high-dimensional models and local nonlinearities relate to friction, hysteresis, or nonlinear actuator dynamics. Applications in which such models arise can, e.g., be found in high-speed milling or drilling applications. A benefit of such an approach in which model reduction is applied to the linear subsystem only is the fact that a wide range of computationally efficient model reduction methods for linear systems exist.

Assumption 3.1 Now, we adopt the following assumptions on the system Σ:

- Σ_{lin} is asymptotically stable (i.e., A is Hurwitz), implying that Σ_{lin} is input-to-state convergent,
- Σ_{nl} is input-to-state convergent.

By the grace of the first bullet in Assumption 3.1, we have that there exist steady-state operators defined as $\mathscr{F}(u, v) := \bar{x}_{u,v}$, with $\bar{x}_{u,v}$ being the steady-state solutions of the convergent system Σ_{lin}, and $\mathscr{F}_i(u, v) = C_i\bar{x}_{u,v}$, $i \in \{y, w\}$, where the latter define the steady-state output operators of Σ_{lin} for outputs y and w. These steady-state output operators are (by linearity) incrementally bounded as

$$\|\mathscr{F}_i(u_2, v_2) - \mathscr{F}_i(u_1, v_1)\|_\infty \le \chi_{ix}(\gamma_{xu}\|u_2 - u_1\|_\infty + \gamma_{xv}\|v_2 - v_1\|_\infty), \quad (3.68)$$

for $i \in \{y, w\}$. In (3.68), γ_{xu}, γ_{xv} denote the gain functions of the steady-state operator $\mathscr{F}(u, v)$, whereas χ_{ix} represent incremental bounds on the output equations of Σ_{lin}. The assumption in the second bullet of Assumption 3.1 implies that there exists a steady-state operator $\mathscr{G}w := \bar{z}_w$, which satisfies $\|\mathscr{G}w_2 - \mathscr{G}w_1\| \le \gamma_{zw}\|w_2 - w_1\|_\infty$. If additionally, there exists an incremental bound for the output function h of Σ_{nl} such that $\|h(z_2) - h(z_1)\|_\infty \le \chi_{vz}\|z_2 - z_1\|_\infty$, then we have that the steady-state output operator $\mathscr{G}_v w := h(\bar{z}_w)$ of Σ_{nl} satisfies

$$\|\mathscr{G}_v(w_2) - \mathscr{G}_v(w_1)\|_\infty \leq \chi_{vz} \circ \gamma_{zw} \|w_2 - w_1\|_\infty. \tag{3.69}$$

All the gain functions above are class $\mathscr{K}_\infty$ functions.

Next, the total nonlinear system Σ is assumed to satisfy the following small-gain condition.

Assumption 3.2 There exist class $\mathscr{K}_\infty$ functions ρ_1 and ρ_2 such that Σ satisfies the small-gain condition

$$(id + \rho_1) \circ \gamma_{xv} \circ \chi_{vz} \circ (id + \rho_2) \circ \gamma_{zw} \circ \chi_{wx}(s) \leq s, \tag{3.70}$$

for all $s \geq 0$.

Assumption 3.2 implies, see [4, 6], that the feedback interconnection Σ is input-to-state convergent.

As a next step, we assume that the reduced-order linear system $\hat{\Sigma}_{lin}$, see Fig. 3.3, given by

$$\hat{\Sigma}_{lin}: \begin{cases} \dot{\hat{x}} &= \hat{A}\hat{x} + \hat{B}_u u + \hat{B}_v \hat{v} \\ \hat{y} &= \hat{C}_y \hat{x} \\ \hat{w} &= \hat{C}_w \hat{x}, \end{cases} \tag{3.71}$$

where $\hat{x} \in \mathbb{R}^k$, with $k < n$ establishing the order reduction, $\hat{y} \in \mathbb{R}^p$, $\hat{v} \in \mathbb{R}^s$ and $\hat{w} \in \mathbb{R}^q$, is asymptotically stable. This implies that there exist steady-state output operators for $\hat{\Sigma}_{lin}$: $\hat{\mathscr{F}}_i(u, \hat{v})$, $i \in \{y, w\}$, that are incrementally bounded, i.e.,

$$\|\hat{\mathscr{F}}_i(u_2, \hat{v}_2) - \hat{\mathscr{F}}_i(u_1, \hat{v}_1)\|_\infty \leq \hat{\chi}_{ix} \circ \hat{\gamma}_{xv} \|\hat{v}_2 - \hat{v}_1\|_\infty, \tag{3.72}$$

for $i \in \{y, w\}$. Moreover, we assume that there exists an error bound for the reduction of the linear part of the system according to

$$\|\mathscr{E}_i(u_2, v_2) - \mathscr{E}_i(u_1, v_1)\|_\infty \leq \varepsilon_{iu} \|u_2 - u_1\|_\infty + \varepsilon_{iv} \|v_2 - v_1\|_\infty, \tag{3.73}$$

where $\mathscr{E}_i(u, v) := \mathscr{F}_i(u, v) - \hat{\mathscr{F}}_i(u, v)$, for $i \in \{y, w\}$, and ε_{ij}, for $i \in \{y, w\}$ and $j \in \{u, v\}$, are positive constants.

In fact, the above assumption on the stability of the reduced-order system and the availability of an error bound for the linear reduced-order system can be directly satisfied since reduction techniques exist that guarantee the satisfaction of both assumptions. In fact, an a priori error bound exists when the reduced-order system $\hat{\Sigma}_{lin}$ is obtained by balanced truncation. Namely, an error bound on the norm on the impulse response as in [16, 24] provides a bound on the $\mathscr{L}_\infty$-induced system norm. Alternatively, an error bound can be computed a posteriori using results from [47], typically leading to a tighter bound.

Now, the following result can be formulated, which guarantees input-to-state convergence of the nonlinear reduced-order system $\hat{\Sigma} = (\hat{\Sigma}_{lin}, \Sigma_{nl})$ and provides and error bound for $\hat{\Sigma}$.

Theorem 3.9 *Let $\Sigma = (\Sigma_{lin}, \Sigma_{nl})$ satisfy Assumptions 3.1 and 3.2. Furthermore, let $\hat{\Sigma} = (\hat{\Sigma}_{lin}, \Sigma_{nl})$ be a reduced-order approximation, where $\hat{\Sigma}_{lin}$ is asymptotically stable and let there exist an error bound as in (3.73) on the linear subsystem. Then, the reduced-order nonlinear system $\hat{\Sigma}$ is input-to-state convergent if there exist class $\mathcal{K}_\infty$ functions $\hat{\rho}_1$ and $\hat{\rho}_2$ such that the following small-gain condition is satisfied:*

$$(id + \hat{\rho}_1) \circ \chi_{vz} \circ (id + \hat{\rho}_2) \circ (\varepsilon_{wv} + \chi_{wx} \circ \gamma_{xv})(s) \leq s, \tag{3.74}$$

for all $s \geq 0$.

When (3.74) holds, then the steady-state error $\|\bar{y}_u - \bar{\hat{y}}_u\|_\infty$ is bounded as $\|\bar{y}_u - \bar{\hat{y}}_u\|_\infty \leq \varepsilon \|u\|_\infty$, where $\varepsilon(r)$ is an error bound function.

For the proof and a detailed expression for the error bound function $\varepsilon(r)$ we refer to [6]. This error bound function $\varepsilon(r)$ depends on the properties (gain functions) of the original system Σ and the error bounds for the linear reduction (3.73). As the latter error bound can be obtained a priori (i.e., before the actual reduction is performed), the error bound in Theorem 3.9 also represents an a priori error bound. Note, moreover, that if the small-gain condition on the original system in Assumption 3.2 is satisfied with some margin, the small-gain condition in (3.74) can be satisfied by making the reduction of Σ_{lin} accurate enough, i.e., making ε_{wv} small enough.

Finally, we note that with this convergence-based approach to model reduction we obtain an error bound on the $\mathcal{L}_\infty$-norm of the reduction error. Alternative approaches exist, see [4, 5], exploiting incremental $\mathcal{L}_2$-gain or incremental passivity properties, instead of convergence properties, to obtain reduced-order systems (for a class of nonlinear systems of the same form as considered here) preserving such incremental system properties and complying with an $\mathcal{L}_2$ error bound.

3.10 Conclusions

In this chapter, we have reviewed the notion of convergent systems and its applications to a wide range of design and analysis problems for nonlinear (control) systems. It appears that nonlinear convergent systems inherit certain simplicity from asymptotically stable linear systems. This simplicity is not common to generic nonlinear systems. It allows one to solve a number of analysis and design problems for nonlinear systems in a nonlocal setting and extend previously known local results to nonlocal cases. For Lur'e-type systems it provides a powerful tool for optimization of steady-state performance. Open problems for further work relate to convergence properties for hybrid systems, to investigating how convergence and the existence of FRFs can be used to support system identification for certain classes of nonlinear convergent systems, and to applications of convergent systems to filtering.

References

1. Angeli, D.: Further results on incremental input-to-state stability. IEEE Trans. Autom. Control **54**(6), 1386–1391 (2002)
2. Angeli, D.: A Lyapunov approach to incremental stability properties. IEEE Trans. Autom. Control **47**, 410–421 (2002)
3. Andrieu, V., Jayawardhana, B., Praly, L.: On transverse exponential stability and its use in incremental stability, observer and synchronization. In: Proceedings of IEEE Conference on Decision and Control, pp. 5915–5920. Florence (2013)
4. Besselink, B.: Model reduction for nonlinear control systems: with stability preservation and error bounds. Ph.D. thesis, Eindhoven University of Technology, Eindhoven (2012)
5. Besselink, B., van de Wouw, N., Nijmeijer, H.: Model reduction for nonlinear systems with incremental gain or passivity properties. Automatica **49**(4), 861–872 (2000)
6. Besselink, B., van de Wouw, N., Nijmeijer, H.: Model reduction for a class of convergent nonlinear systems. IEEE Trans. Autom. Control **57**(4), 1071–1076 (2012)
7. Byrnes, C.I., Delli Priscoli, F., Isidori, A.: Output Regulation of Uncertain Nonlinear Systems. Birkhäuser, Boston (1997)
8. Camlibel, K., van de Wouw, N.: On the convergence of linear passive complementarity systems. In: Proceedings of IEEE Conference Decision and Control, pp. 5886–5891 (2007)
9. Chaillet, A., Pogromsky, A., Rüffer, B.: A Razumikhin approach for the incremental stability of delayed nonlinear systems. In: Proceedings of IEEE Conference Decision and Control, pp. 1596–1602 (2013)
10. Demidovich, B.P.: Lectures on Stability Theory (in Russian). Nauka, Moscow (1967)
11. Devasia, S., Paden, B.: Stable inversion for nonlinear nonminimum-phase time-varying systems. IEEE Trans. Autom. Control **43**, 283–288 (1998)
12. Devasia, S., Chen, D., Paden, B.: Nonlinear inversion-based output tracking. IEEE Trans. Autom. Control **41**, 930–942 (1996)
13. Forni, F., Sepulchre, R.: A differential Lyapunov framework for contraction analysis. IEEE Trans. Autom. Control **59**(3), 614–628 (2014)
14. Francis, B.A., Wonham, W.M.: The internal model principle of control theory. Automatica **12**, 457–465 (1976)
15. Fromion, V., Monaco, S., Normand-Cyrot, D.: Asymptotic properties of incrementally stable systems. IEEE Trans. Autom. Control **41**, 721–723 (1996)
16. Glover, K., Curtain, R.F., Partington, J.R.: Realisation and approximation of linear infinite-dimensional systems with error bounds. SIAM J. Control Optim. **26**(4), 863–898 (1988)
17. Haring, M., van de Wouw, N., Nešić, D.: Extremum-seeking control for nonlinear systems with periodic steady-state outputs. Automatica **49**(6), 1883–1891 (2013)
18. Heemels, W.M.P.H., Lazar, M., van de Wouw, N., Pavlov, A.: Observer-based control of discrete-time piecewise affine systems: exploiting continuity twice. In: Proceedings of IEEE Conference Decision and Control, pp. 4675–4680 (2008)
19. Heertjes, M.F., Steinbuch, M.: Stability and performance of variable gain controllers with application to a DVD storage drive. Automatica **40**, 591–602 (2004)
20. Isidori, A., Byrnes, C.I.: Output regulation of nonlinear systems. IEEE Trans. Autom. Control **35**, 131–140 (1990)
21. Isidori, A., Byrnes, C.I.: Limit sets, zero dynamics and internal models in the problem of nonlinear output regulation. IEEE Trans. Autom. Control **48**(10), 1712–1723 (2003)
22. Jouffroy, J.: Some ancestors of contraction analysis. In: Proceedings of IEEE Conference Decision and Control, pp. 5450–5455 (2005)
23. Krstić, M., Wang, H.-H.: Stability of extremum seeking feedback for general nonlinear dynamic systems. Automatica **36**(4), 595–601 (2000)
24. Lam, J., Anderson, B.D.O.: L1 impulse response error bound for balanced truncation. Syst. Control Lett. **18**(2), 129–137 (1992)

25. Leine, R.I., van de Wouw, N.: Stability and Convergence of Mechanical Systems with Unilateral Constraints. Lecture Notes in Applied and Computational Mechanics, vol. 36. Springer, Berlin (2008)
26. Leine, R.I., van de Wouw, N.: Uniform convergence of monotone measure differential inclusions: with application to the control of mechanical systems with unilateral constraints. Int. J. Bifurc. Chaos **18**(5), 1435–1457 (2008)
27. Li, Y., Phillips, S., Sanfelice, R.G.: Results on incremental stability for a class of hybrid systems. In: Proceedings of IEEE Conference Decision and Control (2014)
28. Li, Y., Sanfelice, R.G.: On necessary and sufficient conditions for incremental stability of hybrid systems using the graphical distance between solutions. In: Proceedings of IEEE Conference Decision and Control (2015)
29. Lohmiller, W., Slotine, J.-J.E.: On contraction analysis for nonlinear systems. Automatica **34**, 683–696 (1998)
30. Mees, A.: The describing function matrix. IMA J. Appl. Math. **10**(1), 49–67 (1972)
31. Nijmeijer, H., Mareels, I.M.Y.: An observer looks at synchronization. IEEE Trans. Circuits Syst. –I: Fundam. Theory Appl. **44**(10), 882–890 (1997)
32. Pavlov, A., Marconi, L.: Incremental passivity and output regulation. Syst. Control Lett. **57**(5), 400–409 (2008)
33. Pavlov, A., Pettersen, K.Y.: Stable inversion of non-minimum phase nonlinear systems: a convergent systems approach. In: Proceedings of IEEE Conference Decision and Control (2007)
34. Pavlov, A., Pogromsky, A., van de Wouw, N., Nijmeijer, H.: Convergent dynamics, a tribute to Boris Pavlovich Demidovich. Syst. Control Lett. **52**, 257–261 (2004)
35. Pavlov, A., van de Wouw, N., Nijmeijer, H.: Uniform Output Regulation of Nonlinear Systems: A Convergent Dynamics Approach. Birkhäuser, Boston (2005)
36. Pavlov, A., Pogromsky, A., van de Wouw, N., Nijmeijer, H.: On convergence properties of piecewise affine systems. Int. J. Control **80**(8), 1233–1247 (2007)
37. Pavlov, A., Pogromsky, A.Y., van de Wouw, N., Nijmeijer, H.: On convergence properties of piecewise affine systems. Int. J. Control **80**(8), 1233–1247 (2007)
38. Pavlov, A., van de Wouw, N., Nijmeijer, H.: Frequency response functions for nonlinear convergent systems. IEEE Trans. Autom. Control **52**(6), 1159–1165 (2007)
39. Pavlov, A., van de Wouw, N.: Steady-state analysis and regulation of discrete-time nonlinear systems. IEEE Trans. Autom. Control **57**(7), 1793–1798 (2013)
40. Pavlov, A., Hunnekens, B.G.B., van de Wouw, N., Nijmeijer, H.: Steady-state performance optimization for nonlinear control systems of Lure type. Automatica **49**(7), 2087–2097 (2013)
41. Pliss, V.A.: Nonlocal Problems of the Theory of Oscillations. Academic Press, London (1966)
42. Pogromsky, A., Matveev, A.S.: Input-dependent incremental stability criterion for piecewise linear analogs of Van der Pol systems. In Proceedings of IEEE Conference Decision and Control (2015)
43. Pogromsky, A.Y., Matveev, A.S.: A non-quadratic criterion for stability of forced oscillations. IEEE Trans. Autom. Control **62**(5), 408–412 (2013)
44. Postoyan, R., Biemond, J.J.B., Heemels, W.M.P.H., van de Wouw, N.: Definitions of incremental stability for hybrid systems. In: Proceedings of IEEE Conference Decision and Control (2015)
45. Russo, G., Di Bernardo, M., Sontag, E.D.: Global entrainment of transcriptional systems to periodic inputs. PLoS Comput. Biol. **6**(4), e1000739 (2010)
46. Rüffer, B.S., van de Wouw, N., Mueller, M.: Convergent systems versus incremental stability. Syst. Control Lett. **62**, 277–285 (2013)
47. Scherer, C., Gahinet, P., Chilali, M.: Multiobjective output-feedback control via LMI optimization. IEEE Trans. Autom. Control **42**(7), 896–911 (1997)
48. Sontag, E.D.: On the input-to-state stability property. Eur. J. Control **1**(1), 24–36 (1995)
49. Sontag, E.D.: Contractive systems with inputs. In: Willems, J., Hara, S., Ohta, Y., Fujioka, H. (eds.) Perspectives in Mathematical System Theory, Control, and Signal Processing, pp. 217–228. Springer, Berlin (2010)

50. Stan, G.B., Sepulchre, R.: Analysis of interconnected oscillators by dissipativity theory. IEEE Trans. Autom. Control **52**(2), 256–270 (2007)
51. Tan, Y., Nešić, D., Mareels, I.M.Y.: On non-local stability properties of extremum seeking control. Automatica **42**(6), 889–903 (2006)
52. Tan, Y., Moase, W.H., Manzie, C., Nešić, D., Mareels, I.M.Y.: Extremum seeking from 1922 to 2010. In: Proceedings of the 29th Chinese Control Conference, Beijing, China (2010)
53. van de Wouw, N., Pavlov, A.: Tracking and synchronisation for a class of PWA systems. Automatica **44**(11), 2909–2915 (2008)
54. van den Berg, R.A., Pogromsky, A.Y., Leonov, G.A., Rooda, J.E.: Design of convergent switched systems. In: Pettersen, K.Y., Gravdahl, J.T., Nijmeijer, H. (eds.) Group Coordination and Cooperative Control. Springer, Berlin (2006)
55. van de Wouw, N., Pastink, H.A., Heertjes, M.F., Pavlov, A., Nijmeijer, H.: Performance of convergence-based variable-gain control of optical storage drives. Automatica **44**(1), 15–27 (2008)
56. Yakubovich, V.A.: Matrix inequalities method in stability theory for nonlinear control systems: I. Absolute stability of forced vibrations. Autom. Remote Control **7**, 905–917 (1964)
57. Zamani, M., van de Wouw, N., Majumdar, R.: Backstepping design for incremental stability. IEEE Trans. Autom. Control **56**(9), 2184–2189 (2013)

Part II
Synchronization in Networked Systems

Chapter 4
Synchronisation and Emergent Behaviour in Networks of Heterogeneous Systems: A Control Theory Perspective

Elena Panteley and Antonio Loría

Abstract Generally speaking, for a network of interconnected systems, synchronisation consists in the mutual coordination of the systems' motions to reach a common behaviour. For homogeneous systems that have identical dynamics this typically consists in asymptotically stabilising a common equilibrium set. In the case of heterogeneous networks, in which systems may have different parameters and even different dynamics, there may exist no common equilibrium but an emergent behaviour arises. Inherent to the network, this is determined by the connection graph but it is independent of the interconnection strength. Thus, the dynamic behaviour of the networked systems is fully characterised in terms of two properties whose study may be recast in the domain of stability theory through the analysis of two interconnected dynamical systems evolving in orthogonal spaces: the emergent dynamics and the synchronisation errors relative to the common behaviour. Based on this premise, we present some results on robust stability by which one may assess the conditions for practical asymptotic synchronisation of networked systems. As an illustration, we broach a brief case-study on mutual synchronisation of heterogeneous chaotic oscillators.

4.1 Introduction

As its etymology suggests, synchronisation may be defined as the adjustment of rhythms of repetitive events (phenomena, processes, ...) through sufficiently strong interaction. In dynamical systems theory, we also speak of synchronised systems if their movements are coordinated in time and/or space. It can be of several types: if one

---

Chapter gladly devoted to our dear friend H. Nijmeijer who introduced the authors to the topic of synchronisation 19 years ago.

E. Panteley (✉) · A. Loría
Laboratoire des signaux et systèmes, CNRS, Gif sur Yvette, France
e-mail: Elena.Panteley@lss.supelec.fr

A. Loría
e-mail: Antonio.Loria@lss.supelec.fr

E. Panteley
ITMO University, Kronverkskiy av. 49, St. Petersburg 197101, Russia

N. van de Wouw et al. (eds.), *Nonlinear Systems*, Lecture Notes in Control and Information Sciences 470, DOI 10.1007/978-3-319-30357-4_4

system "dominates" over the rest, we speak of *master–slave* synchronisation; in this case, the motion of the so-called master system becomes a reference for the motion of the so-called slave system(s). Alternatively, synchronisation may be *mutual*, in which case a set of systems synchronise their movements without specified hierarchy. *Controlled* synchronisation of dynamical systems consists, generally speaking, in ensuring that two or more systems coordinate their motions in a desired manner.

Synchronisation has been a subject of intense research in several disciplines before control theory: it was introduced in the 1970s in the USSR in the field of mechanical vibration by Professor Blekhman. Ever since, research on synchronisation has been popular among physicists, e.g. in the context of synchronisation of chaotic systems since the early 1990s, but also among engineers, especially on automatic control. In this community, the paradigm of synchronisation was largely popularised by H. Nijmeijer. His seminal paper [12] is a landmark tutorial on master–slave synchronisation and his pioneer work [13] on mutual synchronisation (of mechanical systems) preceeds the bulk of literature on a paradigm that is nowadays better known in our comunity under the name of consensus—see [19].

Consensus pertains to the case in which a (large) group of interconnected systems mutually synchronise their behaviours. In this case, we speak of *networks* of systems. These are not *just* large-scale and complex systems but they are characterised by decentralised, distributed, networked compositions of (semi)autonomous elements. These new systems are, in fact, *systems of systems*. The complexity of network interconnected systems may not be overestimated. For instance, in neuronal networks, experimental evidence shows that inhibition/excitation unbalance may result in excessive neuronal synchronisation, which, in turn, may be linked to neuro-degenerative diseases such as Parkinson and epilepsy. In energy transformation networks, the improper management of faults, overloads or simply adding to or subtracting a generator from the transportation network may result in power outages or even in large-scale (continent-wide) blackouts.

In this chapter, we briefly describe a framework, which was originally and recently introduced in [15], for analysis and control of synchronisation of networked heterogeneous systems that is, with different parameters or even completely distinct dynamic models.

We limit our study to the *analysis* paradigm, as opposed to that of *controlled* synchronisation. At the expense of technological and dynamical aspects related directly to the network communication (delays, noise, etc.), we focus on *structural* properties of the network that affect the synchronisation of the agents' motions in one way or another. More precisely, the following issues play a key role in analysis and control of synchronisation of networked systems:

- the coupling strength;
- the network topology;
- the type of coupling between the nodes, i.e. how the units are interconnected;
- the dynamics of the individual units.

Among the latter, our technical results establish how the coupling strength affects synchronisation. Our analysis is carried out from a dynamical systems and stability theory viewpoint.

4.2 The Networked Systems Synchronisation Paradigm

Let us consider a network of dynamical systems modelled via ordinary differential equations,

$$\dot{\boldsymbol{x}}_i = f_i(\boldsymbol{x}_i) + \boldsymbol{B}\boldsymbol{u}_i, \qquad i \in \mathscr{I} := \{1, \ldots, N\} \tag{4.1a}$$

$$\boldsymbol{y}_i = \boldsymbol{C}\boldsymbol{x}_i, \tag{4.1b}$$

where $\boldsymbol{x}_i \in \mathbb{R}^n$, $\boldsymbol{u}_i \in \mathbb{R}^m$ and $\boldsymbol{y}_i \in \mathbb{R}^m$ denote the state, the input and the output of the ith unit, respectively. The network's topology is usually described via graph theory: a network of N nodes is defined by its $N \times N$ adjacency matrix $D = [d_{ij}]$ whose (i, j) element, denoted by d_{ij}, specifies an interconnection between the ith and jth nodes. See [19].

The interaction among nodes depends, in general, on the strength of the coupling and on the nodes' state variables or on functions of the latter, i.e. outputs which define the coupling terms. The interaction is also determined by the form of coupling, i.e. the way how the output of one node affects another; this can be linear, as it is fairly common to assume, but it may also be nonlinear, as in the well-known example of Kuramoto's oscillator model in which the interconnection is made via sinusoids—see [3].

Here, we consider a network composed of N heterogeneous diffusively coupled nonlinear dynamical systems in normal form:

$$\dot{y}_i = f_i^1(y_i, z_i) + u_i \tag{4.2a}$$

$$\dot{z}_i = f_i^2(y_i, z_i). \tag{4.2b}$$

As it may be clear from the notation, each unit possesses one input u_i and one output y_i of the same dimension, i.e. $u_i, y_i \in \mathbb{R}^m$. The state z_i corresponds to that of the ith agent's zero dynamics—see [7]. The functions $f_i^1 : \mathbb{R}^m \times \mathbb{R}^{n-m} \to \mathbb{R}^m$, $f_i^2 : \mathbb{R}^m \times \mathbb{R}^{n-m} \to \mathbb{R}^{n-m}$ are assumed to be locally Lipschitz.

It is convenient to remark that there is little loss of generality in considering systems in normal form, these are equivalent to systems of the form (4.1) under the assumption that the matrices $B \in \mathbb{R}^{n\times m}$ and $C \in \mathbb{R}^{m\times n}$ satisfy a similarity condition for CB that is, if there exists U such that $U^{-1}CBU = \Lambda$ where Λ is diagonal positive—see e.g. [17, 18].

We also assume that the units possess certain physical properties reminiscent of energy dissipation and propagation. Notably, one of our main hypotheses is that the solutions are ultimately bounded; we recall the definition below.

Definition 4.1 (*Ultimate boundedness*) The solutions of the system $\dot{x} = f(x)$, defined by absolutely continuous functions $(t, x_\circ) \mapsto x$, are said to be ultimately bounded if there exist positive constants $\Delta_\circ$ and B_x such that for every $\Delta \in (0, \Delta_\circ)$, there exists a positive constant $T(\Delta)$ such that, for all $x_\circ \in B_\Delta = \{x \in \mathbb{R}^n : |x| \leq \Delta\}$ they satisfy

$$|x(t, x_\circ)| \leq B_x \quad \text{for all} \quad t \geq T.$$

If this bound holds for any arbitrarily large Δ then the solutions are globally ultimately bounded.

Ultimate boundedness is a reasonable assumption for the class of systems of interest here, such as oscillators. In a more general context, boundedness holds, for instance, if the units are strictly semi-passive—cf. [14].

Our second main assumption concerns the zero dynamics.

Assumption 4.1 For any compact sets $B_z \subset \mathbb{R}^{n-m}$, $B_y \subset \mathbb{R}^m$ there exist N continuously differentiable positive definite functions $V_{\circ k} : B_z \to \mathbb{R}_+$ with $k \leq N$, class $\mathcal{K}_\infty$ functions γ_{1k}, γ_{2k} and constants $\bar{\alpha}_k$, $\beta_k > 0$ such that

$$\gamma_{1k}(|z|) \leq V_{\circ k}(z) \leq \gamma_{2k}(|z|)$$

$$\nabla V_{\circ k}(z)\left[f_k^2(y, z) - f_k^2(y, z')\right] \leq -\bar{\alpha}_k|z - z'|^2 + \beta_k$$

where $\nabla V_{\circ k} := \frac{\partial V_{\circ k}}{\partial z}$, for all $z, z' \in B_z$ and $y \in B_y$.

Assumption 4.1 may be interpreted as a condition of incremental stability of the zero dynamics in a practical sense. Note that when $\beta_k = 0$, we recover the characterisation provided in [1].

4.2.1 *Network Model*

We assume that the network units are connected via *diffusive coupling*, i.e. for the ith unit the coupling is given by

$$u_i = -\sigma \sum_{j=1}^{N} d_{ij}(y_i - y_j), \tag{4.3}$$

where the scalar σ corresponds to the coupling gain between the units and the individual interconnections weights, d_{ij}, satisfy the property $d_{ij} = d_{ji}$. Assuming that the network graph is connected and undirected, the interconnections amongst the nodes are completely defined by the adjacency matrix, $D = [d_{ij}]_{i,j\in\mathscr{I}}$, which is used to construct the corresponding Laplacian matrix,

$$L = \begin{bmatrix} \sum_{i=2}^{N} d_{1i} & -d_{12} & \dots & -d_{1N} \\ -d_{21} & \sum_{i=1,i\neq 2}^{N} d_{2i} & \dots & -d_{2N} \\ \vdots & \vdots & \ddots & \vdots \\ -d_{N1} & -d_{N2} & \dots & \sum_{i=1}^{N-1} d_{Ni}. \end{bmatrix}.$$

By construction, all row-sums of L are equal to zero. Moreover, since L is symmetric and the network is connected it follows that all eigenvalues of the Laplacian matrix are real and, moreover, L has exactly one eigenvalue (say, λ_1) equal to zero, while others are positive, i.e. $0 = \lambda_1 < \lambda_2 \leq \cdots \leq \lambda_N$.

Next, we introduce a compact notation that is convenient for our purposes of analysis. We introduce the following vectors of outputs, inputs and states, respectively:

$$y = \begin{bmatrix} y_1 \\ \vdots \\ y_N \end{bmatrix} \in \mathbb{R}^{mN}, \quad u = \begin{bmatrix} u_1 \\ \vdots \\ u_N \end{bmatrix} \in \mathbb{R}^{mN}, \quad x = \begin{bmatrix} x_1 \\ \vdots \\ x_N \end{bmatrix} \in \mathbb{R}^{nN}, \quad x_i = \begin{bmatrix} y_i \\ z_i \end{bmatrix} \in \mathbb{R}^{n}$$

as well as the function $F : \mathbb{R}^{nN} \to \mathbb{R}^{nN}$, defined as

$$F(x) = \begin{bmatrix} F_1(x_1) \\ \vdots \\ F_N(x_N) \end{bmatrix}, \quad F_i(x_i) = \begin{bmatrix} f_i^1(y_i, z_i) \\ f_i^2(y_i, z_i) \end{bmatrix}_{i\in\mathscr{I}}. \tag{4.4}$$

With this notation, the diffusive coupling inputs u_i, defined in (4.3), can be re-written in the compact form

$$u = -\sigma[L \otimes I_m]y,$$

where the symbol $\otimes$ stands for the right Kronecker product.[1] Then, the network dynamics becomes

$$\dot{x} = F(x) - \sigma[L \otimes E_m]y \tag{4.5a}$$
$$y = [I_N \otimes E_m^\top]x, \tag{4.5b}$$

where $E_m^\top = [I_m, 0_{m\times(n-m)}]$. The qualitative analysis of the solutions to the latter equations is our main subject of study.

[1]For two matrices A and B of any dimension, $A \otimes B$ consists in a block-matrix in which the ijth block corresponds to $a_{ij}B$.

4.2.2 Dynamic Consensus and Practical Synchronisation

In a general setting, as for instance that of [13], for the purpose of analysis, synchronisation may be qualitatively measured by equating a functional of the trajectories to zero and measuring the distance of the latter to a synchronisation manifold, e.g.

$$\mathscr{S} = \{\boldsymbol{x} \in \mathbb{R}^{nN} : \boldsymbol{x}_1 = \boldsymbol{x}_2 = \cdots = \boldsymbol{x}_N\}. \tag{4.6}$$

For networks of homogeneous systems, i.e. if $f_i = f_j$ for all $i,\ j \in \mathscr{I}$, synchronisation is often described in terms of the asymptotically identical evolution of the units, i.e. $x_i \to x_j$. This is especially clear in the classical consensus paradigm of simple integrators, in which we have $x_i \to x_j \to$ const. In more complex cases, as for instance in problem of formation tracking control, we may have that each unit follows a (possibly unique) reference trajectory, that is, $x_i \to x_j \to x^*(t)$. What is more, controlled synchronisation is sometimes assimilated to a problem of "collective" tracking control—see e.g. [5, 13].

Hence, whether a set-point equilibrium or a reference trajectory, it seems natural to formulate the consensus problem as one of asymptotic stability (or stabilisation for that effect) of the synchronisation manifold $\mathscr{S}$. Such stability problem may be approached, for instance, using tools developed for semi-passive, incrementally passive or incrementally input-output stable systems—see [6, 8, 9, 16, 17, 21]. If the manifold $\mathscr{S}$ is stabilised, one says that the networked units are synchronised. For networks of non-identical units, the paradigm is much more complex due to the fact that the synchronisation manifold $\mathscr{S}$ does not necessarily exist. Yet, it may also be recast in terms of stability analysis.

To that end, we generalise the consensus paradigm by introducing what we call *dynamic consensus*. We shall say that this property is achieved by the systems interconnected in a network if their motions converge to one generated by what we shall call *emergent dynamics*. In the case of undirected graphs, for which the corresponding Laplacian is symmetric, the emergent dynamics is naturally defined as the average of the units' drifts, that is, the functions $f_s^1 : \mathbb{R}^m \times \mathbb{R}^{n-m} \to \mathbb{R}^m$, $f_s^2 : \mathbb{R}^m \times \mathbb{R}^{n-m} \to \mathbb{R}^{n-m}$ defined as

$$f_s^1(y_e, z_e) := \frac{1}{N}\sum_{i=1}^{N} f_i^1(y_e, z_e),$$

$$f_s^2(y_e, z_e) := \frac{1}{N}\sum_{i=1}^{N} f_i^2(y_e, z_e)$$

hence, the emergent dynamics may be written in the compact form

$$\dot{x}_e = f_s(x_e) \qquad x_e = [y_e^\top\ z_e^\top]^\top, \quad f_s := [f_s^{1\top}\ f_s^{2\top}]^\top. \tag{4.7}$$

For the sake of comparison, in the classical (set-point) consensus paradigm, all systems achieving consensus converge to a common equilibrium *point*, that is, $f_s \equiv 0$ and x_e is constant. In the case of formation tracking *control*, Eq. (4.7) can be seen as the reference dynamics to the formation. In the general case of dynamic consensus, the motions of all the units converge to a *motion* generated by the emergent dynamics (4.7).

Then, to study the behaviour of the individual network-interconnected systems, relative to that of the emergent dynamics, we introduce the *average* state (also called mean-field) and its corresponding dynamics. Let

$$x_s = \frac{1}{N}\sum_{i=1}^{N} x_i, \tag{4.8}$$

which comprises an average output, $y_s \in \mathbb{R}^m$, defined as $y_s = E_m^\top x_s$ and the state of the average zero dynamics, $z_s \in \mathbb{R}^{n-m}$, that is, $x_s = [y_s^\top, z_s^\top]^\top$. Now, by differentiating on both sides of (4.8) and after a direct computation in which we use (4.2), (4.3) and the fact that the sums of the elements of the Laplacian's rows equal to zero, i.e.

$$\frac{1}{N}\sum_{i=1}^{N} -\sigma\big[d_{i1}(y_i - y_1) + \cdots + d_{iN}(y_i - y_N)\big] = 0,$$

we obtain

$$\dot{y}_s = \frac{1}{N}\sum_{i=1}^{N} f_i^1(y_i, z_i), \tag{4.9a}$$

$$\dot{z}_s = \frac{1}{N}\sum_{i=1}^{N} f_i^2(y_i, z_i). \tag{4.9b}$$

Then, in order to write the latter in terms of the average state x_s, we use the functions f_s^1 and f_s^2 defined above so, from (4.9), we derive the *average dynamics*

$$\dot{y}_s = f_s^1(y_s, z_s) + \frac{1}{N}\sum_{i=1}^{N}\big[f_i^1(y_i, z_i) - f_i^1(y_s, z_s)\big], \tag{4.10a}$$

$$\dot{z}_s = f_s^2(y_s, z_s) + \frac{1}{N}\sum_{i=1}^{N}\big[f_i^2(y_i, z_i) - f_i^2(y_s, z_s)\big]. \tag{4.10b}$$

It is to be remarked that this model is intrinsic to the diffusively interconnected network. Indeed, since the row-sums of the Laplacian equals to zero, the interconnection strength σ does not appear in (4.10). Another interesting feature of Eq. (4.10) is that they may be regarded as composed of the nominal part

$$\dot{y}_s = f_s^1(y_s, z_s)$$
$$\dot{z}_s = f_s^2(y_s, z_s)$$

and the perturbation terms $\left[f_i^1(y_i, z_i) - f_i^1(y_s, z_s)\right]$ and $\left[f_i^2(y_i, z_i) - f_i^2(y_s, z_s)\right]$. The former corresponds exactly to (4.7), only re-written with another state variable. In the case that dynamic consensus is achieved (that is, in the case of complete synchronisation) and the graph is balanced and connected, we have $(y_i, z_i) \to (y_s, z_s)$. Nonetheless, in the case of a heterogeneous network, asymptotic synchronisation is in general hard to achieve hence, $y_i \not\to y_s$ and, consequently, the terms $\left[f_i^1(y_i, z_i) - f_i^1(y_s, z_s)\right]$ and $\left[f_i^2(y_i, z_i) - f_i^2(y_s, z_s)\right]$ do not vanish.

Thus, from a dynamical systems' viewpoint, the average dynamics may be considered as a perturbed variant of the emergent dynamics. Consequently, it appears natural to study the problem of dynamic consensus, recast in that of *robust* stability analysis, in a broad sense. On one hand, in contrast to the more commonly studied case of state synchronisation, we shall admit that synchronisation may be established with respect to part of the variables only, i.e. with respect to the outputs y_i. More precisely, for the former case, similarly to (4.6), we introduce the *state* synchronisation manifold

$$\mathscr{S}_x = \{x \in \mathbb{R}^{nN} : x_1 - x_s = x_2 - x_s = \cdots = x_N - x_s = 0\} \tag{4.11}$$

and, for the study of *output* synchronisation, we analyse the stability of the manifold

$$\mathscr{S}_y = \{y \in \mathbb{R}^{mN} : y_1 - y_s = y_2 - y_s = \cdots = y_N - y_s = 0\}. \tag{4.12}$$

Since, in the general case of heterogeneous networks, the perturbation terms may prevail it becomes natural to study synchronisation in a practical sense, that is, by seeking to establish stability of the output or state synchronisation manifolds $\mathscr{S}_y$ or $\mathscr{S}_x$ in a *practical* sense only. This is precisely defined next.

Consider a parameterised system of differential equations

$$\dot{x} = f(x, \varepsilon), \tag{4.13}$$

where $x \in R^n$, the function $f : \mathbb{R}^n \to \mathbb{R}^n$ is locally Lipschitz and ε is a scalar parameter such that $\varepsilon \in (0, \varepsilon_\circ]$ with $\varepsilon_\circ < \infty$. Given a closed set $\mathscr{A}$, we define the norm $|x|_{\mathscr{A}} := \inf_{y \in \mathscr{A}} |x - y|$.

Definition 4.2 For the system (4.13), we say that the closed set $\mathscr{A} \subset \mathbb{R}^n$ is practically uniformly asymptotically stable if there exists a closed set $\mathscr{D}$ such that $\mathscr{A} \subset \mathscr{D} \subset \mathbb{R}^n$ and

(1) the system is forward complete for all $x_\circ \in \mathscr{D}$;
(2) for any given $\delta > 0$ and $R > 0$, there exist $\varepsilon^* \in (0, \varepsilon_\circ]$ and a class $\mathscr{K}\mathscr{L}$ function $\beta_{\delta R}$ such that, for all $\varepsilon \in (0, \varepsilon^*]$ and all $x_\circ \in \mathscr{D}$ such that $|x_\circ|_{\mathscr{A}} \le R$, we have

$$|x(t, x_\circ, \varepsilon)|_{\mathscr{A}} \leq \delta + \beta_{\delta R}\big(|x_\circ|_{\mathscr{A}}, t\big).$$

Remark 4.1 Similarly, to the definition of uniform global asymptotic stability of a set, the previous definition includes three properties: uniform boundedness of the solutions with respect to the set, uniform stability of the set and uniform practical convergence to the set.

The following statement, which establishes practical asymptotic stability of sets, may be deduced along the lines of the proof of the main result in [4].

Proposition 4.1 *Consider the system $\dot{x} = f(x)$, where $x \in \mathbb{R}^n$ and f is continuous and locally Lipschitz. Assume that the system is forward complete, there exists a closed set $\mathscr{A} \subset \mathbb{R}^n$ and a C^1 function $V : \mathbb{R}^n \to \mathbb{R}_+$ as well as functions $\alpha_1, \alpha_2 \in \mathscr{K}_\infty$, $\alpha_3 \in \mathscr{K}$ and a constant $c > 0$, such that, for all $x \in \mathbb{R}^n$,*

$$\alpha_1(|x|_{\mathscr{A}}) \leq V(x) \leq \alpha_2(|x|_{\mathscr{A}})$$
$$\dot{V} \leq -\alpha_2(|x|_{\mathscr{A}}) + c.$$

Then, for any $R, \varepsilon > 0$ there exists a constant $T = T(R, \varepsilon)$ such that for all $t \geq T$ and all $x_\circ$ such that $|x_\circ|_{\mathscr{A}} \leq R$

$$|x(t, x_\circ)|_{\mathscr{A}} \leq r + \varepsilon,$$

where $r = \alpha_1^{-1} \circ \alpha_2 \circ \alpha_3^{-1}(c)$.

4.3 Network Dynamics

In the previous section, we motivated, albeit intuitively, the study of dynamic consensus and practical synchronisation as a stability problem of the attractor of the emergent dynamics as well as of the synchronisation manifold. In this section, we render this argument formal by showing that the networked dynamical systems model (4.5) is equivalent, up to a coordinate transformation, to a set of equations composed of the average system dynamics (4.10) with average state x_s and a synchronisation errors equation with state $e = [e_1^\top \; \dots \; e_N^\top]^\top$, where $e_i = x_i - x_s$ for all $i \in \mathscr{I}$. It is clear that $x \in \mathscr{S}_x$ if and only if $e = 0$; hence, the general synchronisation problem is recast in the study of stability of the dynamics of e and x_s.

4.3.1 New Coordinates

Let us formally justify that the choice of coordinates x_s and e completely and appropriately describe the networked systems' behaviours.

Considering a network with an undirected and connected graph, the Laplacian matrix $L = L^\top$ has a single zero eigenvalue $\lambda_1 = 0$ and its corresponding right and left eigenvectors v_{r1}, v_{l1} coincide with $v = \frac{1}{\sqrt{N}}\mathbf{1}$ where $\mathbf{1} \in \mathbb{R}^N$ denotes the vector $[1\ 1\ \dots\ 1]^\top$. Moreover, since L is symmetric and non-negative definite, there exists (see [2, Chap. 4, Theorems 2 and 3]) an orthogonal matrix U (i.e. $U^{-1} = U^\top$) such that $L = U\Lambda U^\top$ with $\Lambda = \text{diag}\{[0\ \lambda_2\ \dots\ \lambda_N]\}$, where $\lambda_i > 0$ for all $i \in [2, N]$, are eigenvalues of L. Furthermore, the ith column of U corresponds to an eigenvector of L related to the ith eigenvalue, λ_i. Therefore, recognising v as the first eigenvector, we decompose the matrix U as:

$$U = \left[\frac{1}{\sqrt{N}}\mathbf{1}\quad U_1\right], \tag{4.14}$$

where $U_1 \in \mathbb{R}^{N\times N-1}$ is a matrix composed of $N-1$ eigenvectors of L related to $\lambda_2, \dots, \lambda_N$ and, since the eigenvectors of a real symmetric matrix are orthogonal, we have

$$\frac{1}{\sqrt{N}}\mathbf{1}^\top U_1 = 0, \qquad U_1^\top U_1 = I_{N-1}.$$

Based on the latter observations, we introduce the coordinate transformation

$$\bar{x} = \mathscr{U}^\top x, \tag{4.15}$$

where the block diagonal matrix $\mathscr{U} \in \mathbb{R}^{nN\times nN}$ is defined as

$$\mathscr{U} = U \otimes I_n,$$

which, in view of (4.15), is also orthogonal. Then, we use (4.14) to partition the new coordinates $\bar{x}$, i.e.

$$\bar{x} = \begin{bmatrix}\bar{x}_1\\ \bar{x}_2\end{bmatrix} = \begin{bmatrix}\frac{1}{\sqrt{N}}\mathbf{1}_N^\top \otimes I_n\\ U_1^\top \otimes I_n\end{bmatrix} x.$$

The coordinates $\bar{x}_1$ and $\bar{x}_2$ thus obtained are equivalent to the average x_s and the synchronisation errors e, respectively. Indeed, observing that the state of the average unit, defined in (4.8), may be re-written in the compact form

$$x_s = \frac{1}{N}(\mathbf{1}^\top \otimes I_n)x, \tag{4.16}$$

we obtain $\bar{x}_1 = \sqrt{N}x_s$. On the other hand, $\bar{x}_2 = 0$ if and only if $e = 0$. To see the latter, let $\mathscr{U}_1 = U_1 \otimes I_n$, then, using the expression

$$(A \otimes B)(C \otimes D) = AC \otimes BD, \tag{4.17}$$

we obtain

$$\mathscr{U}_1 \mathscr{U}_1^\top = (U_1 U_1^\top) \otimes I_n$$

and, observing that

$$U_1 U_1^\top = I_N - \frac{1}{N}\mathbf{1}\mathbf{1}^\top,$$

we get

$$\mathscr{U}_1 \mathscr{U}_1^\top = \left(I_N - \frac{1}{N}\mathbf{1}\mathbf{1}^\top\right) \otimes I_n. \tag{4.18}$$

Therefore, multiplying $\bar{x}_2 = \mathscr{U}_1^\top x$ by $\mathscr{U}_1$ and using (4.18), we obtain

$$\begin{aligned} \mathscr{U}_1 \bar{x}_2 &= \left[\left(I_N - \frac{1}{N}\mathbf{1}\mathbf{1}^\top\right) \otimes I_n\right] x \\ &= x - \frac{1}{N}(\mathbf{1}\mathbf{1}^\top \otimes I_n)x, \end{aligned}$$

which, in view of (4.17), is equivalent to

$$\begin{aligned} \mathscr{U}_1 \bar{x}_2 &= x - \frac{1}{N}\big(\mathbf{1} \otimes I_n\big)\big(\mathbf{1}^\top \otimes I_n\big)x \\ &= x - \big(\mathbf{1} \otimes I_n\big)x_s = e. \end{aligned}$$

Since $\mathscr{U}_1$ has column rank equal to $(N-1)n$, which corresponds to the dimension of $\bar{x}_2$, we see that $\bar{x}_2$ is equal to zero if and only if so is e.

Even though the state space of (x_s, e) is of higher dimension than that of the original networked system (4.1), only both together, the synchronisation error dynamics and the average dynamics, may give a complete characterisation of the network behaviour. Thus, the states x_s and e are *intrinsic* to the network and not the product of an artifice with purely theoretical motivations.

We proceed to derive the differential equations in terms of the average state x_s and the synchronisation errors e.

4.3.2 Dynamics of the Average Unit

Using the network dynamics Eq. (4.5a), as well as (4.16), we obtain

$$\dot{x}_s = \frac{1}{N}(\mathbf{1}^\top \otimes I_n)F(x) - \frac{1}{N}\sigma(\mathbf{1}^\top \otimes I_n)[L \otimes E_m]y. \tag{4.19}$$

Now, using the property of the Kronecker product, (4.17), and in view of the identity $\mathbf{1}^\top L = 0$, we obtain

$$(\mathbf{1}^\top \otimes I_n)(L \otimes E_m) = (\mathbf{1}^\top L) \otimes (I_n E_m) = 0.$$

This reveals the important fact that the average dynamics, i.e. the right-hand side of (4.19), is independent of the interconnections gain σ, even though the solutions $x_s(t)$ are, certainly, affected by the synchronisation errors; hence, by the coupling strength.

Now, using (4.4) and defining

$$f_s(x_s) := \frac{1}{N} \sum_{i=1}^{N} F_i(x_s) \tag{4.20}$$

we obtain

$$\dot{x}_s = f_s(x_s) + \frac{1}{N} \sum_{i=1}^{N} \Big[F_i(x_i) - F_i(x_s) \Big].$$

Therefore, defining

$$G_s(e, x_s) := \frac{1}{N} \sum_{i=1}^{N} \Big[F_i(e_i + x_s) - F_i(x_s) \Big],$$

we see that we may write the average dynamics in the compact form,

$$\dot{x}_s = f_s(x_s) + G_s(e, x_s). \tag{4.21}$$

Furthermore, since the functions F_i, with $i \in \mathscr{I}$, are locally Lipschitz so is the function G_s and, moreover, there exists a continuous, positive, non-decreasing function $k : \mathbb{R}_+ \times \mathbb{R}_+ \to \mathbb{R}_+$, such that

$$|G_s(e, x_s)| \le k\big(|e|, |x_s|\big)|e|.$$

In summary, the average dynamics is described by the Eq. (4.21). That is, it consists in the nominal system (4.7), which corresponds to the emergent dynamics, perturbed by the synchronisation error of the network via the term G_s.

4.3.3 Dynamics of the Synchronisation Errors

To study the effect of the synchronisation errors, $e(t)$, on the emergent dynamics, we start by introducing the vectors

$$F_s(x_s) := \left[F_1(x_s)^\top \quad \dots \quad F_N(x_s)^\top\right]^\top \tag{4.22}$$
$$\tilde{F}(e, x_s) := \begin{bmatrix} F_1(x_1) - F_1(x_s) \\ \vdots \\ F_N(x_N) - F_N(x_s) \end{bmatrix} = \begin{bmatrix} F_1(e_1 + x_s) - F_1(x_s) \\ \vdots \\ F_N(e_N + x_s) - F_N(x_s) \end{bmatrix}$$

i.e. $\tilde{F}(e, x_s) = F(x) - F_s(x_s)$. Then, differentiating on both sides of

$$e = x - (\mathbf{1} \otimes I_n)x_s$$

and using (4.5a) and (4.21), we obtain

$$\begin{aligned} \dot{e} &= -\sigma[L \otimes E_m]y + F(x) - (\mathbf{1} \otimes I_n)\left[f_s(x_s) + G_s(e, x_s)\right] \\ &= -\sigma[L \otimes E_m]y + [F(x) - F_s(x_s)] + F_s(x_s) - (\mathbf{1} \otimes I_n)\left[f_s(x_s) + G_s(e, x_s)\right] \\ &= -\sigma[L \otimes E_m]y + \left[F_s(x_s) - (\mathbf{1} \otimes I_n)f_s(x_s)\right] + \left[\tilde{F}(e, x_s) - (\mathbf{1} \otimes I_n)G_s(e, x_s)\right]. \end{aligned} \tag{4.23}$$

Next, let us introduce the output synchronisation errors $e_{yi} = y_i - y_s$, that is, $e_y = [e_{y1}^\top, \dots, e_{yN}^\top]^\top$, which may also be written as

$$e_y = y - \mathbf{1} \otimes y_s, \tag{4.24}$$

and let us consider the first term and the two groups of bracketed terms on the right-hand side of (4.23), separately. For the term $\left(L \otimes E_m\right)y$ we observe, from (4.24), that

$$[L \otimes E_m]\, y = [L \otimes E_m]\left[e_y + \mathbf{1} \otimes y_s\right]$$

and we use (4.17) and the fact that $L\mathbf{1} = 0$ to obtain

$$[L \otimes E_m]\, y = [L \otimes E_m]\, e_y.$$

Second, concerning the first bracket on the right-hand side of (4.23), we observe that, in view of (4.20) and (4.22),

$$f_s(x_s) = \frac{1}{N}(\mathbf{1}^\top \otimes I_n)F_s(x_s).$$

Therefore,

$$F_s(x_s) - (\mathbf{1} \otimes I_n)f_s(x_s) = F_s(x_s) - \frac{1}{N}(\mathbf{1} \otimes I_n)(\mathbf{1}^\top \otimes I_n)F_s(x_s).$$

Then, using (4.17) we see that

$$\frac{1}{N}(\mathbf{1} \otimes I_n)(\mathbf{1}^\top \otimes I_n) = \frac{1}{N}(\mathbf{1}\mathbf{1}^\top) \otimes I_n. \tag{4.25}$$

So, introducing

$$P = I_{nN} - \frac{1}{N}(\mathbf{1}\mathbf{1}^\top) \otimes I_n,$$

we obtain

$$F_s(x_s) - (\mathbf{1} \otimes I_n) f_s(x_s) = P F_s(x_s). \tag{4.26}$$

Finally, concerning the term $\tilde{F}(e, x_s) - (\mathbf{1} \otimes I_n) G_s(e, x_s)$ on the right-hand side of (4.23), we see that, by definition, $G(e, x_s) = \frac{1}{N}\big(\mathbf{1}^\top \otimes I_n\big)\tilde{F}(e, x_s)$, hence, from (4.25), we obtain

$$(\mathbf{1} \otimes I_n) G_s(e, x_s) = \frac{1}{N}\big[(\mathbf{1}\mathbf{1}^\top) \otimes I_n\big]\tilde{F}(e, x_s)$$

and

$$\begin{aligned} \tilde{F}(e, x_s) - (\mathbf{1} \otimes I_n) G_s(e, x_s) &= \left(I_{nN} - \frac{1}{N}(\mathbf{1}\mathbf{1}^\top) \otimes I_n \right) \tilde{F}(e, x_s) \\ &= P\tilde{F}(e, x_s). \end{aligned} \tag{4.27}$$

Using (4.26) and (4.27) in (4.23), we see that the latter may be expressed as

$$\dot{e} = -\sigma\big[L \otimes E_m\big]e_y + P\big[\tilde{F}(e, x_s) + F_s(x_s)\big].$$

The utility of this equation is that it clearly exhibits three terms: a term linear in the output e_y which reflects the synchronisation effect of diffusive coupling between the nodes, the term $P\tilde{F}(e, x_s)$ which vanishes with the synchronisation errors, i.e. if $e = 0$, and the term

$$P F_s(x_s) = \begin{bmatrix} F_1(x_s) - \frac{1}{N}\sum_{i=1}^N F_i(x_s) \\ \vdots \\ F_N(x_s) - \frac{1}{N}\sum_{i=1}^N F_i(x_s) \end{bmatrix} = \begin{bmatrix} F_1(x_s) - f_s(x_s) \\ \vdots \\ F_N(x_s) - f_s(x_s) \end{bmatrix},$$

which represents the variation between the dynamics of the individual units and the average unit. This term equals to zero, e.g. when the nominal dynamics, f_i in (4.1a), of all the units are identical that is, in the case of a homogeneous network (Fig. 4.1).

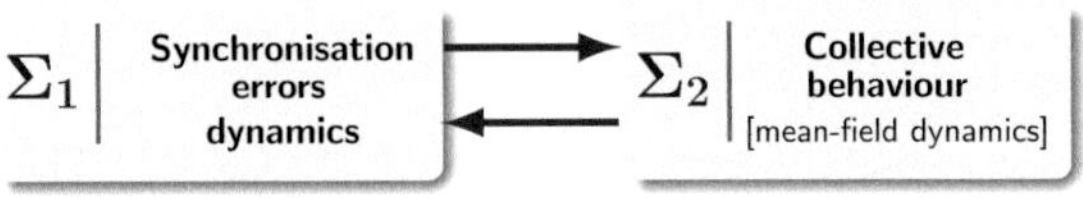

Fig. 4.1 Interaction between synchronisation and the collective dynamics

4.4 Stability analysis

All is in place to present our main statements on stability of the networked systems model (4.5). For the purpose of analysis, we use the equations previously developed, in the coordinates e and x_s, which we recall here for convenience:

$$\dot{x}_s = f_s(x_s) + G_s(e, x_s), \tag{4.28a}$$
$$\dot{e} = -\sigma\big[L \otimes E_m\big]e_y + P\big[\tilde{F}(e, x_s) + F_s(x_s)\big]. \tag{4.28b}$$

These equations correspond to those of two feedback interconnected systems, as it is illustrated in Fig. 4.1. For the system (4.28a), we study the stability with respect to a compact attractor which is proper to the emergent dynamics and we establish conditions under which the average of the trajectories of the interconnected units remains close to this attractor. For the system (4.28b) we study robust stability of the synchronisation manifolds $\mathscr{S}_y$ and $\mathscr{S}_x$.

4.4.1 *Practical Synchronisation Under Diffusive Coupling*

We formulate conditions that ensure practical global asymptotic stability of the sets $\mathscr{S}_x$ and $\mathscr{S}_y$—see (4.11), (4.12). This implies practical state and output synchronisation of the network, respectively. Furthermore, we show that the upper bound on the state synchronisation error depends on the mismatches between the dynamics of the individual units of the network.

Theorem 4.1 (Output synchronisation) *Let the solutions of the system (4.5) be globally ultimately bounded. Then, the set $\mathscr{S}_y$ is practically uniformly globally asymptotically stable with $\varepsilon = 1/\sigma$. If, moreover, Assumption 4.1 holds, then there exists a function $\beta \in \mathscr{K}_\infty^3$ such that for any $\varepsilon \geq 0$ and $R > 0$ there exist $T^* > 0$ and $\sigma^* > 0$ such that the solutions of (4.28b) with $\sigma = \sigma^*$ satisfy*

$$|e(t, x_\circ)| \leq \beta(\bar{\alpha}_k, \beta_k, \Delta_f) + \varepsilon, \qquad \forall\, t \geq T^*,\ x_\circ \in B_R := \{x_\circ : |x_\circ| \leq R\}$$

where

$$\Delta_f = \max_{|x|\leq B_x}\ \max_{k,i\in N} \Big\{ |f_k^2(x_k) - f_i^2(x_k)| \Big\}. \tag{4.29}$$

The bound on the synchronisation errors, β, is a function of the constants $\bar{\alpha}_k$, β_k defined in Assumption 4.1 as well as on the degree of heterogeneity of the network, characterised by Δ_f. In the definition of the latter, B_x corresponds to a compact set to which the solutions ultimately converge by assumption. That is, Theorem 4.1 guarantees, in particular, that the perturbing effect of heterogeneity in the network may be diminished at will by increasing the interconnection strength.

The proof of this theorem is provided in [15]. Roughly speaking, the first statement (synchronisation) follows from two properties of the networked system—namely, negative definiteness of the second smallest eigenvalue of the Laplacian metric L and global ultimate boundedness. Now, global ultimate boundedness holds, e.g. under the following assumption.

Assumption 4.2. For each i, the system (4.2) defines a strictly semi-passive map $u_i \mapsto y_i$ with continuously differentiable and radially unbounded storage functions $V_i : \mathbb{R}^n \to \mathbb{R}_+$, where $i \in \mathscr{I}$. That is, there exist positive definite and radially unbounded storage functions V_i, positive constants ρ_i, continuous functions H_i and positive continuous functions ϱ_i such that

$$\dot{V}_i(x_i) \leq y_i^\top u_i - H_i(x_i)$$

and $H_i(x_i) \geq \varrho_i(|x_i|)$ for all $|x_i| \geq \rho_i$.

Indeed, the following statement is reminiscent of [16, Corollary 1].

Proposition 4.2. *Consider a network of N diffusively coupled units (4.5). Let the graph of interconnections be undirected and connected and assume that all the units of the network are strictly semi-passive (i.e. Assumption 4.2 holds). Then, the solutions of the system (4.5) are ultimately bounded.*

Proof. We proceed as in the proof of [16, Lemma 1] and [22, Proposition 2.1]. Let Assumption 4.2 generate positive definite storage functions V_i, as well as functions ϱ_i, H_i and constants ρ_i, defined as above and let

$$V_\Sigma(x) := \sum_{i=1}^{N} V_i(x_i).$$

Then, taking the derivative of $V_\Sigma(x)$ along trajectories of the system (4.5), we obtain

$$\begin{aligned} \dot{V}_\Sigma(x) &\leq -\sigma y^\top \big[L \otimes I_m\big] y - \sum_{i=1}^{N} H_i(x_i) \\ &\leq -\sum_{i=1}^{N} H_i(x_i), \end{aligned} \tag{4.30}$$

where for the last inequality we used the fact that Laplacian matrix is semi-positive definite. Next, let $\bar{\rho} = \max_{1\leq i\leq N}\{\rho_i\}$ and consider the function $\bar{\varrho} : [\bar{\rho}, +\infty) \to \mathbb{R}_{\geq 0}$ as $\bar{\varrho}(s) = \min_{1\leq i\leq N}\{\varrho_i(s)\}$. Note that $\bar{\varrho}$ is continuous and $\bar{\varrho}(s)$ positive for all $s \geq \bar{\rho}$. Furthermore, for any $|x| \geq N\bar{\rho}$ there exists $k \in \mathscr{I}$ such that $|x_k| \geq \frac{1}{N}|x| \geq \bar{\rho}$. Therefore, for all $|x| \geq N\bar{\rho}$,

$$\sum_{i=1}^{N} H_i(x_i) \geq H_k(x_k) \geq \bar{\varrho}(|x_k|) \geq \bar{\varrho}\left(\frac{1}{N}|x|\right).$$

Using the last bound in (4.30) we obtain, for all $|x| \geq N\bar{\rho}$,

$$\dot{V}_\Sigma(x) \leq -\bar{\varrho}\left(\frac{1}{N}|x|\right).$$

Hence, invoking [23, Theorem 10.4] we conclude that the solutions of the system (4.5) are ultimately bounded. □

Some interesting corollaries, on state synchronisation, follow from Theorem 4.1, for instance, if the interconnections among the network units depend on the whole state, that is, if $y = x$.

Corollary 4.1. *Consider the system (4.5). Let Assumptions 4.1 and 4.2 be satisfied and let $y = x$. Then, the system is forward complete and the set $\mathscr{S}_x$ is practically, uniformly, globally and asymptotically stable with $\varepsilon = 1/\sigma$.*

The constant Δ_f defined in (4.29) represents the maximal possible mismatch between the dynamics of any individual unit and that of the averaged unit, on a ball of radius B_x. The more heterogeneous is the network, the bigger is the constant Δ_f. Conversely, in the case that all the zero dynamics of the units are identical, we have $\Delta_f = 0$. In this case, we obtain the following statement.

Corollary 4.2. *Consider the system (4.5) under Assumptions 4.1 and 4.2. Assume that the functions f_i^2, which define zero dynamics of the network units, are all identical, i.e. $f_i^2(x) = f_j^2(x)$ for all $i, j \in \mathscr{I}$ and all $x \in \mathbb{R}^n$. Then the set $\mathscr{S}_x$ is practically uniformly globally asymptotically stable with $\varepsilon = 1/\sigma$.*

4.4.2 *On Practical Stability of the Collective Network Behaviour*

Now we analyse the behaviour of the average unit, whose dynamics is given by the Eq. (4.28a). We assume that the nominal dynamics of average unit (i.e. with $e = 0$) has a stable compact attractor $\mathscr{A}$ and we establish that the stability properties of this attractor are preserved under the network interconnection, albeit, slightly weakened.

Assumption 4.3. For the system (4.7), there exists a compact invariant set $\mathscr{A} \subset \mathbb{R}^n$ which is asymptotically stable with a domain of attraction $\mathscr{D} \subset \mathbb{R}^n$. Moreover, we assume that there exists a continuously differentiable Lyapunov function $V_\mathscr{A} : \mathbb{R}^n \to \mathbb{R}_{\geq 0}$ and functions $\alpha_i \in \mathscr{K}_\infty$, $i \in \{1, \ldots, 4\}$ such that for all $x_e \in \mathscr{D}$ we have

$$\begin{aligned}
\alpha_1(|x_e|_\mathscr{A}) \leq V_\mathscr{A}(x_e) &\leq \alpha_2(|x_e|_\mathscr{A}) \\
\dot{V}_\mathscr{A}(x_e) &\leq -\alpha_3(|x_e|_\mathscr{A}) \\
\left|\frac{\partial V_\mathscr{A}}{\partial x_e}\right| &\leq \alpha_4(|x_e|).
\end{aligned}$$

The second part of the assumption (the existence of V) is purely technical whereas the first part is essential to analyse the emergent synchronised behaviour as well as the synchronisation properties of the network, recast as a (robust) stability problem. The following statement applies to the general case of diffusively coupled networks.

Theorem 4.2. *For the system (4.5), assume that the solutions are globally ultimately bounded and Assumptions 4.1, 4.3 hold. Then, there exist a non-decreasing function* $\gamma : \mathbb{R}_+ \times \mathbb{R}_+ \to \mathbb{R}_+$ *and, for any* $R, \varepsilon > 0$ *there exists* $T^* = T^*(R, \varepsilon)$, *such that for all* $t \geq T^*$ *and all* $x_\circ$ *such that* $|x_\circ|_{\mathscr{A}} \leq R$,

$$|x_s(t, x_\circ)|_{\mathscr{A}} \leq \gamma(\Delta_f, R) + \varepsilon.$$

In the case that the network is state practically synchronised, it follows that the set $\mathscr{A}$ is practically stable for the network (4.5).

Corollary 4.3. *Consider the system (4.5) under Assumption 4.3. If the set* $\mathscr{S}_x$ *is practically uniformly globally asymptotically stable for this system, then the attractor* $\mathscr{A}$ *defined in Assumption 4.3 is practically asymptotically stable for the average unit (4.21).*

4.5 Example

To illustrate our theoretical findings we present a brief case-study of analysis of interconnected heterogeneous systems via diffusive coupling. We consider three of the best known chaotic oscillators: the Rössler [20], the Lorenz [10] and the Lü system [11]. The dynamics equations of these *forced* oscillators are the following:

$$\text{LORENZ OSCILLATOR:} \quad \left\{ \begin{aligned} \begin{bmatrix} \dot{x}_\ell \\ \dot{y}_\ell \end{bmatrix} &= \begin{bmatrix} \gamma(y_\ell - x_\ell) \\ r x_\ell - y_\ell - x_\ell z_\ell \end{bmatrix} + u_\ell \\ \dot{z}_\ell &= x_\ell y_\ell - b z_\ell \end{aligned} \right.$$

$$\text{LÜ OSCILLATOR:} \quad \left\{ \begin{aligned} \begin{bmatrix} \dot{x}_m \\ \dot{y}_m \end{bmatrix} &= \begin{bmatrix} -\dfrac{\alpha\beta}{\alpha+\beta} x_m - y_m z_m + c \\ \alpha y_m + x_m z_m \end{bmatrix} + u_m \\ \dot{z}_m &= \beta z_m + x_m y_m . \end{aligned} \right.$$

$$\text{RÖSSLER OSCILLATOR:} \quad \left\{ \begin{aligned} \begin{bmatrix} \dot{x}_r \\ \dot{y}_r \end{bmatrix} &= \begin{bmatrix} a_r x_r + y_r \\ -(x_r - y_r) \end{bmatrix} + u_r \\ \dot{z}_r &= b_r + y_r z_r - c_r z_r . \end{aligned} \right.$$

The values of the parameters of the three systems are fixed in order for them to exhibit a chaotic behaviour when unforced. These are collected in Table 4.1.

Table 4.1 Parameter values of chaotic oscillators

Rössler	Lorenz	Lü 3rd order
$a_r = 0.15$	$\gamma = 16$	$\alpha = -10$
$b_r = 0.2$	$r_2 = 45.6$	$\beta = -4$
$c_r = 10$	$b = 4$	$\delta = 1$

Since the three chaotic systems are *oscillators* their trajectories are globally ultimately bounded, they converge to the strange attractors depicted in Fig. 4.2. In this figure we also show the phase portrait for the average solutions $x_s(t)$ for the three unforced oscillators (with $u_\ell = u_m = u_r = 0$). Then, we apply the respective inputs

$$\begin{aligned} u_\ell &= -\sigma\big[d_{13}(\mathrm{y}_\ell - \mathrm{y}_m) + d_{12}(\mathrm{y}_\ell - \mathrm{y}_r)\big], \quad d_{12} = 2,\ d_{13} = 4, \\ u_r &= -\sigma\big[d_{12}(\mathrm{y}_r - \mathrm{y}_\ell) + d_{23}(\mathrm{y}_r - \mathrm{y}_m)\big], \quad d_{23} = 3, \\ u_m &= -\sigma\big[d_{13}(\mathrm{y}_m - \mathrm{y}_\ell) + d_{23}(\mathrm{y}_m - \mathrm{y}_r)\big], \end{aligned}$$

where $\mathrm{y}_{(\cdot)}$ are measurable outputs. We have simulated two scenarios: in the first case, we assume that only the $x_{(\cdot)}$ coordinates of each oscillator are measured hence,

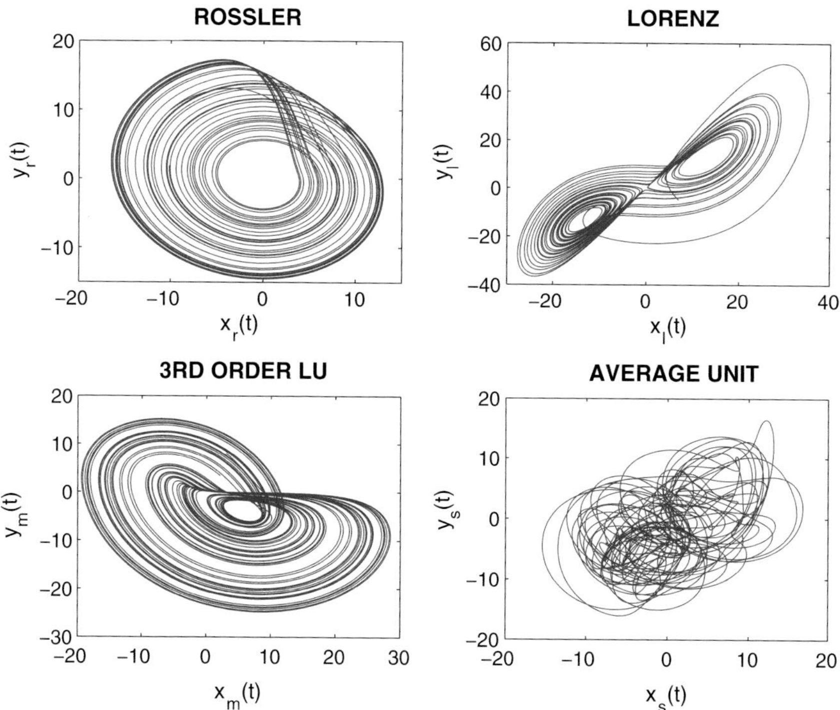

Fig. 4.2 Phase portraits of the three chaotic oscillators, Rössler, Lorenz and Lü, as well as that of the average dynamics, in the absence of interconnection, i.e. with $\sigma = 0$

$$\mathrm{y}_\ell := [x_\ell \; 0]^\top \quad \mathrm{y}_m := [x_m \; 0]^\top \quad \mathrm{y}_r := [x_r \; 0]^\top$$

In Fig. 4.3, we depict the phase portrait for the Rössler system overlapped with that of the average solutions. It is appreciated, on one hand, that the synchronisation errors diminish as the interconnection gain is increased. On the other hand, the behaviour of the oscillators' solutions also changes: for relatively large values of σ (50 and 80), the chaotic behaviour is lost and the systems stabilise.

In the second scenario, we assume that both $x_{(\cdot)}$ and $y_{(\cdot)}$ are measured hence,

$$\mathrm{y}_\ell := [x_\ell \; y_\ell]^\top \quad \mathrm{y}_m := [x_m \; y_m]^\top \quad \mathrm{y}_r := [x_r \; y_r]^\top.$$

The simulation results in this case, for different values of the interconnection gain σ, are showed in Figs. 4.4 and 4.5. With two inputs, the systems "loose" the chaotic response and stabilise to an equilibrium. In Fig. 4.4 we plot the norm of the output synchronisation errors $|e_y(t)| = |\mathrm{y}(t) - \mathbf{1} \otimes y_s(t)|$; it is clearly appreciated that the errors diminish as the interconnection gain increases. Finally, in Fig. 4.5 we show the phase portraits for four different values of σ; it is clear that output synchronisation occurs.

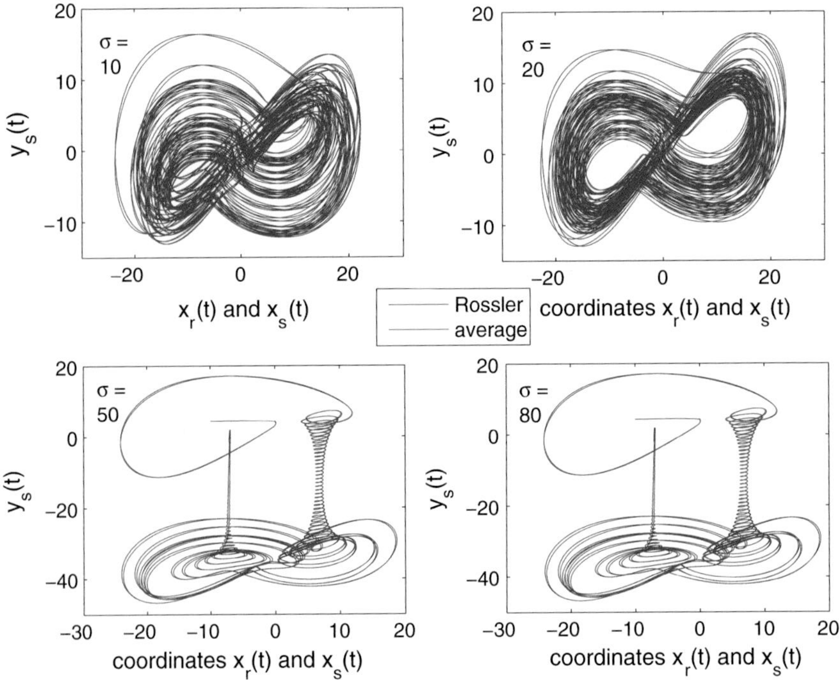

Fig. 4.3 Phase portraits of the Rössler oscillator compared to that of the average unit, for different values of the interconnection gain σ

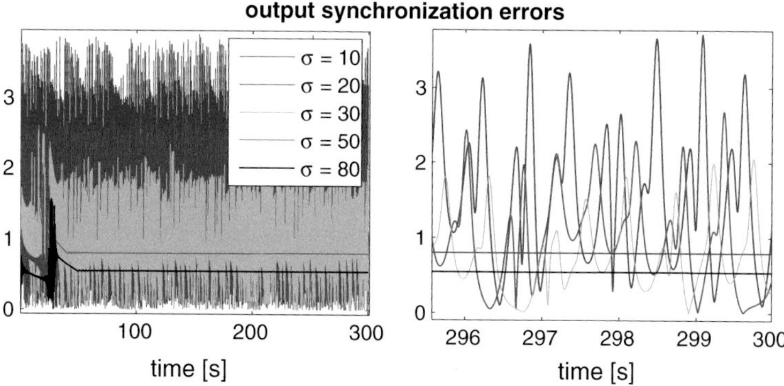

Fig. 4.4 Norms of the output synchronisation errors, $|e_y(t)|$, for different values of the interconnection gain σ

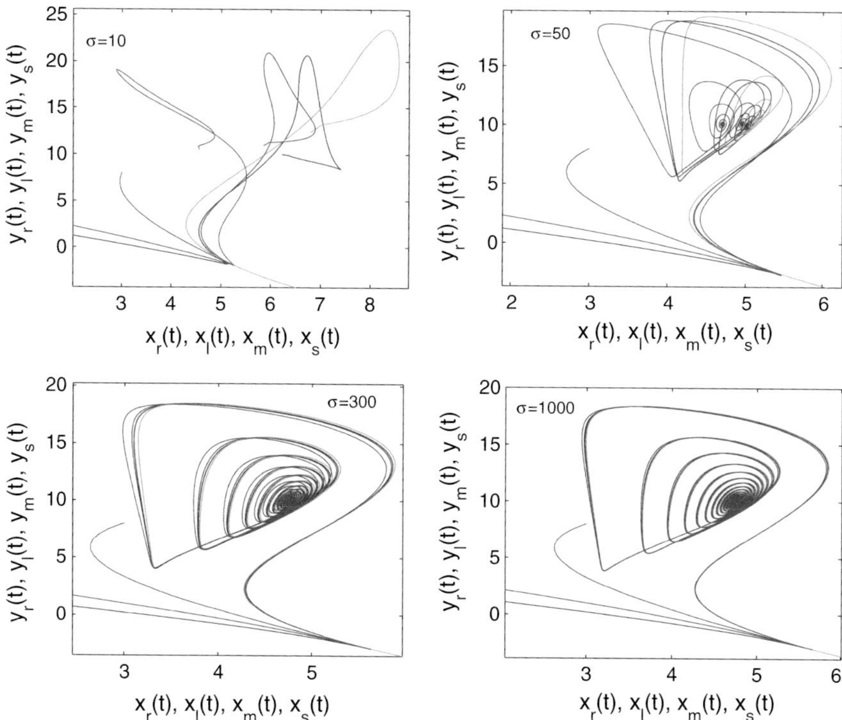

Fig. 4.5 Phase portraits of the three chaotic oscillators, Rössler, Lorenz and Lü, as well as that of the average dynamics, with two inputs and for different values of the interconnection gain σ

References

1. Angeli, D.: A Lyapunov approach to incremental stability properties. IEEE Trans. Autom. Cont. **47**(3), 410–421 (2002)
2. Bellman, R.E.: Introduction to Matrix Analysis, vol. 10. McGraw-Hill, New York (1970)
3. Belykh, I., de Lange, E., Hasler, M.: Synchronization of bursting neurons: what matters in the network topology. Phys. Rev. Lett. **94**, 188101 (2005)
4. Corless, M., Leitmann, G.: Continuous state feedback guaranteeing uniform ultimate boundedness for uncertain dynamic systems. IEEE Trans. Autom. Cont. **26**(5), 1139–1144 (1981)
5. Efimov, D., Panteley, E., Loría, A.: Switched mutual–master–slave synchronisation: Application to mechanical systems. In: Proceedings of the 17th IFAC World Congress, pp. 11508–11513, Seoul, Korea (2008)
6. Franci, A., Scardovi, L., Chaillet, A.: An input–output approach to the robust synchronization of dynamical systems with an application to the Hindmarsh–Rose neuronal model. In: Proceedings of the 50th IEEE Conference on Decision and Control and European Control Conference (CDC-ECC), pp. 6504–6509 (2011)
7. Isidori, A.: Nonlinear Control Systems II. Springer, London (1999)
8. Jouffroy, J., Slotine, J.J.: Methodological remarks on contraction theory. In: Proceedings of the 43rd IEEE Conference on Decision and Control, vol. 3, pp. 2537–2543 (2004)
9. Lohmiller, W., Slotine, J.-J.: Contraction analysis of non-linear distributed systems. Int. J. Control **78**, 678–688 (2005)
10. Lorenz, E.N.: Deterministic nonperiodic flow. J. Atmos. Sci. **20**, 130–141 (1963)
11. Lü, J.H., Chen, G.: A new chaotic attractor coined. Int. J. Bifurc. Chaos **12**(3), 659–661 (2002)
12. Nijmeijer, H., Mareels, I.: An observer looks at synchronization. IEEE Trans. Circuits Sys. I: Fundam. Theory Appl. **44**(10), 882–890 (1997)
13. Nijmeijer, H., Rodríguez-Angeles, A.: Synchronization of Mechanical Systems. Nonlinear Science, Series A, vol. 46. World Scientific, Singapore (2003)
14. Oguchi, T., Yamamoto, T., Nijmeijer, H.: Synchronization of bidirectionally coupled nonlinear systems with time-varying delay. Topics in Time Delay Systems, pp. 391–401. Springer, Berlin (2009)
15. Panteley, E.: A stability-theory perspective to synchronisation of heterogeneous networks. Habilitation à diriger des recherches (DrSc Dissertation). Université Paris Sud, Orsay, France(2015)
16. Pogromski, A.Y., Glad, T., Nijmeijer, H.: On difffusion driven oscillations in coupled dynamical systems. Int. J. Bifurc. Chaos Appl. Sci. Eng. **9**(4), 629–644 (1999)
17. Pogromski, A.Y., Nijmeijer, H.: Cooperative oscillatory behavior of mutually coupled dynamical systems. IEEE Trans. Circuits Sys. I: Fundam. Theory Appl. **48**(2), 152–162 (2001)
18. Pogromski, A.Y., Santoboni, G., Nijmeijer, H.: Partial synchronization: from symmetry towards stability. Phys. D: Nonlinear Phenom. **172**(1–4), 65–87 (2002)
19. Ren, W., Beard, R.W.: Distributed consensus in multivehicle cooperative control. Springer, Heidelberg (2005)
20. Rössler, O.E.: An equation for hyperchaos. Phys. Lett. A **71**(2–3), 155–157 (2007)
21. Scardovi, L., Arcak, M., Sontag, E.D.: Synchronization of interconnected systems with an input–output approach. Part I: main results. In: Proceedings of the 48th IEEE Conference on Decision and Control, pp. 609–614 (2009)
22. Steur, E., Tyukin, I., Nijmeijer, H.: Semi-passivity and synchronization of diffusively coupled neuronal oscillators. Phys. D: Nonlinear Phenom. **238**(21), 2119–2128 (2009)
23. Yoshizawa, T.: Stability Theory by Lyapunov's Second Method. The Mathematical Society of Japan, Tokyo (1966)

Chapter 5
Anticipating Synchronization and State Predictor for Nonlinear Systems

Toshiki Oguchi

Abstract This chapter discusses a synchronization problem, anticipating synchronization, and its application in control design for nonlinear systems with delays. The anticipating synchronization phenomena was initially reported by Voss (2000), and Oguchi and Nijmeijer (2005) then generalized it from the framework of control theory. This chapter revisits the anticipating synchronization problem and introduces a state predictor based on synchronization. Furthermore, we discuss recent progress on predictor design for nonlinear systems with delays.

5.1 Introduction

It is a great pleasure to contribute this chapter to the festschrift of Prof. Nijmeijer on the occasion of his 60th birthday. I had the opportunity to collaborate with Prof. Nijmeijer while working at Eindhoven University of Technology as a visiting researcher in 2003. Since then, we have continued to work together on the subject of synchronization and control of multi-agent systems with delays. This chapter revisits anticipating synchronization that occurs in master–slave systems, which was my first joint work project with Henk. In addition, I discuss the use of anticipating synchronization in control.

Recently, synchronization phenomena have been the subject of growing interest in various fields, including applied physics, biology, applied mathematics, social sciences, control engineering, and so on. As an extension of the synchronization of coupled systems, Voss [1] discovered the occurrence of anticipating synchronization of chaotic systems with time-delay in a unidirectional coupling configuration. Anticipated synchronization can be recognized as a state prediction method that does not require numerical integration. These phenomena are also observed in physical systems. Voss [2] demonstrated an experimental realization of such a phenomenon in an

T. Oguchi (✉)
Department of Mechanical Engineering, Graduate School of Science and Engineering, Tokyo Metropolitan University, 1-1, Minami-Osawa, Hachioji-shi, Tokyo 192-0397, Japan
e-mail: t.oguchi@tmu.ac.jp

N. van de Wouw et al. (eds.), *Nonlinear Systems*, Lecture Notes in Control and Information Sciences 470, DOI 10.1007/978-3-319-30357-4_5

electronic circuit. Masoller [3] considered anticipating synchronization of chaotic external-cavity semiconductor lasers with numerical simulation. Sivaprakasam et al. [4] observed the occurrence of anticipating synchronization using chaotic semiconductor diode lasers. These chaotic systems are described by one-dimensional systems with time-delay in the state, but anticipating synchronization can occur in more general systems. In a previous study [5], we generalized the phenomenon from the framework of control theory. Subsequently, we also considered the use of this synchronization phenomenon as a predictor of nonlinear systems. More recently, we have been attempting to use this synchronization-based predictor to extend the applicability of the finite spectrum assignment [6] for nonlinear retarded systems [7]. In last half of this chapter, we introduce a prediction control scheme combined with a synchronization-based predictor.

5.2 Anticipating Synchronization

This section revisits anticipating synchronization of coupled chaotic systems and introduces a recent advance in this direction.

In the original study concerning anticipating synchronization [1], Voss first considered the following equation, which is a scalar system and behaves chaotically.

$$\dot{x}(t) = -\alpha x(t) - \beta f(x(t-\tau)), \tag{5.1}$$

where $x \in \mathbb{R}$, and α and β are constants. For this system, we consider the following response system.

$$\dot{z}(t) = -\alpha z(t) - \beta f(x(t)), \tag{5.2}$$

where $z \in \mathbb{R}$. The dynamics of the prediction error $e(t) \triangleq z(t-\tau) - x(t)$ is given by

$$\dot{e}(t) = -\alpha e(t)$$

and a necessary and sufficient condition for the error e to converge to 0 is that $\alpha > 0$. Then, the synchronization manifold $x = z(t-\tau)$ is globally asymptotically stable and this phenomenon is called anticipating synchronization or anticipatory synchronization. Therefore, if $\alpha > 0$, $z(t)$ estimates the future value $x(t+\tau)$. In this case, the availability of anticipating synchronization is not only independent of the nonlinear function f but also of the length of time-delay τ. This prediction method is simple, and the convergence property of the error dynamics is globally guaranteed, but the convergence property is dependent on the dynamics of the master system.

As a simple modification, we consider the following scalar retarded system [8].

$$\dot{x}(t) = -ax(t-\tau) + bf(x(t-\tau)), \tag{5.3}$$

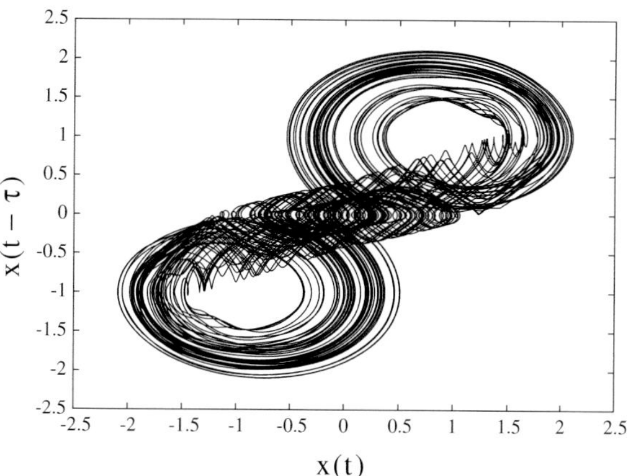

Fig. 5.1 Chaotic attractor for $a = 0.2$, $b = 0.2$, and $\tau = 10$

where $f(x) = \tanh(10x)$. Choosing $a = 0.2$, $b = 0.2$, and $\tau = 10$, the system (5.3) has a chaotic attractor depicted in Fig. 5.1. For this system, we design the slave system as follows.

$$\dot{z}(t) = -az(t-\tau) + bf(x(t)) + k(z(t-\tau) - x(t)). \tag{5.4}$$

The prediction error dynamics for this case is given by

$$\dot{e}(t) = -(a-k)e(t-\tau), \tag{5.5}$$

where $e(t) = z(t-\tau) - x(t)$. Then, a necessary and sufficient condition for asymptotic stability of (5.5) is that the following inequality holds

$$0 < (a-k)\tau < \frac{\pi}{2}.$$

If $k = 0$, i.e., there is no correction term, this inequality does not hold for the parameters stated above. For k satisfying $a - \frac{\pi}{2/\tau} < k < a$, however, $e = 0$ is asymptotically stable, and then e asymptotically converges to zero, which means that $z(t)$ estimates the future value $x(t+\tau)$ of the master system (Figs. 5.2 and 5.3).

These prediction schemes can be generalized into multidimensional systems as follows. Consider the following multidimensional difference-differential equation system,

$$\begin{cases} \dot{x}(t) = A_0 x(t) + F(y(t-\tau)) \\ y(t) = h(x(t)), \end{cases} \tag{5.6}$$

where $x \in \mathbb{R}^n$, $y \in \mathbb{R}^m$, $A_0 \in \mathbb{R}^{n\times n}$, $F : \mathbb{R}^m \to \mathbb{R}^n$ and $h : \mathbb{R}^n \to \mathbb{R}^m$. For the system (5.6), we consider the following prediction scheme,

$$\dot{z}(t) = A_0 z(t) + F(y(t)), \tag{5.7}$$

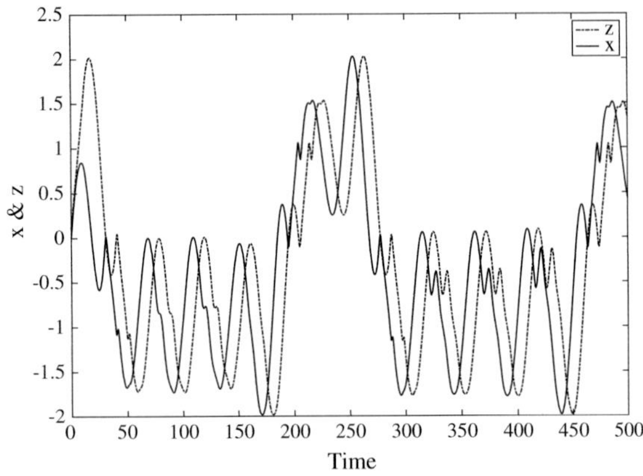

Fig. 5.2 The behaviors of $x(t)$ and $z(t)$ with $x(\theta) = 0.1$ and $z(\theta) = 0$ for $\theta \in [-\tau, 0]$

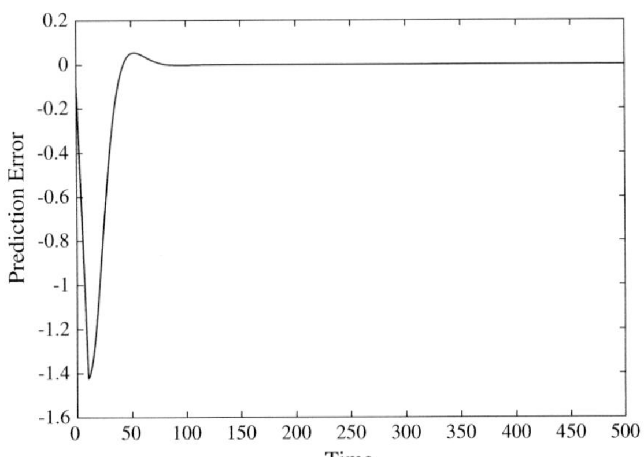

Fig. 5.3 Prediction error: $e(t) = z(t - \tau) - x(t)$

in which $z \in \mathbb{R}^n$. Since the dynamics of the prediction error $e(t) = z(t - \tau) - x(t)$ is described by

$$\dot{e}(t) = A_0 e(t),$$

the error converges to zero if and only if matrix A_0 is Hurwitz, i.e., A_0 has all eigenvalues in the open left half-plane. Therefore, if A_0 is a Hurwitz matrix, the output $z(t)$ of the predictor (5.7) estimates the future value $x(t + \tau)$ of the state $x(t)$.

However, to achieve anticipating synchronization, the master system is required to satisfy the following conditions:

(i) A_0 is a Hurwitz matrix, and
(ii) the time-delayed term $F(\cdot)$ in the dynamics (5.6) depends on the output only.

These restrictions can be relaxed by adding a coupling term into the slave system as follows. Consider the following possibly chaotic nonlinear system:

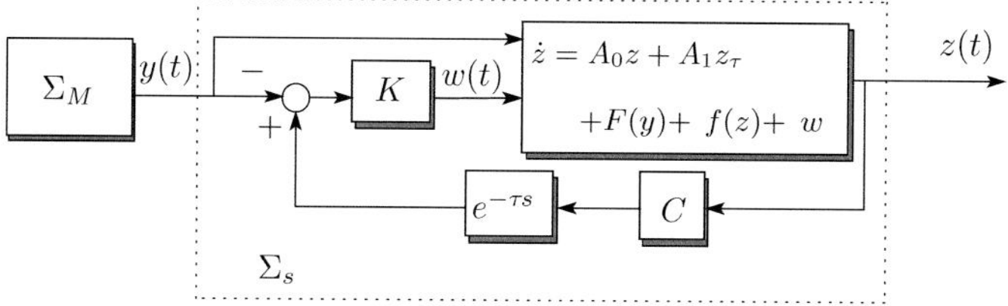

Fig. 5.4 Configuration of the prediction scheme

$$\Sigma_M : \begin{cases} \dot{x}(t) = A_0 x(t) + A_1 x(t-\tau) + F(y(t-\tau)) + f(x(t)) \\ y(t) = Cx(t) \\ x(0) = x_0, \end{cases} \tag{5.8}$$

where $x \in \mathbb{R}^n$, A_0 and A_1 are $n \times n$ matrices, and $f : \mathbb{R}^n \to \mathbb{R}^n$ is a smooth vector field. For the system (5.8), we construct the predictor

$$\Sigma_s : \begin{cases} \dot{z}(t) = A_0 z(t) + A_1 z(t-\tau) + F(y(t)) + f(z(t)) + K\{Cz(t-\tau) - y(t)\} \\ z(t) = z_0, \ t \in [-\tau, 0], \end{cases} \tag{5.9}$$

where $z \in \mathbb{R}^n$ and $K \in \mathbb{R}^{n\times m}$ is a constant matrix. For simplicity of notation, we will often suppress notation of the explicit dependence of time t when no confusion can arise. The configuration of the prediction scheme is shown in Fig. 5.4.

Then, the dynamics of the prediction error $e \triangleq z_\tau(t) - x(t)$, where $z_\tau(t) \triangleq z(t-\tau)$, is given by

$$\begin{aligned} \dot{e} =& A_0 e + (A_1 + KC)e_\tau + f(e+x) - f(x) \\ \triangleq& A_0 e + A_d e_\tau + \phi(x,e), \end{aligned} \tag{5.10}$$

with $\phi(x,e) = f(e+x) - f(x)$. Clearly, $e \equiv 0$ is a solution of (5.10), and the system (5.9) acts as a predictor for (5.8) if the error dynamics (5.10) has $e = 0$ as an asymptotically stable equilibrium. Consequently, the anticipating synchronization problem can be reduced to the stability problem of the prediction error dynamics.

If the master system does not have a nonlinear term $f(x)$, the prediction error dynamics is simply

$$\dot{e} = A_0 e + (A_1 + KC)e_\tau,$$

and the stability of the system is determined by the location of the roots of the following characteristic equation.

$$\det(sI - A_0 - (A_1 + KC)e^{-s\tau}) = 0.$$

Therefore, if the most-right root has a negative real part, e converges to zero, and $z(t)$ predicts the future value of $x(t)$.

Next, we consider the case in which the nonlinear term $f(x)$ in the system (5.8) exists. By applying the Lyapunov stability theorem, several sufficient conditions for the zero solution for Eq. (5.10) to be asymptotically stable have been proposed. In the following section, we introduce sufficient conditions derived under several different assumptions.

5.3 Synchronization Conditions

This section introduces some sufficient conditions for anticipating synchronization.

5.3.1 General Approach

We define the Jacobian of $\phi(x, e)$ with respect to e at $e = 0$ by

$$D(x) \triangleq \left(\frac{\partial \phi(x, e)}{\partial e}\right)_{e=0},$$

whose components are functions of x. Then, the linearization of (5.10) about $e = 0$ is given by

$$\dot{e} = A_0 e + A_d e_\tau + D(x)e. \tag{5.11}$$

It is well known that if $e = 0$ of (5.11) is asymptotically stable, then $e = 0$ of (5.10) is also asymptotically stable [9]. Therefore, using this linearized system, we obtain the following lemma concerning the stability of the trivial solution $e = 0$ of (5.10).

Lemma 5.1 ([5]) *Assume the nonempty set $\Omega \subset \mathbb{R}^n$ is such that all trajectories of the system Σ_M converge to it. If there exist symmetric and positive-definite matrices P, Γ_0, Γ_1 and Γ_2 of dimension $n \times n$ satisfying*

$$(A_0 + A_d + D(x))^T P + P(A_0 + A_d + D(x)) + \tau(A_0^T \Gamma_0 A_0 + A_d^T \Gamma_1 A_d + D(x)^T \Gamma_2 D(x)) + \tau P A_d(\Gamma_0^{-1} + \Gamma_1^{-1} + \Gamma_2^{-1}) A_d^T P < 0$$

for all $x \in \Omega$, then the zero solution of (5.10) is asymptotically stable.

Here, we assume that each entry of matrix $D(x(t))$ is bounded, and that the bound is known a priori, because x is a trajectory of the master system. The approximated error system (5.11) can then be rewritten by the following polytopic system:

$$\dot{e}(t) = A_0 e(t) + A_d e(t-\tau) + \sum_{j=1}^{k} p_j(x(t)) D_j e(t), \tag{5.12}$$

where D_j are constant matrices and $p_j(x(t)) \in [0, 1]$ are polytopic coordinates satisfying the convex sum property $\sum_{j=1}^{k} p_j(x(t)) = 1$.

Lemma 5.2 ([5]) *If there exist a symmetric and positive-definite matrix $P > 0$ and positive-definite matrices Γ_k, $0 \leq k \leq m$, of dimension $n \times n$ satisfying*

$$\begin{aligned}&(\bar{A} + D_j)^T P + P(\bar{A} + D_j)\\&\quad + \tau\Big\{(\bar{A} + D_j)^T \Gamma_0 (\bar{A} + D_j) + A_d^T \Gamma_1 A_d + P A_d (\Gamma_0^{-1} + \Gamma_1^{-1}) A_d^T P\Big\} < 0\end{aligned} \tag{5.13}$$

for $j = 1, \ldots, k$, where $\bar{A} = A_0 + A_d$, then the zero solution of (5.11) is asymptotically stable.

Using Lemma 5.2 we obtain the following theorem concerning how to choose the gain K stabilizing system (5.10).

Theorem 5.1 *Let $\tau > 0$. If there exist a symmetric and positive-definite matrix $P > 0$ and a matrix Y such that the following LMI holds for each $j = 1, \ldots, k$:*

$$\begin{bmatrix} M_{11}^j + M_{11}^{jT} & A_0^T P & A_1^T P + C^T Y^T & P A_1 + Y C \\ P A_0 & \frac{-1}{\tau} P & 0 & 0 \\ P A_1 + Y C & 0 & \frac{-1}{\tau} P & 0 \\ A_1^T P + C^T Y^T & 0 & 0 & \frac{-1}{2\tau} P \end{bmatrix} < 0, \tag{5.14}$$

where $M_{11}^j = P(A_0 + A_1 + D_j) + Y C$, then the zero solution of (5.10) is asymptotically stable, and $K = P^{-1} Y$ is a stabilizing gain matrix.

5.3.2 Sector Condition

In this section, we consider the case where $\phi(x, e)$ in (5.10) is a scalar function given by $B\phi(y(t), Ce(t))$. For simplicity, we assume that the master system is a single output system, i.e., $m = 1$. In addition, we assume that the function $\phi(y, y_e)$ satisfies the following sector condition: there exist a constant $\alpha > 0$ such that $0 \leq \phi(y, y_e) \leq \alpha y_e$ for all $y \in \{y = Cx | x \in \Omega\}$.

A possible criterion was derived by Huijberts et al. [10] in order to derive a sufficient condition for anticipating synchronization of chaotic Lur'e systems. The criterion described below is an extension of the improved delay- dependent stability criterion by Xu et al. [11] for systems with a nonlinearity satisfying a sector condition.

Theorem 5.2 ([10]) *Let $\bar{\tau}$ be given and assume that there exist matrices $P > 0$, $Q > 0$, $Z > 0$, Y, and W and a constant $\gamma > 0$ such that the following LMI is satisfied.*

$$\begin{bmatrix} PA_0 + A_0^T P + Y + Y^T + Q & PA_1 - Y + W^T & PB + \gamma\lambda C^T & -\bar{\tau}Y & \bar{\tau}A_0^T Z \\ A_1^T P - Y^T + W & -Q - W - W^T & 0 & -\bar{\tau}W & \bar{\tau}A_1^T Z \\ B^T P + \gamma\lambda C & 0 & -2\gamma & 0 & \bar{\tau}B^T Z \\ -\bar{\tau}Y^T & -\bar{\tau}W^T & 0 & -\bar{\tau}Z & 0 \\ \bar{\tau}ZA_0 & \bar{\tau}ZA_1 & \bar{\tau}ZB & 0 & -\bar{\tau}Z \end{bmatrix} < 0. \tag{5.15}$$

Then, the origin is an asymptotically stable equilibrium point for every $0 < \tau \le \bar{\tau}$.

Remark 5.1 The LMI condition (5.15) in Theorem 5.2 is derived using the Lyapunov–Krasovskii functional

$$V_1(e_t) = e^T(t)Pe(t) + \int_{t-\tau}^{t} e^T(a)Qe(a)da + \int_{-\tau}^{0}\int_{t+b}^{t} \dot{e}^T(a)Z\dot{e}(a)dadb, \tag{5.16}$$

where e_t denotes $e_t = e(t+\theta)$ for $\theta \in [-\tau, 0]$. On the other hand, to solve the master–slave synchronization problem, Han [12] also proposed a synchronization condition for Lur'e systems with a delay coupling using a Lur'e–Postnikov Lyapunov functional candidate. Adjusting the functional for our problem as

$$\begin{aligned} V_2(e_t) =& e(t)^T Pe(t) + \int_{t-\tau}^{t} e^T(\xi)Qe(\xi)d\xi \\ &+ \int_{t-\tau}^{t} \tau(\tau - t + \xi)\dot{e}^T(\xi)R\dot{e}(\xi)d\xi + 2\int_{0}^{Ce(t)} \gamma\Psi(x^1, \xi)d\xi, \end{aligned} \tag{5.17}$$

where P, Q, and R are positive-definite matrices and γ is a positive constant, this functional can cause the same stability condition as the LMI condition by replacing τR with Z.

In this section, we introduced some LMI-based condition for anticipating synchronization. Since these results are based on the Lyapunov–Krasovskii functional approach, the key point is how to evaluate nonlinear terms remaining in the prediction error dynamics. Apart from those mentioned, synchronization conditions can be derived by assuming the QUAD condition. Meanwhile, if the value of the remaining nonlinear term is relatively small, it can be treated as a perturbation that vanishes at $e = 0$ (see Appendix).

5.4 Predictor-Based Control

When we consider control systems in practical situations, there are greater or lesser time-delays in real systems. This section considers use of the slave system in anticipating synchronization as a predictor. For the past ten years, with the spread of networked systems, the study of time-delay systems has attracted a growing interest in control engineering and science. In particular, a large number of useful results have been obtained for linear systems. The study on delay compensation for linear systems with input delays has a relatively long history, and the Smith predictor [13] is one of the most popular delay compensators. Kravaris and Wright [14] proposed an extension of the predictor for nonlinear systems by combining it with the input–output linearization. Further extensions and modifications have been proposed by many researchers. In recent years, input delay systems have attracted attention for a delay compensation method via the backstepping transformation approach proposed by Krstić [15]. In this section, as a completely different approach to these predictors, we introduce the synchronization-based predictor.

To explain the fundamental idea, we consider the following linear system with input delay.

$$\begin{cases} \dot{x}(t) = Ax(t) + bu(t-\tau) \\ y(t) = cx(t), \end{cases} \tag{5.18}$$

where $x \in \mathbb{R}^n$, u, $y \in \mathbb{R}$, and $\tau > 0$ is a time-delay. Here we assume that if $\tau = 0$, (A, b) is controllable and (c, A) is observable. In addition, we introduce the pure delay operator σ that shifts the time from t to $t - \tau$ and is defined as

$$\sigma\lambda(t) = \lambda(t-\tau),$$

where $\lambda(t)$ is a function defined on the interval $[t-\tau, t]$.

Using the pure delay operator σ, the system (5.18) can be rewritten as

$$\dot{x}(t) = Ax(t) + b(\sigma u(t)).$$

For this system, the synchronization-based predictor is designed as follows.

$$\dot{z}(t) = Az(t) + bu(t) + H(y(t) - c(\sigma x(t))). \tag{5.19}$$

The prediction error dynamics then is given by

$$\dot{e}(t) = Ae(t) - Hc\sigma e(t), \tag{5.20}$$

where $e = z_\tau - x(t)$. If we choose H such that the zero solution of (5.20) is asymptotically stable, $z(t-\tau)$ converges to $x(t)$.

Now we design a feedback law as $u(t) = -Kz(t)$ using $z(t)$ instead of $x(t)$. Then, substituting it into (5.18) and (5.19), the total dynamics can be written as

$$\begin{bmatrix} \dot{x} \\ \dot{e} \end{bmatrix} = \begin{bmatrix} A - bK & -bK \\ 0 & A - Hc\sigma \end{bmatrix} \begin{bmatrix} x(t) \\ e(t) \end{bmatrix}. \tag{5.21}$$

The characteristic equation is given by

$$\det\begin{bmatrix} sI-(A-bK) & bK \\ 0 & sI-(A-Hc\sigma) \end{bmatrix}$$
$$= \det(sI-(A-bK))\det(sI-(A-Hc\sigma)) = 0,$$

where $\sigma := e^{-s\tau}$. From this equation, we know that the stability of the prediction error dynamics and system with feedback are independent. This result corresponds to a counterpart of the so-called "separation principle" of controller and observer. From the above, we know that this predictor is an extension of the full state observer. Furthermore, Nijmeijer and Mareels [16] stated the relationship between synchronization and observer theory, and the relationship between anticipating synchronization and this predictor is comparable to the relation between synchronization and the full state observer.

5.4.1 System Configuration

Consider the following input delay system

$$\begin{cases} \dot{x}(t) = f(x(t), u(t-\tau)) \\ y(t) = Cx(t), \end{cases} \tag{5.22}$$

where f is smooth with respect to x and u. We assume that when $\tau = 0$ holds, i.e., the system is delay-free, a feedback $u = \gamma(x(t))$ is so designed that the closed-loop system achieves satisfactory stability. However, if $\tau \neq 0$, the closed-loop system becomes $\dot{x}(t) = f(x(t), \gamma(x(t-\tau)))$ and may be destabilized by the existence of the delay. If we could obtain the future value $x(t+\tau)$ of x, by applying a feedback $u(t) = \gamma(x(t+\tau))$, the closed-loop system would be given by $\dot{x}(t) = f(x(t), \gamma(x(t)))$.

From such a viewpoint, we adopt a state predictor based on anticipating synchronization to estimate the future value $x(t+\tau)$ of x, and the output of the predictor is used in place of the actual state of the system in the feedback. The synchronization-based predictor is described by

$$\dot{z}(t) = f(z(t), u(t)) + K\{Cz(t-\tau) - y(t)\},$$

and the controller is given by $u(t) = \gamma(z(t))$.

The configuration of the proposed control scheme is shown in Fig. 5.5.

In this case, the dynamics of the prediction error defined by $e(t) = z(t-\tau) - x(t)$ is given by

$$\dot{e}(t) = f(e(t) + x(t), u(t-\tau)) - f(x(t), u(t-\tau)) + KCe(t-\tau),$$

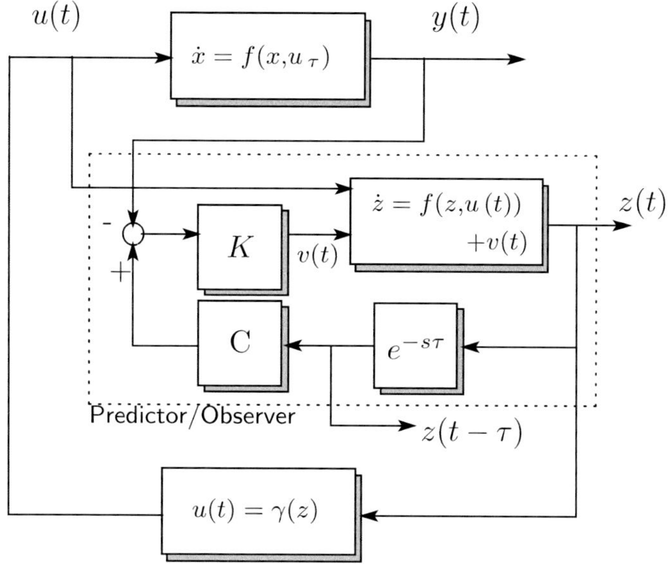

Fig. 5.5 Configuration of the proposed control scheme

and we know that the stability of this dynamics also depends on u. Therefore, under additional assumptions, we consider this problem.

First, we consider the following system.

$$\begin{cases} \dot{x}(t) = f(x(t)) + gu(t-\tau) \\ y(t) = Cx(t), \end{cases} \tag{5.23}$$

where $x \in \mathbb{R}^n$, u, $y \in \mathbb{R}$, $f(0) = 0$, $h(0) = 0$, and g is a constant vector.

The proposed synchronization-based predictor is then defined by

$$\dot{z}(t) = f(z(t)) + gu(t) + K(Cz(t-\tau) - y(t)).$$

In addition, the prediction error dynamics is given by

$$\dot{e}(t) = f(e(t) + x(t)) - f(x(t)) + KCe(t-\tau),$$

where $e(t) = z(t-\tau) - x(t)$. Therefore, by applying the synchronization criteria discussed in Sect. 5.3, we can judge the stability of $e = 0$ or design the coupling gain K.

5.4.2 Linearizable Systems

Next, as a special case, we consider the following affine system.

$$\begin{cases} \dot{x}(t) = f(x(t)) + g(x)u(t-\tau) \\ y(t) = h(x(t)), \end{cases} \tag{5.24}$$

where $x \in \mathbb{R}^n$, u, $y \in \mathbb{R}$, f and g are smooth vector fields with $f(0) = 0$, and h is a smooth function with $h(0) = 0$.

For the system (5.24), we assume that the following conditions hold for any $x \in \mathbb{R}^n$:

1. $\dim(\mathrm{span}\,\{dh, dL_f h, \dots, dL_f^{n-1}h\}) = n$,
2. $[ad_f^i r, ad_f^j r] = 0, 0 \le i, j \le n-1$,
3. $[g, ad_f^j r] = 0, 0 \le j \le n-1$,
4. $ad_f^i r$ for $0 \le i \le n-1$ are complete vector fields,
   where r is the vector field solution of

$$\begin{bmatrix} \langle dh, r\rangle \\ \vdots \\ \left\langle d(L_f^{n-1}h), r\right\rangle \end{bmatrix} = \begin{bmatrix} 0 \\ \vdots \\ 1 \end{bmatrix}.$$

Here $\langle \cdot, \cdot \rangle$ denotes an inner product on $\mathbb{R}^n$.

Then, there exists a global diffeomorphism

$$\bar{x}(t) = T(x(t))$$

such that the system (5.24) is globally transformable into the following linear observable system [22]:

$$\begin{aligned} \dot{\bar{x}}(t) =& \begin{bmatrix} 0\;0 \cdots 0 & -\alpha_0 \\ 1\;0 \cdots 0 & -\alpha_1 \\ \ddots\; 0 & \vdots \\ 0\;0 \cdots 1 & -\alpha_{n-1} \end{bmatrix} \bar{x}(t) + Bu(t-\tau) \\ :=& A\bar{x}(t)(t) + Bu(t-\tau) \qquad (5.25) \\ y(t) =& \begin{bmatrix} 0\;0 \cdots 0\;1 \end{bmatrix} \bar{x}(t) := C\bar{x}(t). \qquad (5.26) \end{aligned}$$

where B is given by $B = \frac{\partial T(x)}{\partial x} g \circ T^{-1}(z)$. As a result, the predictor is constructed by

$$\dot{\bar{z}}(t) = A\bar{z}(t) + Bu(t) + K(C\bar{z}(t-\tau) - y(t)). \tag{5.27}$$

Then, the prediction error dynamics in the transformed coordinates is given by

$$\dot{\bar{e}}(t) = A\bar{e}(t) + KC\bar{e}(t-\tau),$$

where $e(t) = \bar{z}(t-\tau) - \bar{x}(t)$, and the predicted value of x is given by

$$z(t) = T^{-1}(\bar{z}(t)).$$

Therefore, if the most-right root of the corresponding characteristic equation

$$\det(sI - A - KCe^{-s\tau}) = 0$$

has a negative real part, (5.27) works as a predictor for the system (5.24).

Conditions 1–3 guarantee that the system is globally linearized by a coordinate transformation, and they are extremely restrictive.

5.4.3 An Example

We consider a simplified boiler subsystem described by the following equation.

$$\begin{cases} \dot{x}(t) = A_0 x(t) + f(x(t)) + g_1 u_1(t-\tau) + g_2 u_2(t-\tau) \\ y(t) = \begin{bmatrix} h_1(x(t)) \\ h_2(x(t)) \end{bmatrix} = \begin{bmatrix} x_2(t) \\ x_5(t) \end{bmatrix}, \end{cases} \tag{5.28}$$

where $x \in \mathbb{R}^5$, $u_i \in \mathbb{R}$, $y \in \mathbb{R}^2$,

$$A_0 = \begin{pmatrix} -\frac{1}{T_1} & 0 & 0 & 0 & 0 \\ \frac{1}{T_2} & -\frac{1}{T_2} & 0 & 0 & 0 \\ 0 & 0 & -\frac{1}{T_3} & 0 & 0 \\ 0 & 0 & \frac{1}{T_4} & -\frac{1}{T_4} & 0 \\ 0 & -\frac{1}{T_5} & 0 & \frac{a_3}{T_5} & 0 \end{pmatrix},$$

$$f(x) = \left(0, \frac{1}{a_1 T_2} x_5 x_1(t), 0, 0, \frac{a_2}{T_5} x_4^2(t)\right)^T,$$

$$g_1 = \left(\frac{1}{T_1}, 0, 0, 0, 0\right)^T, g_2 = \left(0, 0, \frac{1}{T_3}, 0, 0\right)^T,$$

where T_i and a_i for $i = 1, 2, 3$ are constants, and τ is a constant time-delay. Furthermore, the state variables x_i for $i = 1, \ldots, 5$ are normalized with percentage. Since x_5 is shifted the operating point γ into zero, the bound of x_5 is given by $-\gamma \le x_5 \le 100 - \gamma$. Since this system has time-delays at the inputs, it is called an input time-delay system.

If $\tau = 0$ holds, then this system is exactly linearizable [17, 18] by the following coordinate transformation and feedback.

$$
\begin{aligned}
\xi(t) &= \left(h_1(x), L_f h_1(x), h_2(x), L_f h_2(x), L_f^2 h_2(x)\right)^T \\
\begin{pmatrix} u_1 \\ u_2 \end{pmatrix} &= G(x)^{-1} \begin{pmatrix} -L_f^2 h_1(x) + v_1(t) \\ -L_f^3 h_2(x) + v_2(t) \end{pmatrix},
\end{aligned}
\tag{5.29}
$$

where the non-singular matrix $G(x)$ is given by

$$
G(x) = \begin{pmatrix} L_{g_1} L_f h_1(x) & L_{g_2} L_f h_1(x) \\ L_{g_1} L_f^2 h_2(x) & L_{g_2} L_f^2 h_2(x) \end{pmatrix}.
$$

However, when τ is not 0, all nonlinear terms cannot be canceled out by applying the same feedback (5.29). In order to accomplish exact linearization for nonzero τ, the nonlinear terms at time $t + \tau$ must be evaluated at time t. Therefore, we attempt to estimate the future value of the state x by using a predictor based on the synchronization discussed above.

The state predictor based on synchronization is given by

$$
\dot{z}(t) = A_0 z(t) + f(z) + g_1 u_1(t) + g_2 u_2(t) + K\{Cz(t-\tau) - y(t)\}, \tag{5.30}
$$

where $z(\theta) = z_0$ for $-\ell \le \theta \le 0$. The dynamics of the prediction error $e(t) = z(t - \tau) - x(t)$ is given by

$$
\begin{aligned}
\dot{e}(t) &= A_0 e(t) + KCe(t-\tau) + \{f(e+x) - f(x)\} \\
&\triangleq A_0 e(t) + KCe(t-\tau) + F(e, x),
\end{aligned}
\tag{5.31}
$$

where $F(e, x) \triangleq f(e+x) - f(x)$, and this term satisfies

$$
\|F(e, 0)\| = 0 \text{ and } \|F(e, x)\| < \gamma \|e(t)\|.
$$

The coupling gain K is designed as follows.

If there exist a symmetric and positive-definite matrix $P > 0$ and a matrix Y such that the following LMI holds:

$$
\begin{bmatrix}
M_{11} + M_{11}^T & A_0^T P & C^T Y^T & YC \\
P A_0 & \frac{-1}{\tau} P & 0 & 0 \\
YC & 0 & \frac{-1}{\tau} P & 0 \\
C^T Y^T & 0 & 0 & \frac{-1}{2\tau} P
\end{bmatrix} < 0,
\tag{5.32}
$$

where $M_{11} = P(A_0 + \gamma I) + YC$, then the zero solution of system (5.31) is asymptotically stable and $K = P^{-1} Y$ is a stabilizing gain matrix. Consequently, it is guaranteed that the delayed output $z(t-\tau)$ of the predictor (5.30) converges to $x(t)$.

By combining this state predictor with the linearizing feedback (5.29), we can obtain the linearizing feedback with the state prediction as follows.

$$\begin{pmatrix} u_1 \\ u_2 \end{pmatrix} = G(z)^{-1} \begin{pmatrix} -L_f^2 h_1(z) + v_1(t) \\ -L_f^3 h_2(z) + v_2(t) \end{pmatrix}$$
$$v_1 \triangleq -s_1 h_1(z) - s_2 L_f h_1(z) + y_{1,ref}(t)$$
$$v_2 \triangleq -s_3 h_2(z) - s_4 L_f h_2(z) - s_5 L_f^2 h_2(x) + y_{2,ref}(t),$$

where s_i are coefficients of Hurwitz polynomials:

$$\left.\begin{array}{l} \lambda^2 + s_2\lambda + s_1 \\ \lambda^3 + s_5\lambda^2 + s_4\lambda + s_3 \end{array}\right\}$$

Figures 5.6 and 5.7 show a simulation result. In this simulation, the time-delay is $\tau = 2.0$, the initial condition is given by $y_1(t) = 20$ and $y_5(t) = 0$ for $t \leq 0$, the stabilizing gain matrix by

$$K = \begin{pmatrix} -0.0560 & -0.0983 & 0 & 0 & 0.0019 \\ 0.0001 & 0.0019 & 0 & 0.0005 & -0.1952 \end{pmatrix}^T,$$

and the coefficients s_i are given by

$$(s_1, s_2, s_3, s_4, s_5) = (1.0, 2.0, 1.0, 3.0, 3.0).$$

In Fig. 5.6, the dashed line represents the reference signal $y_{1,ref}$, and $y_{2,ref}(t) = 0$ for $t \geq 0$. This figure shows that the boiler system is exactly linearized by application of the obtained feedback. The prediction error e of the predictor (5.30) for $t \geq \tau$ is illustrated in Fig. 5.7. This figure shows that the prediction error converges to zero and the predictor based on anticipating synchronization works well as the predictor of $x(t)$.

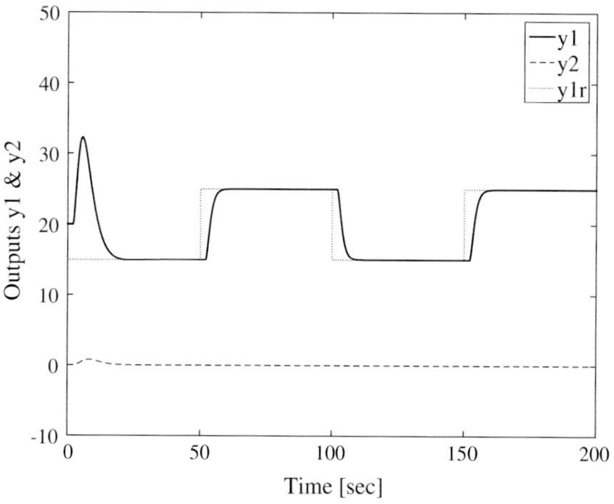

Fig. 5.6 Outputs y_1 and y_2

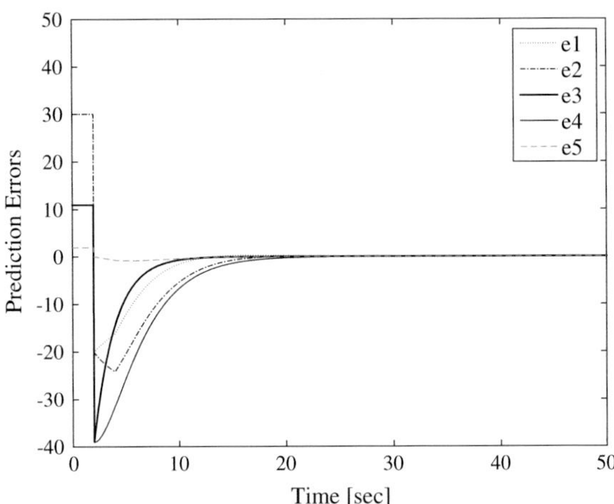

Fig. 5.7 Prediction error $e(t) = z(t - \tau) - x(t)$

5.5 Conclusions

In this chapter, we revisited anticipating synchronization of master–slave type coupled systems. After introducing some synchronization conditions derived in the form of LMIs, we proposed to use this scheme as a state predictor. Today, the major concern in the field of chaos synchronization has shifted from master–slave synchronization to network synchronization; anticipating synchronization, however, can only be observed in the master–slave type system or unidirectional path networks. For example, anticipating synchronization cannot appear in two bidirectional coupled systems because $x_1(t - \tau) = x_2(t)$ or $x_2(t - \tau) = x_1(t)$ are not synchronization manifolds. However, by using anticipating synchronization as the prediction scheme of the system, the consensus problem of multi-agent systems with input delays is also investigated [19].

Appendix: Stability of Systems with Uncertainties

We consider the following system with uncertainties and/or perturbation.

$$\begin{cases} \dot{x}(t) = f(t, x_t) + g(t, x_t), & \forall t \geq 0 \\ x_{t_0} = \phi(t) & \forall t \in [-\tau, 0], \end{cases} \tag{5.33}$$

where $x \in \mathbb{R}^n$, $x_t(\theta) = x(t + \theta)$ for $\theta \in [-\tau, 0]$, $\phi \in \mathscr{C}[-\tau, 0]$, and $g(t, x_t)$ denotes the uncertainty and/or perturbation.

For a function $\phi \in \mathscr{C}([a, b], \mathbb{R}^n)$, define the continuous norm $\|\cdot\|_c$ by $\|\phi\|_c := \sup_{\theta \in [a,b]} \|\phi(\theta)\|$, where $\|\cdot\|$ denotes the Euclidean norm of a vector.

Suppose now that the nominal system

$$\dot{x}(t) = f(t, x_t) \tag{5.34}$$

has a delay-dependent asymptotically stable equilibrium point at the origin. Then the following theorem is well-known.

Theorem 5.3 ([9]) *Let there exist a continuous functional* $V(t, \phi) : \mathbb{R} \times \mathscr{C}[-\tau, 0] \to \mathbb{R}$ *such that*

$$w_1(\|\phi(0)\|) \leq V(t, \phi) \leq w_2(\|\phi\|_c),$$

$w_1(r) \to \infty$ *as* $r \to \infty$ *and* $\dot{V}(x_t) \leq 0$.

Let Z *be the set of those elements from* $\mathscr{C}[-\tau, 0]$ *for which* $\dot{V} = 0$ *and* Q *is the largest invariant set situated in* Z*. Then all solutions of (5.34) tend to* Q *as* $t \to \infty$*. In particular, if the set* Q *has the only zero element, then the trivial solution of (5.34) is asymptotically stable.*

Here $\dot{V}$ is the right derivative of V along the solutions of (5.34), i.e.,

$$\dot{V}(x_t) := \lim_{h \to 0+} \sup \frac{1}{h}(V(x_{t+h}) - V(x_t)).$$

To derive a sufficient condition for robust stability of perturbed systems, we use the following type of Lyapunov–Krasovskii functional throughout this section.

$$V(x_t) = V_0(x(t)) + V_1(x_t), \tag{5.35}$$

where V_0 is a positive-definite function of $x(t)$, and $V_1(x_t)$ consists of the sum of the integrals of the functional $x(t+\theta)$, such as

$$V_1(x_t) := \int_{t-\tau}^{t} \bar{V}_1(x(\theta))d\theta + \int_{-\tau}^{0} \int_{t+s}^{t} \bar{V}_2(x(\theta))d\theta ds \\ + \cdots .$$

Note that the derivative of $V_1(x_t)$ with respect to time does not contain $\dot{x}$.

Then we obtain the following result.

Theorem 5.4 *Let* $x = 0$ *be a delay-dependent asymptotically stable equilibrium point of the nominal system (5.34). Let* $V(t, x_t) : \mathbb{R} \times \mathscr{C} \to \mathbb{R}$ *be a Lyapunov–Krasovskii functional (5.35) that satisfies the following conditions:*

(5.4a) $\alpha_1\|\psi(0)\|^2 \leq V(t, \psi) \leq \alpha_2\|\psi\|_c^2$

(5.4b) the time derivative of V *along the trajectories of the unperturbed system (5.34) satisfies* $\dot{V}_{(5.34)}(t, \psi) \leq -\alpha_3\|\psi(0)\|^2$

where $\dot{V}_{(5.34)}$ *means the derivation of* V *along the solution of (5.34) and* $\psi \in Q_H := \{\psi \in \mathscr{C}[-\tau, 0] : \|\psi\|_c < H\}$. *In addition, suppose that*

(5.4c) $\left\| \frac{\partial V_0(x)}{\partial x} \right\| \leq \alpha_4 \|x\|$

(5.4d) $\|g(t, 0)\| = 0$

(5.4e) $\|g(t, \psi)\| < \gamma \|\psi(0)\|$

(5.4f) $\gamma < \frac{\alpha_3}{\alpha_4}$

where α_i, $i = 1, \ldots, 4$ *and* γ *are positive constants. Then, the origin is an asymptotically stable equilibrium point of the perturbed system (5.33). Moreover, if all the assumptions hold globally, then the origin is globally asymptotically stable.*

Proof From the condition (5.4d), $x = 0$ is also an equilibrium point of the perturbed system (5.33). The existence of V satisfying condition (5.4a) and (5.4b) guarantees that the zero solution of the unperturbed system (5.34) is asymptotically stable from the Lyapunov–Krasovskii theorem [9, 20]. The time derivative of V along the trajectories of (5.33) satisfies

$$\begin{aligned}\dot{V}(x_t) &\leq -\alpha_3 \|x(t)\|^2 + L_g V_0 \\ &\leq -\alpha_3 \|x(t)\|^2 + \alpha_4 \gamma \|x(t)\|^2 \\ &\leq -(\alpha_3 - \alpha_4 \gamma) \|x(t)\|^2.\end{aligned}$$

From the Lyapunov–Krasovskii theorem, if $\gamma < \frac{\alpha_3}{\alpha_4}$, the zero solution of the system (5.33) is also asymptotically stable.

Example 5.1 We consider the following Chua's circuit with time-delay [21].

$$\begin{aligned}\dot{x} &= \begin{bmatrix} -\alpha(1+b) & \alpha & 0 \\ 1 & -1 & 1 \\ 0 & -\beta & -\gamma \end{bmatrix} x + \begin{bmatrix} -\alpha(a-b) \\ 0 \\ 0 \end{bmatrix} \varphi(x_1) + \begin{bmatrix} 0 \\ 0 \\ -\beta \end{bmatrix} \varepsilon \sin(\eta x_1(t-\tau)) \\ &:= A_0 x + B\varphi(x_1) + B_1 \xi(x_1(t-\tau)) \qquad (5.36) \\ y(t) &= \begin{bmatrix} 1 & 0 & 0 \end{bmatrix} x := Cx,\end{aligned}$$

where $\varphi(x_1) = \frac{1}{2}(|x_1 + 1| - |x_1 - 1|)$ and $\xi(x_1(t)) = \sin(\eta x_1(t))$ with $\alpha = 10$, $\beta = 19.53$, $\gamma = 0.1636$, $a = -1.4325$, $b = -0.7831$, $\eta = 0.5$, $\varepsilon = 0.2$ and $\tau = 0.019$.

If we construct the following predictor,

$$\Sigma_s : \begin{cases} \dot{z} = A_0 z + B\varphi(z_1) + B_1 \xi(y) + K\{Cz(t-\tau) - y\} \\ z(t) = z_0, \ t \in [-2\tau, 0], \end{cases}$$

then the dynamics of the prediction error $e(t) = z(t-\tau) - x(t)$ is given by

$$\begin{aligned}\dot{e} &= A_0 e + KCe(t-\tau) + B\{\varphi(e_1 + x_1) - \varphi(x_1)\} \\ &\triangleq A_0 e + KCe(t-\tau) + B\phi(x_1, e_1). \qquad (5.37)\end{aligned}$$

By choosing a Lyapunov–Krasovskii functional as

$$V(e_t) = e(t)^T Pe(t) + \int_0^{\tau}\int_{t-s}^{t} \phi^T B^T RB\phi duds + \sum_{j=0}^{1}\int_{j\tau}^{(j+1)\tau}\int_{t-s}^{t} e^T(u)Q_j e(u)duds,$$

and a coupling gain as $K = \left[-12.1, -2.25, 3.71\right]^T$, anticipating synchronization of the unperturbed systems can be accomplished. Now, we consider the effect of a perturbation on anticipating synchronization. We assume that A_0 in both the master and slave systems contains uncertainties as follows.

$$A_0 = \begin{bmatrix} -\alpha(1+b)+\Delta & \alpha & 0 \\ 1 & -1+\Delta & 1 \\ 0 & -\beta & -\gamma+\Delta \end{bmatrix}.$$

The prediction error dynamics is then obtained by

$$\dot{e} := A_0 e + KCe(t-\tau) + B\phi(x_1, e_1) + g,$$

where $g = \Delta Ie(t)$ In this case, if Δ is bounded, the perturbation satisfies the conditions (5.4d) and (5.4e). In addition, we can obtain the corresponding parameters of condition (5.4f) of Theorem 5.4 as follows: $\gamma < 0.0594$. Considering $x(0) = [1, -2, 0]^T$, $x(\theta) = 0$ for $\theta < 0$ as the initial condition of the master system, $z(\theta) = 0$ for $\theta < 0$ as that of the slave system, and the perturbation $\Delta = 0.05$, the behavior of the prediction error $e(t) = z(t - \tau_1) - x(t)$ is depicted in Fig. 5.8. Since the prediction error converges to the origin, we know that the effect of the perturbation vanishes, and the slave system works effectively as a predictor of the master system.

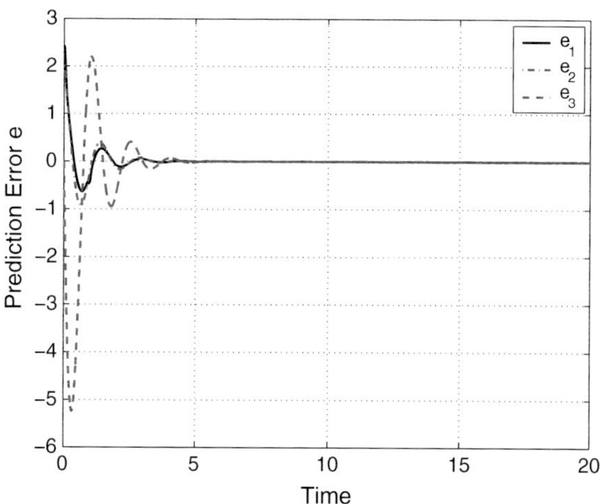

Fig. 5.8 Prediction Error (Theorem 5.4)

References

1. Voss, H.U.: Anticipating chaotic synchronization. Phys. Rev. E **61**, 5115–5119 (2000)
2. Voss, H.U.: Real-time anticipation of chaotic states of an electronic circuit. Int. J. Bifurc. Chaos **12**, 1619–1625 (2002)
3. Masoller, C.: Anticipation in the synchronization of chaotic semiconductor laser with optical feedback. Phys. Rev. Lett. **86**, 2782–2785 (2001)
4. Sivaprakasam, S., et al.: Experimental demonstration of anticipating synchronization in chaotic semiconductor lasers with optical feedback. Phys. Rev. Lett. **87**, 154101 (2001)
5. Oguchi, T., Nijmeijer, H.: Prediction of chaotic behavior. IEEE Trans. Circuits Syst. I: Regular paper **52**, 2464–2472 (2005)
6. Oguchi, T.: A finite spectrum assignment for retarded nonlinear systems and its solvability condition. Int. J. Control **80**(6), 898–907 (2007)
7. Oguchi, T.: Finite spectrum assignment for nolinear time-delay systems. In: Malisoff, M., et al. (eds.) Recent Results on Nonlinear Delay Control Systems, pp. 291–310. Springer, Cham (2015)
8. Lakshmanan, M., Senthilkumar, D.V.: Dynamics of Nonlinear Time-Delay Systems. Springer, Heidelberg (2011)
9. Kolmanovskii, V., Myshkis, A.: Introduction to the Theory and Applications of Functional Differential Equations. Kluwer Academic, Berlin (1999)
10. Huijberts, H., Nijmeijer, H., Oguchi, T.: Anticipating synchronization of chaotic Lur'e systems. Chaos: Interdiscip. J. Nonlinear Sci. **17**, 013117 (2007)
11. Xu, S., Lam, J.: Improved delay-dependent stability results for time-delay systems. IEEE Trans. Autom. Control **50**(3), 384–387 (2005)
12. Han, Q.-L.: New delay-dependent synchronization criteria for Lur'e systems using time delay feedback control. Phys. Lett. A **360**, 563–569 (2007)
13. Smith, O.J.: A controller to overcome dead time. ISA J. **6**(2), 28–33 (1959)
14. Kravaris, C., Wright, R.A.: Deadtime compensation for nonlinear processes. AIChE J. **35**(9), 1535–1542 (1989)
15. Krstić, M.: Delay Compensation for Nonlinear, Adaptive, and PDE Systems. Birkhäuser, Boston (2009)
16. Nijmeijer, H., Mareels, I.M.Y.: An observer looks at synchronization. IEEE Trans. Circuits Syst. **I**(44), 882–890 (1997)
17. Nijmeijer, H., van der Schaft, A.: Nonlinear Dynamical Control Systems. Springer, Heidelberg (1990)
18. Isidori, A.: Nonlinear Control Systems -An Introduction. Springer, Heidelberg (1995)
19. Cao, Y., Oguchi, T., Verhoeckx, P.B., Nijmeijer, H.: Predictor-based consensus control of a multi-agent system with time-delays. In: Proceedings of 2015 IEEE Conference on Control Application (CCA), pp. 113–118 (2015)
20. Hale, J.K., Verduin Lunel, S.M.: Introduction to Functional Differential Equations. Springer, Heidelberg (1993)
21. Cruz-Hernández, C.: Synchronization of time-delay Chua's oscillator with application to secure communication. Nonlinear Dyn. Syst. Theory **4**, 1–13 (2004)
22. Marino, R., Tomei, P.: Nonlinear Control Design- Geometric, Adaptive and Robust. Prentice Hall, Upper Saddle River (1995)

Chapter 6
Delays Effects in Dynamical Systems and Networks: Analysis and Control Interpretations

Wim Michiels

Abstract Time-delays are important components of many systems from engineering, economics and the life sciences, due to the fact that the transfer of material, energy and information is mostly not instantaneous. They appear for instance as computation and communication lags, they model transport phenomena and hereditary effects and they arise as feedback delays in control loops. The aim of the chapter is to present a guided tour on stand-alone and interconnected systems with delays, thereby explaining some important qualitative properties. The focus rather lies on the main ideas as technical details are avoided. Different mechanisms with which delays can interact with the system are outlined, with the emphasis on the effects of delays on stability. It is clarified how these mechanisms affect control design problems. Not only limitations induced by delays in control loops are discussed, but also opportunities to use delays in the construction of controllers. Finally, extensions of these results toward networks of interconnected dynamical systems are discussed, with the focus on relative stability problems, in particular the synchronization problem.

6.1 Introduction

Time-delays are important components of many systems from engineering, economics, and the life sciences, due to the fact that the transfer of material, energy and information is mostly not instantaneous. They appear, for instance, as computation and communication lags, they model transport phenomena and heredity effects and they arise as feedback delays in control loops. An overview of applications, ranging from traffic flow control and lasers with phase-conjugate feedback, over (bio)chemical reactors and cancer modeling, to control of communication networks and control via networks is included in [12].

The presence of time-delays in dynamical systems may induce complex behavior, and this behavior is not always intuitive. Even if a system's equation is scalar,

W. Michiels (✉)
Department of Computer Science, KU Leuven, Celestijnenlaan 200A, 3001 Heverlee, Belgium
e-mail: Wim.Michiels@cs.kuleuven.be

N. van de Wouw et al. (eds.), *Nonlinear Systems*, Lecture Notes in Control and Information Sciences 470, DOI 10.1007/978-3-319-30357-4_6

oscillations may occur. Time-delays in control loops are usually associated with degradation of performance and robustness, but, at the same time, there are situations where time-delays are used as controller parameters.

The aim of this chapter is to describe some important properties of control systems subjected to time-delays and to outline principles behind analysis and synthesis methods. Throughout the text, the results will be illustrated by means of the scalar system

$$\dot{x}(t) = u(t - \tau), \tag{6.1}$$

which, controlled with instantaneous state feedback, $u(t) = -kx(t)$, leads to the closed-loop system

$$\dot{x}(t) = -kx(t - \tau). \tag{6.2}$$

Although this didactic example is extremely simple, we shall see that its dynamics are already very rich and shed a light on delay effects in control loops.

In some works, the analysis of (6.2) is called the *hot shower problem*, as it can be interpreted as a (over)simplified model for a human adjusting the temperature in a shower: $x(t)$ then denotes the difference between the water temperature and the desired temperature as felt by the person, the term $-kx(t)$ models the reaction of the person by further opening or closing taps, and the delay is due to the propagation with finite speed of the water in the ducts.

The structure of the chapter is as follows. In Sect. 6.2, we outline fundamental properties of time-delays systems. In Sect. 6.3, we discuss spectral properties of linear time-delay systems. In Sect. 6.4, we discuss limitations induced by delays in control loops, but also opportunities of using delays for control purposes, and, in Sect. 6.5, we make the leap to networks of interconnected systems, focusing on the synchronization problem. A short version of Sects. 2–4 appeared in [7].

6.2 Basic Properties of Time-Delay Systems

6.2.1 Functional Differential Equation

We focus on a model for a time-delay system described by

$$\dot{x}(t) = A_0x(t) + A_1x(t - \tau), \quad x(t) \in \mathbb{R}^n. \tag{6.3}$$

This is an example of a *functional differential equation* (FDE) of *retarded type*. The term FDE stems from the property that the right-hand side can be interpreted as a functional evaluated at a piece of trajectory. The term retarded expresses that the right-hand side does not explicitly depend on $\dot{x}$.

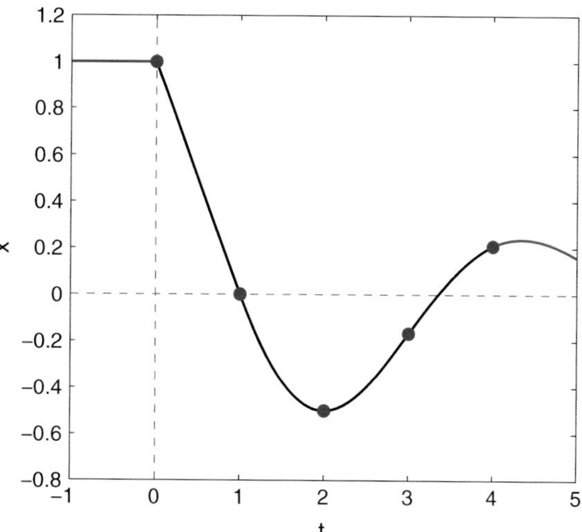

Fig. 6.1 Solution of (6.2) for $\tau = 1$, $k = 1$ and initial condition $\phi \equiv 1$

As a first difference with an ordinary differential equation, the initial condition of (6.3) at $t = 0$ is a function ϕ from $[-\tau,\ 0]$ to $\mathbb{R}^n$. For all $\phi \in \mathscr{C}([-\tau,\ 0], \mathbb{R}^n)$, where $\mathscr{C}([-\tau,\ 0],\ \mathbb{R}^n)$ is the space of continuous functions mapping the interval $[-\tau,\ 0]$ into $\mathbb{R}^n$, a forward solution $x(\phi)$ exists and is uniquely defined. In Fig. 6.1, a solution of the scalar system (6.2) is shown. The discontinuity in the derivative at $t = 0$ stems from $A_0\phi(0) + A_1\phi(-\tau) \neq \lim_{\theta\to 0-} \dot{\phi}(\theta)$. Due to the smoothing property of an integrator, however, at $t = n \in \mathbb{N}$, the discontinuity will only be present in the $(n + 1)-$th derivative. This illustrates a second property of functional differential equations of retarded type: solutions become smoother as time evolves. As a third major difference with ODEs, backward continuation of solutions is not always possible [8].

The extension of methods for time-integration (time stepping, simulation) from ordinary to delay differential equations naturally follows from the properties of solutions sketched above: the "history" of the solution should be taken into account in the time-stepper, and special attention should be paid to so-called break points, where the solution is non-smooth. A key reference is the book [1].

6.2.2 Reformulation in a First-Order Form

The state of system (6.3) at time t is the minimal information needed to continue the solution, which, once again, boils down to a function segment $x_t(\phi)$ where $x_t(\phi)(\theta) = x(t + \theta),\ \theta \in [-\tau,\ 0]$ (in Fig. 6.1 the function x_t is shown in red for $t = 5$). This suggests that (6.3) can be reformulated as a standard ordinary differential equation over the infinite-dimensional space $\mathscr{C}([-\tau,\ 0], \mathbb{R}^n)$. This equation takes the form

$$\frac{d}{dt}z(t) = \mathscr{A}z(t),\ z(t) \in \mathscr{C}([-\tau,\ 0], \mathbb{R}^n) \tag{6.4}$$

where operator $\mathscr{A}$ is given by

$$\begin{aligned} \mathscr{D}(\mathscr{A}) &= \left\{ \begin{array}{r} \phi \in \mathscr{C}([-\tau_m,\ 0],\ \mathbb{R}^n) : \dot{\phi} \in \mathscr{C}([-\tau_m,\ 0],\ \mathbb{R}^n), \\ \dot{\phi}(0) = A_0\phi(0) + A_1\phi(-\tau) \end{array} \right\}, \\ \mathscr{A}\,\phi &= \frac{d\phi}{d\theta}. \end{aligned} \tag{6.5}$$

The relation between solutions of (6.3) and (6.4) is given by $z(t)(\theta) = x(t+\theta),\ \theta \in [-\tau,\ 0]$. Note that all system information is concentrated in the non-local boundary condition describing the domain of $\mathscr{A}$. The representation (6.4) is closely related to a description by an advection PDE with a non-local boundary condition [5].

6.2.3 *Asymptotic Growth Rate of Solutions and Stability*

The reformulation of (6.3) into the standard form (6.4) allows us to define stability notions and to generalize the stability theory for ordinary differential equations in a straightforward way, with the main change that the state space is $\mathscr{C}([-\tau,\ 0], \mathbb{R}^n)$. For example, the null solution of (6.3) is exponentially stable if and only if there exist constants $C > 0$ and $\gamma > 0$ such that

$$\forall\phi \in \mathscr{C}([-\tau_m,\ 0], \mathbb{R}^n) \quad \|x_t(\phi)\|_s \leq Ce^{-\gamma t}\|\phi\|_s,$$

where $\|\cdot\|_s$ is the supremum norm, $\|\phi\|_s = \sup_{\theta\in[-\tau,\ 0]} \|\phi(\theta)\|_2$. As the system is linear, asymptotic and exponential stability are equivalent. A direct generalization of Lyapunov's second method yields the following theorem:

Theorem 6.1 *The null solution of linear system (6.3) is asymptotically stable if there exist a continuous functional $V : \mathscr{C}([-\tau,\ 0], \mathbb{R}^n) \to \mathbb{R}$ (also-called Lyapunov-Krasovskii functional) and continuous non-decreasing functions $u, v, w : \mathbb{R}^+ \to \mathbb{R}^+$ with*

$$u(0) = v(0) = w(0) = 0 \text{ and } u(s) > 0, v(s) > 0, w(s) > 0 \text{ for } s > 0,$$

such that for all $\phi \in \mathscr{C}([-\tau,\ 0], \mathbb{R}^n)$

$$u(\|\phi\|_s) \leq V(\phi) \leq v(\|\phi(0)\|_2),\ \dot{V}(\phi) \leq -w(\|\phi(0)\|_2),$$

where

$$\dot{V}(\phi) = \limsup_{h \to 0+} \frac{1}{h} \left[V(x_h(\phi)) - V(\phi) \right].$$

Converse Lyapunov theorems and the construction of so-called complete-type Lyapunov-Krasovskii functionals are discussed in [4].

Imposing a particular structure on the functional, e.g., a form depending only a finite number of free parameters, often leads to easy-to-check stability criteria (for instance, in the form of Linear Matrix Inequalities (LMIs)), yet as price to pay the obtained results may be conservative in the sense that the sufficient stability conditions might not be close to necessary conditions. As an alternative to Lyapunov functionals, Lyapunov functions can be used as well, provided that the condition $\dot{V} < 0$ is relaxed (the so-called Lyapunov-Razhumikhin approach), see, for example, [2].

More recent contributions on stability of systems with time-varying delay originate from a similar perturbation point of view, where the system is seen as a perturbation of a system with constant (possibly non-zero) delay.

6.2.4 Delay Differential Equations as Perturbation of ODEs

Many results on stability, robust stability, and control of time-delay systems are explicitly or implicitly based on a perturbation point of view, where delay differential equations are seen as perturbations of ordinary differential equations. For instance, in the literature a classification of stability criteria is often presented in

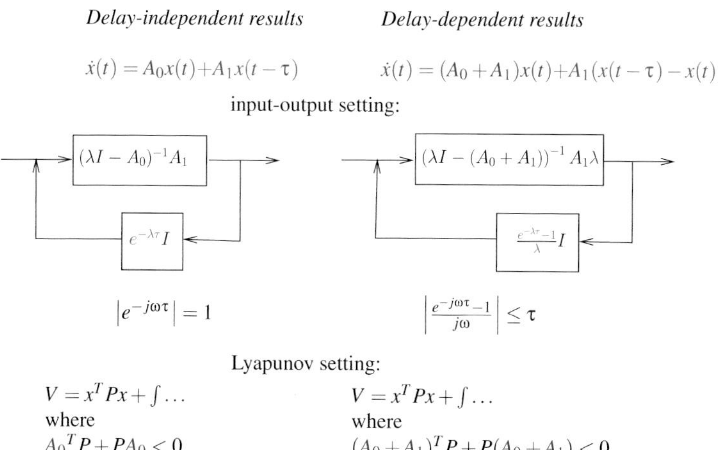

Fig. 6.2 The classification of stability criteria in delay-independent results and delay-dependent results stems from two different perturbation viewpoints. Here, perturbation terms are printed in *red*

terms of *delay-independent* criteria (conditions holding for all values of the delays) and *delay-dependent* criteria (usually holding for all delays smaller than a bound). This classification has its origin at two different ways of seeing (6.3) as a perturbation of an ODE, with as nominal system $\dot{x}(t) = A_0x(t)$ and $\dot{x}(t) = (A_0 + A_1)x(t)$ (system for zero delay), respectively. This observation is illustrated in Fig. 6.2 for results based on input-output and Lyapunov based approaches.

6.3 The Spectrum of Linear Time-Delay Systems

6.3.1 Two Eigenvalue Problems

The substitution of an exponential solution in (6.3) leads us to the *nonlinear eigenvalue problem*

$$\left(\lambda I - A_0 - A_1 e^{-\lambda\tau}\right) v = 0,\ \lambda \in \mathbb{C},\ v \in \mathbb{C}^n,\ v \neq 0. \tag{6.6}$$

The solutions of the equation $\det(\lambda I - A_0 - A_1 e^{-\lambda\tau}) = 0$ are called characteristic roots. Similarly, formulation (6.4) leads to the equivalent *infinite-dimensional linear eigenvalue problem*

$$(\lambda I - \mathscr{A})\, u = 0,\ \lambda \in \mathbb{C},\ u \in \mathscr{C}([-\tau,\ 0], \mathbb{C}^n),\ u \not\equiv 0. \tag{6.7}$$

The combination of these two viewpoints lay at the basis of most methods for computing characteristic roots, see [6]. On the one hand, discretizing (6.7), i.e., approximating $\mathscr{A}$ with a matrix, and solving the resulting standard eigenvalue problems allows to obtain global information, for example, estimates of *all* characteristic roots in a given compact set or in a given right-half plane. On the other hand, the (finitely many) nonlinear equations (6.6) allow to make *local corrections* on characteristic root approximations up to the desired accuracy, e.g., using Newton's method or inverse residual iteration.

Linear time-delay systems satisfy spectrum-determined growth properties of solutions. For instance, the zero solution of (6.3) is asymptotically stable if and only if all characteristic roots are in the open left-half plane.

In Fig. 6.3 (left), the rightmost characteristic roots of (6.2) are depicted for $k\tau = 1$. Note that, since the characteristic equation can be written as $\lambda\tau + k\tau e^{-\lambda\tau} = 0$, k and τ can be combined into one parameter. In Fig. 6.3 (right), we show the real parts of the characteristic roots as a function of $k\tau$. The plots illustrate some important spectral properties of retarded type FDEs. First, even though there are in general infinitely many characteristic roots, their number in *any* right-half plane is always finite. Second, the individual characteristic roots, as well as the *spectral abscissa*, i.e., the supremum of the real parts of all characteristic roots, continuously depend on parameters. Related to this, a loss or gain of stability is always associated with

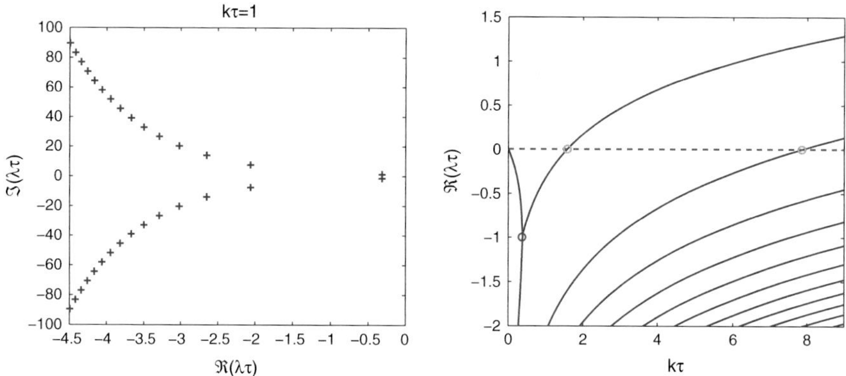

Fig. 6.3 (*Left*) rightmost characteristic roots of (6.2) for $k\tau = 1$. (*Right*) real parts of rightmost characteristic roots as a function of $k\tau$

characteristic roots crossing the imaginary axis. Figure 6.3 (right) also illustrates the transition to a delay-free system as $k\tau \to 0^+$.

6.3.2 *Critical Delays: A Finite Dimensional Characterization*

Assume that, for a given value of k, we are looking for values of the delay τ_c for which (6.2) has a characteristic root $j\omega_c$ on the imaginary axis. From $j\omega = -ke^{-j\omega\tau}$ we get

$$\omega_c = k,\ \tau_c = \frac{\frac{\pi}{2} + l2\pi}{\omega_c},\quad l = 0, 1, \ldots,\quad \Re\left\{ \left.\frac{d\lambda}{d\tau}\right|_{(\tau_c, j\omega_c)} \right\}^{-1} = \frac{1}{\omega_c^2}. \tag{6.8}$$

Critical delay values τ_c are indicated with green circles on Fig. 6.3 (right). The above formulas first illustrate an *invariance property* of imaginary axis roots and their crossing direction with respect to delay shifts of $2\pi/\omega_c$. Second, the number of possible values of ω_c is one and thus *finite*. More generally, substituting $\lambda = j\omega$ in (6.6) and treating τ as a free parameter leads to a *two-parameter eigenvalue problem*

$$(j\omega I - A_0 - A_1 z)v = 0, \tag{6.9}$$

with ω on the real axis and $z := \exp(-j\omega\tau)$ on the unit circle. Most methods to solve such a problem boil down to an elimination of one of the independent variables ω or z. As an example of an elimination technique, we directly get from (6.9),

$$\begin{aligned}&j\omega \in \sigma(A_0 + A_1 z),\ -j\omega \in \sigma(A_0^* + A_1^* z^{-1})\\ &\Rightarrow \det\left((A_0 + A_1 z) \oplus (A_0^* + A_1^* z^{-1})\right) = 0\\ &\Rightarrow \det\left((A_0 z + A_1 z^2) \oplus (A_0^* z + A_1^*)\right) = 0.\end{aligned}$$

where $\sigma(\cdot)$ denotes the spectrum and $\oplus$ the Kronecker sum. Clearly, the resulting quadratic eigenvalue problem in z is finite-dimensional.

6.4 Control of Time-Delay Systems

6.4.1 Limitations Induced by Delays

It is well known that delays in controls loop may lead to a significant degradation of performance and robustness and even to instability [9, 11]. Let us return to example (6.2). As illustrated with Fig. 6.3 and expressions (6.8), the system loses stability if τ reaches the value $\pi/2k$, while stability cannot not be recovered for larger delays. The maximum achievable exponential decay rate of the solutions, which corresponds to the minimum of the spectral abscissa, is given by $-1/\tau$, hence, large delays can only be tolerated at the price of a degradation of the rate of convergence. It should be noted that the limitations induced by delays are even more stringent if the uncontrolled systems is exponentially unstable, which is not the case for (6.2).

The analysis in the previous sections gives a hint why control is difficult in the presence of delays: the system is inherently infinite-dimensional. As a consequence, most control design problems which involve determining a finite number of parameters can be interpreted as reduced-order control design problems or as control design problems for under-actuated systems, which are both known to be hard problems.

6.4.2 Fixed-Order Control

Most standard control design techniques lead to controllers whose dimension is larger or equal to the dimension of the system. For infinite-dimensional time-delay system such controllers might have a disadvantage of being complicated and hard to implement. To see this, for a system with delay in the state the generalization of static state feedback, $u(t) = k(x)$ is given by

$$u(t) = \int_{-\tau}^{0} x(t+\theta)d\mu(\theta),$$

where μ is a function of bounded variation. However, in the context of large-scale systems it is known that reduced-order controllers often perform relatively well compared to full-order controllers, while they are much easier to implement.

Recently new methods for the design of controllers with a prescribed order (dimension) or structure have been proposed [6]. These methods rely on a direct optimization of appropriately defined cost functions (spectral abscissa, $\mathscr{H}_2/\mathscr{H}_\infty$ criteria). While $\mathscr{H}_2$ criteria can be addressed within a derivative-based optimization framework, $\mathscr{H}_\infty$ criteria and the spectral abscissa require targeted methods for *non-smooth optimization problems*. To illustrate the need for such methods consider again Fig. 6.3 (right): minimizing the spectral abscissa for a given value of τ as a function of the controller gain k leads to an optimum where the objective function is not differentiable, even not locally Lipschitz, as shown by the red circle. In case of multiple controller parameters, the path of steepest descent in the parameter space typically has phases along a manifold characterized by the non-differentiability of the objective function.

6.4.3 Using Delays as Controller Parameters

In contrast to the detrimental effects of delays, there are situations where delays have a beneficial effect and are even used as controller parameters, see [12]. For instance, delayed feedback can be used to stabilize oscillatory systems where the delay serves to adjust the phase in the control loop. An illustration is given in Fig. 6.4, which depicts the stability regions of oscillator

$$\ddot{x}(t) = -x(t) + u(t), \;\; y(t) = x(t),$$

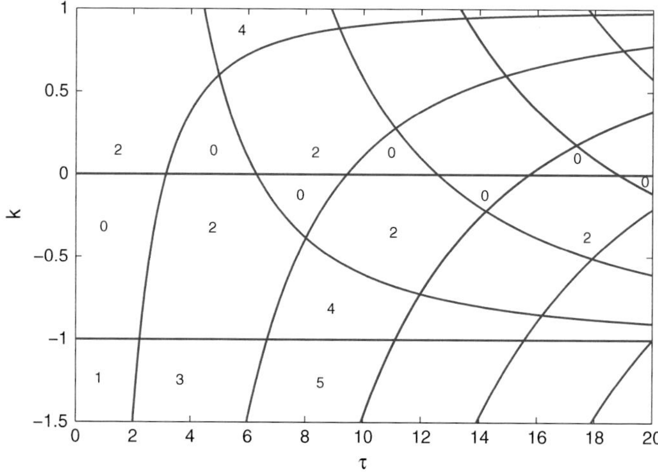

Fig. 6.4 Stability regions in the (k, τ)- parameter space of an oscillator controlled by delayed output feedback. The numbers refers to the number of characteristic roots in the open *right-half* plane

controlled with delayed output feedback, $u(t) = -ky(t-\tau)$. Note that for $\tau = 0$ the (second order) system is not stabilizable by static output feedback. For small k, the sequence of stability—instability regions in the delay parameter space is related to the feedback acting in-phase or anti-phase with respect to the oscillation of the uncontrolled system. Note that, by taking k sufficiently small, systems with arbitrarily large input delay can be stabilized as well.

Delayed terms in control laws can also be used to approximate derivatives in the control action. For example a PD controller can by approximated by a control law of the form

$$u(t) = -k_1 y(t) - k_2 \frac{y(t) - y(t-\tau)}{\tau},$$

for τ sufficiently small. This principle can be extended towards higher-order derivatives (combined with a scaling property it allows, for instance, to derive stabilizing control laws for integrator chains, using control laws of the form $u(t) = \sum_i k_i y(t-\tau_i)$ where the delays can be prescribed [8]). The experimental reconstruction and visualization of attractors of systems of high order via delayed time-series of the output is grounded in a similar idea.

Control laws of the form

$$u(t) = F(y(t) - y(t-\tau))$$

which depend on time-differences of state or output variables, so-called Pyragas type feedback, have the property that the position of equilibria and the shape of periodic orbits with period τ are not affected (since the control law vanishes whenever $y(t) = y(t-\tau)$), in contrary to their stability properties.

Last but not least, delays can be used in control schemes to generate predictions or to stabilize predictors, which allow to compensate delays and improve performance [5, 15]. Let us illustrate the main idea once more with the system (6.1).

System (6.1) has a special structure, in the sense that the delay is only in the input, and it is advantageous to exploit this structure in the context of control. Coming back to the didactic example, the person who is taking a shower is -possibly after some bad experiences- aware about the delay and will take into account his/her prediction of the system's reaction when adjusting the cold and hot water supply. Let us, to conclude, formalize this. The uncontrolled system can be rewritten as $\dot{x}(t) = v(t)$, where $v(t) = u(t-\tau)$. We know u up to the current time t, thus we know v up to time $t+\tau$ and, if $x(t)$ is also known, we can predict the value of x at time $t+\tau$:

$$x_p(t+\tau) = x(t) + \int_t^{t+\tau} v(s)ds = x(t) + \int_{t-\tau}^{t} u(s)ds,$$

and use the predicted state for feedback. With the control law $u(t) = -kx_p(t+\tau)$ there is only one closed-loop characteristic root at $\lambda = -k$, i.e., as long as the model used in the predictor is exact, the delay in the loop is compensated by the prediction. For further reading on prediction based controllers, see, e.g., [5] and the references therein.

6.5 From Stand-Alone Systems to Networks of Interconnected Systems

We briefly discuss the effects of delays in networks of nonlinear coupled systems, focusing on synchronization.

There are many examples of networks of interacting dynamical systems that exhibit collective behavior. The most unambiguous form of collective behavior is that of synchronization, which refers to the state in which all systems in the network behave identically. Fireflies emit their light pulses at the same instants in time; crickets chirp in unison for extended periods of time; and electrons move in synchrony in superconducting Josephson junctions [14].

Synchronization is a form of *relative* stability, i.e., a stability property of the zero solution of the error dynamics, which describe the differences between the state variables of the systems. A special case of synchronization problems are consensus problems, where the state variables of the different systems converge to a common steady state value. Consensus problem are also important in the context of distributed decision making, social networks (opinion dynamics), distributed and cooperative control (adaptive cruise control, formation stabilization), just to mention a few [10].

Networks of identical diffusive delay-coupled systems are studied in depth in [13]. In order to highlight the effects of coupling delays, it suffices to analyze a model problem consisting of two coupled systems,

$$\dot{x}_i(t) = f(x_i(t)) + u_i(t), \quad i = 1, 2,$$

with coupling

$$\begin{cases} u_1(t) = k_1(x_2(t-\tau) - x_1(t)), \\ u_2(t) = k_2(x_1(t-\tau) - x_2(t)). \end{cases} \tag{6.10}$$

The linear coupling functions (possibly with zero delay) as in (6.10) appear in a large number of applications, such as, networks of coupled neurons, networks of biological systems, coupled mechanical systems and electrical systems. The coupling signal $u_i(t)$ of node i is (in general) defined as the sum of weighted differences of time-delayed outputs of connected systems and the node's own output at time instant t, $y_i(t)$. In this type of coupling, the delay models the effect of finite speed of signal transmission.

Most nonlinear oscillators encountered in applications from engineering, physics and biology are semi-passive, meaning that they behave as passive systems sufficiently far from the origin. Semi-passivity implies boundedness of solutions of the coupled systems under mild additional conditions, also in the presence of delay.

The presence of observable synchronized behavior requires two properties to be satisfied: the presence of a forward invariant manifold characterized by synchronized motion, a so-called synchronization manifold, and (2) the satisfaction of a stability property of this manifold, in the sense that it attracts neighboring solutions. To clarify the role of delays, let us consider the above example and make a change of coordinate, $e(t) = x_2(t) - x_1(t)$, resulting in

$$\begin{cases} \dot{x}_1(t) = f(x_1(t)) - k_1(x_1(t) - x_1(t-\tau)) + k_1 e(t-\tau), \\ \dot{e}(t) \;\; = f(x_1(t) + e(t)) - f(x_1(t)) - k_2 e(t) - k_1 e(t-\tau) \\ \qquad\qquad +(k_1 - k_2)(x_1(t) - x_1(t-\tau)). \end{cases} \tag{6.11}$$

In the delay free case (6.11) becomes

$$\begin{cases} \dot{x}_1(t) = f(x_1(t)) + k_1 e(t) \\ \dot{e}(t) \;\; = f(x_1(t) + e(t)) - f(x_1(t)) - (k_1 + k_2) e(t). \end{cases} \tag{6.12}$$

Comparing (6.11) and (6.12) we conclude the following:

1. In the presence of delay, the existence of a partial synchronization manifold (i.e., $e \equiv 0$ being a solution of the second equation) requires $k_1 = k_2$, whereas $e \equiv 0$ always solves the second equation of (6.12). This illustrates that *additional* structural requirements on the coupling may be necessary in the presence of delay.
2. Assuming $k_1 = k_2$, the dynamics on the synchronization manifold of (6.11) are described by
$$\dot{x}_1(t) = f(x_1(t)) - k_1(x_1(t) - x_1(t-\tau)),$$
which can be interpreted as the dynamics of one system, controlled with Pyragas type time-delayed feedback. Since the dynamics on the synchronization manifold are affected by the coupling, we call the type of coupling (6.10) *invasive* for $\tau \neq 0$. On the other hand, the dynamics on the synchronization manifold of (6.12) are described by the dynamics of one individual system.
3. For $k_1 = k_2$, the dynamics transversal to the synchronization manifold of (6.11), and its stability properties, are determined by the stability properties of an equilibrium of a delay differential equation with time-varying coefficients (considering x_1 as an exogenous signal), necessitating targeted analysis and synthesis tools. Local stability is described by the linearization
$$\delta\dot{e}(t) = \frac{\partial f}{\partial x}(x_1(t))\delta e(t) - k_2\delta e(t) - k_1\delta e(t-\tau).$$
In the case of more complex networks, a decoupling of the linearized error dynamic is possible based on the eigenstructure of the graph Laplacian matrix.

Remark 6.1 Besides (6.10) another frequently encountered coupling type is described by

$$\begin{cases} u_1(t) = k_1(x_2(t-\tau) - x_1(t-\tau)), \\ u_2(t) = k_2(x_1(t-\tau) - x_2(t-\tau)), \end{cases} \tag{6.13}$$

that is, both the node's own output and the outputs of connected systems are delayed by an amount of τ. This type of coupling models the effects of sensor/actuator delay. Coupling (6.13) is *non-invasive* since it vanishes on the synchronization manifold and, therefore, it does not affect the dynamics.

For large networks, the emerging behavior in bifurcations may be very complex, and the error dynamics are high dimensional and hard to analyze, but simplifies when the coupling strength is strong. For more information we refer to [13] and the references therein.

Finally, let us take, instead of mutual coupling, a master-slave setting and make a connection with predictive feedback in Sect. 6.4.3. Suppose we want to predict the state of a possibly chaotic system

$$\dot{x}(t) = f(x(t)),$$

over a time-window of length τ. To achieve this one can build an "observer" of the form

$$\dot{z}(t) = f(z(t)) + K(z(t-\tau) - x(t)).$$

The key idea is that the second, correction term vanishes on the moment that x behaves as a delayed version of z. The prediction error $e(t) = z(t-\tau) - x(t)$ satisfies

$$\dot{e}(t) = f(x(t) + e(t)) - f(x(t)) + Ke(t-\tau). \tag{6.14}$$

The predictor is stable and the prediction reliable when K and τ are such that the null solution of (6.14) is asymptotically stable. Hence, eventually the synthesis of the predictor boils down to a stabilization problem with delayed feedback, which is prone to the fundamental limitations sketched in Sect. 6.4.1. The above idea is elaborated in [3] and called anticipating synchronization.

6.6 Conclusions

Time-delay systems, which appear in a large number of applications, in particular in the context of networked systems, are a class of infinite-dimensional systems, resulting in rich dynamics and challenges from a control point of view. The different representations and interpretations and, in particular, the combination of viewpoints lead a wide variety of analysis and synthesis tools.

References

1. Bellen, A., Zennaro, M.: Numerical Methods for Delay Differential Equations. Oxford University Press, Oxford (2003)
2. Gu, K., Kharitonov, V.L., Chen, J.: Stability of Time-delay Systems. Birkhäuser, Boston (2003)
3. Huijberts, H., Nijmeijer, H., Oguchi, T.: Anticipating synchronization of chaotic Lur'e systems. Chaos **17**, 013117 (2007)
4. Kharitonov, V.L.: Time-Delay Systems. Lyapunov Functionals and Matrices. Birkhäuser, Basel (2013)
5. Krstić, M.: Delay Compensation for Nonlinear, Adaptive, and PDE Systems. Birkhäuser, Boston (2009)
6. Michiels, W.: Design of fixed-order stabilizing and $\mathscr{H}_2 - \mathscr{H}_\infty$ optimal controllers: an eigenvalue optimization approach. Lect. Notes Control Inf. Sci. **423**, 201–216 (2012)
7. Michiels, W.: Control of linear systems with delays. In: Baillieul, J., Samad, T. (eds.) The Encyclopedia of Systems and Control. Springer, Berlin (2015)
8. Michiels, W., Niculescu, S.I.: Stability, Control and Computation for Time-Delay Systems. An Eigenvalue Based Approach, 2nd edn. SIAM (2014)
9. Niculescu, S.I.: Delay Effects on Stability. A Robust Control Approach, Volume 269 of Lecture Notes in Control and Information Sciences. Springer, Berlin (2001)
10. Olfati-Saber, R., Murray, R.M.: Consensus problems in networks of agents with switching topology and time-delays. IEEE Trans. Autom. Control **49**(9), 1520–1533 (2004)
11. Richard, J.-P.: Time-delay systems: an overview of recent advances and open problems. Automatica **39**(10), 1667–1694 (2003)
12. Sipahi, R., Niculescu, S., Abdallah, C., Michiels, W., Gu, K.: Stability and stabilization of systems with time delay. IEEE Control Syst. Mag. **31**(1), 38–65 (2011)
13. Steur, E., Oguchi, T., Nijmeijer, H.: Synchronization of Systems with Time-delayed Coupling. World Scientific, Singapore (2016) (In press)
14. Strogatz, S.H.: Sync: The Emerging Science of Spontaneous Order. 1st edn. Hyperion, New York (2003)
15. Zhong, Q.-C.: Robust Control of Time-Delay Systems. Springer, London (2006)

Chapter 7
Emergence of Oscillations in Networks of Time-Delay Coupled Inert Systems

Erik Steur and Alexander Pogromsky

Abstract We discuss the emergence of oscillations in networks of single-input–single-output systems that interact via linear time-delay coupling functions. Although the systems itself are inert, that is, their solutions converge to a globally stable equilibrium, in the presence of coupling, the network of systems exhibits ongoing oscillatory activity. We address the problem of emergence of oscillations by deriving conditions for; 1. solutions of the time-delay coupled systems to be bounded, 2. the network equilibrium to be unique, and 3. the network equilibrium to be unstable. If these conditions are all satisfied, the time-delay coupled inert systems have a nontrivial oscillatory solution. In addition, we show that a necessary condition for the emergence of oscillations in such networks is that the considered systems are at least of second order.

7.1 Introduction

This chapter is concerned with networks of identical single-input-single-output systems that interact via linear time-delay coupling functions. A little bit more precise, the coupling for a system in a network is defined to be the weighted difference of the time-delayed output of its neighbors and its own, non-delayed output. The delay models in this case the time it takes a signal to propagate from its source to its

E. Steur (✉)
Institute for Complex Molecular Systems and Department of Mechanical Engineering, Eindhoven University of Technology, P.O. Box 513, 5600 MB Eindhoven, The Netherlands
e-mail: e.steur@tue.nl

A. Pogromsky
Department of Mechanical Engineering, Eindhoven University of Technology, Eindhoven, The Netherlands
e-mail: a.pogromsky@tue.nl

A. Pogromsky
Department of Control Systems and Informatics, Saint-Petersburg National Research University of Information Technologies Mechanics and Optics, Saint Petersburg, Russia

N. van de Wouw et al. (eds.), *Nonlinear Systems*, Lecture Notes in Control and Information Sciences 470, DOI 10.1007/978-3-319-30357-4_7

destination, and therefore it is reasonable to assume that the systems have immediate access to their own outputs. We consider the case that the systems are inert, that is, in absence of coupling each system has a globally asymptotically stable equilibrium. We address the problem that, nevertheless, oscillations emerge in network of the time-delay coupled systems.

The problem of emergence of oscillations in coupled inert systems goes back to the early fifties of the previous century, starting with Alan Turing's work on morphogenesis [25]. About twenty years later, Steven Smale, being inspired by the work of Turing, proposed a fourth-order model of chemical kinetics that, even though the model is inert or "dead", two identical copies of them in diffusive interaction become "alive", in the sense that they start to oscillate for an infinite amount of time [18]. According to Smale there is a paradoxical aspect to the model:

> One has two dead (mathematically dead) cells interacting by a diffusion process, which has a tendency in itself to equalize the concentrations. Yet in interaction, a state continues to pulse indefinitely.

Because of the importance of the class of equations coupled via diffusion in many fields of science, Smale posed the sharp problem to "axiomatize" the necessary conditions for diffusion-driven oscillations. A partial solution to his problem was proposed in [23],[1] where the dynamics of two Lur'e systems in diffusive interaction was studied using frequency methods. In that paper, it was shown that diffusion-driven oscillations are possible with third-order systems. It was proved in [14] that diffusion-driven oscillations cannot emerge from a unique equilibrium in case the systems are of order lower than three. In that same paper, constructive conditions were presented for the emergence of diffusion-driven oscillations. It is worth mentioning that oscillations may emerge in networks of diffusively coupled systems of order two; In that case the oscillations are born after a secondary bifurcation of equilibria [1].

The above-mentioned studies all considered diffusive coupling, which is (typically) symmetric and delay-free. We introduce a time-delay in the coupling terms. Such time-delay coupling functions appear, among others, in network of neurons [6], electrical circuits [16], and networked control systems [17].

We present conditions for emergence of oscillations in networks of time-delay coupled inert systems. In particular, we present conditions for the solutions of the time-delay coupled systems to be bounded, we discuss when the network equilibrium is unique, and we derive a condition (at the level of the dynamics of the systems that comprise the network) for the network equilibrium to be unstable. If all these conditions are satisfied the coupled system is oscillatory. Our results imply immediately that only if the dimension of the systems is at least two, then in time-delayed interaction one may have oscillatory activity in the network. The results we present in this chapter extend our previous results reported in [22] in the sense that we remove the restriction to undirected networks.

We remark that we will only consider the case that the coupled systems can *not* only be oscillatory for zero time-delay. The reason for this is that the results

[1] A minor flaw in that paper was corrected in [24].

of [14, 15], which consider the delay-free case, remain true for sufficiently small time-delays. A proof of this claim follows almost immediately from Rouché's theorem, cf. [5].

7.2 Preliminaries

Let $\mathbb{R}$ and $\mathbb{C}$ denote the real numbers and complex numbers, respectively. $\mathbb{R}_+$ is the set of positive real numbers and $\overline{\mathbb{R}}_+ = \mathbb{R}_+ \cup \{0\}$ is the set of the nonnegative real numbers. For a number $x = a + bi \in \mathbb{C}$ with $a, b \in \mathbb{R}$ and i being the imaginary unit, $i^2 = -1$, we denote $\Re(x) = a$ and $\Im(x) = b$. Let $\mathbb{C}_+ := \{x \in \mathbb{C} \mid \Re(x) \in \mathbb{R}_+\}$ and $\overline{\mathbb{C}}_+ := \{x \in \mathbb{C} \mid \Re(x) \in \overline{\mathbb{R}}_+\}$. Given positive integers p, q, and r, for $\mathscr{X} \subset \mathbb{R}^p$ and $\mathscr{Y} \subset \mathbb{R}^q$ we denote by $\mathscr{C}^r(\mathscr{X}, \mathscr{Y})$ the space of continuous functions from $\mathscr{X}$ into $\mathscr{Y}$ that are at least r-times continuously differentiable. If $r = 0$ we simply write $\mathscr{C}(\mathscr{X}, \mathscr{Y})$ instead of $\mathscr{C}^0(\mathscr{X}, \mathscr{Y})$. We denote $\mathscr{C} := \mathscr{C}([-\tau, 0], \mathbb{R}^{Nn})$ and we let this space be equipped with the norm

$$\|\phi\| = \sup_{-\tau \leq \theta \leq 0} |\phi(\theta)|, \quad \phi \in \mathscr{C}.$$

Here $|\cdot|$ is the Euclidean norm in $\mathbb{R}^{Nn}$, $|x| = \sqrt{x^\top x}$, where $^\top$ denotes transposition. For a positive integer k we let I_k denote the $k \times k$ identity matrix and $\mathbf{1}_k$ denotes the column vector of length k with all entries equal to 1.

Let $\xi \in \mathscr{C}([0, \infty), \mathbb{R})$ be bounded on the whole interval of definition. Such a function is *oscillatory* (in the sense of Yakubovich) if $\lim_{t\to\infty} \xi(t)$ does not exist. In that spirit we say that a system is oscillatory if it admits the following properties: 1. the solutions of the system are *uniformly (ultimately) bounded* (such that solutions are defined on $[0, \infty)$) and, 2. the system has *a finite number of hyperbolically unstable equilibria*.[2] In other words, if the initial data are not an equilibrium solution or do not belong to a stable manifold of an equilibrium, then at least one state variable of an oscillatory system is an oscillatory function of time.

7.3 Problem Setting

We consider networks consisting of N single-input-single-output systems of the form

$$\begin{cases} \dot{x}^j(t) = f(x^j(t)) + Bu^j(t) \\ y^j(t) = Cx^j(t) \end{cases} \tag{7.1}$$

---

[2]An equilibrium solution of a delay differential equation is called hyperbolic if the roots of its associated characteristic equation have nonzero real part, cf. [9].

with $j = 1, \ldots, N$, states $x^j(t) \in \mathbb{R}^n$, inputs $u^j(t) \in \mathbb{R}$, outputs $y^j(t) \in \mathbb{R}$, $f : \mathbb{R}^n \to \mathbb{R}^n$ is a sufficiently smooth function and matrices B, C of appropriate dimension with CB a positive constant. We shall assume that:

C1. the system (7.1) with $u^j \equiv 0$ has a unique equilibrium x_0, i.e., $f(x_0) = 0$, which is globally asymptotically stable and locally exponentially stable.

Note that local exponential stability of the equilibrium is equivalent to all eigenvalues of the matrix

$$J_0 = J(x_0),$$

with $J(x) = \frac{\partial f}{\partial x^j}(x)$ being the Jacobian matrix of f at x, having strictly negative real part, i.e., J_0 is Hurwitz.

Systems (7.1) interact via linear time-delay coupling functions of the form

$$u^j(t) = \sigma \sum_{\ell} a_{j\ell}[y^\ell(t-\tau) - y^j(t)] \tag{7.2}$$

where positive constant σ is the coupling strength, positive constant τ is the (propagation) delay, and nonnegative constants $a_{j\ell}$ are the interconnection weights. In particular, a_{jl} is positive if and only if there is a connection *from* system ℓ *to* system j. Define the $N \times N$ matrix $A = \left(a_{j\ell}\right)$. Matrix A is the (weighted) adjacency matrix of the graph that specifies the interaction structure. Note that we allow the graph to be *directed*. We shall assume that the matrix A is irreducible and has zero diagonal entries. This is equivalent to saying that the graph is *simple*, i.e., there is at most one edge from node j to node ℓ and self-connections are absent, and *strongly connected*, i.e., every pair of systems can be joined by a sequence of directed edges. In addition, we assume that

C2. each row-sum of A equals 1.

The latter assumption is not strictly necessary but it simplifies notation significantly. Moreover, this assumption ensures that the synchronous (oscillatory) state exists, cf. [12, 19]. We remark that **C2** implies, by the Gershgorin Disc Theorem, that all eigenvalues of A are located in the closed unit disc in $\mathbb{C}$.

7.4 Conditions for Oscillation

Given that **C1** and **C2** hold true we establish conditions for

1. the solutions of the coupled system (7.1), (7.2) to be uniformly bounded and uniformly ultimately bounded;
2. the network equilibrium

$$X_0 = \mathbf{1}_N \otimes x_0$$

to be the unique, but unstable equilibrium.

Clearly, if both points hold true, the coupled system is oscillatory. Uniqueness of the network equilibrium is not necessary for the existence of oscillations. However, the stability properties of additional equilibria are difficult to assess as the locations of these additional equilibrium solutions depend on σ. In addition, it is worth mentioning that for a unique equilibrium the state of the coupled system can be oscillatory only if one of its outputs is an oscillatory function of time. Indeed, in case none of the outputs is an oscillatory function the value of each coupling function is (or converges to) zero such that, by **C1**, the system is not oscillatory.

7.4.1 Bounded Solutions

Consider a single system (7.1) and let $u^j(\cdot)$ be a piece-wise continuous input function being defined on $[0, T)$, $T \in \mathbb{R}_+$, and taking values in a compact set $\mathscr{U} \subset \mathbb{R}$. Let $x^j(\cdot) = x^j(\cdot; x_0^j, u^j[0, T))$ be the solution of system (7.1) corresponding to input $u^j(\cdot)$ being defined on $[0, T]$ and coinciding with x_0^j at $t = 0$. Then we define a (strictly) $\mathscr{C}^r$*-semipassive* system as follows.

Definition 7.1 Suppose that there is a function $S \in \mathscr{C}^r(\mathbb{R}^n, \overline{\mathbb{R}}_+)$, called the *storage function*, such that

$$S(x^j(t)) - S(x^j(0)) \leq \int_0^t \left[(y^j u^j)(s) - H(x^j(s))\right] \mathrm{d}s \tag{7.3}$$

with $H \in \mathscr{C}(\mathbb{R}^n, \mathbb{R})$ and $t \in (0, T]$. If there is a constant $R > 0$ and a nonnegative nondecreasing function $h : \overline{\mathbb{R}}_+ \to \overline{\mathbb{R}}_+$ such that

$$H(s) \geq h(|s|) \tag{7.4}$$

for all $|s| \geq R$, then system (7.1) is called $\mathscr{C}^r$*-semipassive*. If (7.4) holds for all $|s| \geq R$ with a function h that is strictly increasing and such that $h(s) \to \infty$ as $s \to \infty$, then system (7.1) is called *strictly* $\mathscr{C}^r$*-semipassive*.

Remark 7.1 In case the storage function S is continuously differentiable, i.e., $r \geq 1$, then (7.3) can be replaced by the differential inequality

$$\dot{S}_{(7.1)}(x^j(t)) \leq (y^j u^j)(t) - H(x^j(t)),$$

where the subscript (7.1) means that the derivative of S is taken along solutions of (7.1) for given input $u^j(\cdot)$.

Lemma 7.1 (Boundedness) *Let* $w_0, w_1 : [0, \infty) \to [0, \infty)$ *be strictly increasing functions that satisfy* $w_0(0) = w_1(0) = 0$ *and* $w_0(s), w_1(s) \to \infty$ *as* $s \to \infty$. *Suppose that each system* (7.1) *is strictly* $\mathscr{C}^1$*-semipassive with storage function S that satisfies*

$$w_0(|x^j(t)|) \leq S(x^j(t)) \leq w_1(|x^j(t)|).$$

Then for each fixed σ and fixed τ the solutions of the coupled systems (7.1), (7.2) *are uniformly bounded and uniformly ultimately bounded.*

Proof (*Sketch, a full proof is found in* [20]) Let $\phi \in \mathscr{C}$,

$$\phi(\theta) = \begin{pmatrix} \phi^1(\theta) \\ \vdots \\ \phi^N(\theta) \end{pmatrix}, \quad \phi^1(\theta), \ldots, \phi^N(\theta) \in \mathbb{R}^n, \quad \theta \in [-\tau, 0],$$

and consider the functional

$$V(\phi) = \nu_1 S(\phi^1(0)) + \nu_2 S(\phi^2(0)) + \cdots + \nu_N S(\phi^N(0)) + \frac{\sigma}{2} \sum_j \nu_j a_{j\ell} \int_{-\tau}^{0} \left(\phi^{\ell\top}(s) C^\top C \phi^\ell(s) \right) \mathrm{d}s.$$

Here ν_i are positive constants such that

$$\begin{pmatrix} \nu_1 & \nu_2 & \cdots & \nu_N \end{pmatrix} (I_N - A) = \nu^\top (I_N - A) = 0.$$

The existence of the positive vector $\nu \in \mathbb{R}^N$ is implied by the Perron–Frobenius theorem for irreducible matrices, cf. [10]. Note that the matrix $I_N - A$ is irreducible as A is assumed to be irreducible. Then, invoking the strict semipassivity property and after some simple algebraic manipulations, we find that

$$\dot{V}(\phi) \le -\nu_1 H(\phi^1(0)) - \nu_2 H(\phi^2(0)) - \cdots - \nu_N H(\phi^N(0)) \le -W(|\phi(0)|) + M$$

for some strictly increasing function $W : \overline{\mathbb{R}}_+ \to \overline{\mathbb{R}}_+$ and positive constant M. An application of Theorem 4.2.10 of [3] completes the proof. □

7.4.2 Uniqueness and Instability of the Network Equilibrium

We shall start with establishing conditions for instability of the network equilibrium. Using **C2** we can write the coupled system dynamics as

$$\dot{x}(t) = F(x(t)) + \sigma \left[(A \otimes BC)x(t-\tau) - (I_N \otimes BC)x(t) \right] \tag{7.5}$$

where

$$x(t) = \begin{pmatrix} x^1(t) \\ x^2(t) \\ \vdots \\ x^N(t) \end{pmatrix}, \quad F(x(t)) = \begin{pmatrix} f(x^1(t)) \\ f(x^2(t)) \\ \vdots \\ f(x^N(t)) \end{pmatrix},$$

and $\otimes$ denotes the Kronecker (tensor) product. A linearization of (7.5) around the network equilibrium X_0 yields the dynamics

$$\dot{\tilde{x}}(t) = [I_N \otimes (J_0 - \sigma BC)]\,\tilde{x}(t) + (\sigma A \otimes BC)\tilde{x}(t-\tau). \tag{7.6}$$

It is well known that the zero solution of the linear system (7.6) is unstable for some $\sigma > 0$ and $\tau > 0$ if (and only if) its associated characteristic equation

$$\Delta(\lambda; \sigma, \tau) = 0 \tag{7.7}$$

with

$$\Delta(\lambda; \sigma, \tau) := \det\left(\lambda I_{Nn} - I_N \otimes (J_0 - \sigma BC) - (\sigma A \otimes BC)\exp(-\lambda\tau)\right)$$

has a root in $\mathbb{C}_+$ for that σ and τ, cf. [9, 11]. However, computing the roots of the characteristic equation (7.7) for a large number of points in the (σ, τ)-parameter space may be cumbersome. As a solution, we will present (sufficient) conditions for instability of the network equilibrium at the level of the dynamics of the system (7.1). For that purpose we denote

$$\mathscr{H}(s) = C(sI_n - J_0)^{-1}B = \frac{p(s)}{q(s)}$$

the linear transfer function from u^j to y^j of the system (7.1) at its equilibrium. Here $p(s)$ is a polynomial of degree $n-1$ and $q(s)$ is a polynomial of degree n.[3] It is assumed that p and q are co-prime.

Lemma 7.2 (Instability) *Suppose that* **C2** *holds true. Let*

$$\eta = \inf_{\omega > 0} \Re\left(\mathscr{H}(i\omega)\right).$$

If $\eta < 0$, then for each $\sigma \geq \frac{-1}{2\eta}$ there exists a $\tau > 0$ such that the characteristic equation (7.7) *has a root in $\mathbb{C}_+$.*

The proof of the lemma is provided in the Appendix.

It is important to note that the condition for instability in Lemma 7.2 is *delay-dependent*. As we have remarked already in the introduction, we focus in this chapter only on delay-dependent conditions for oscillations.

We continue with conditions for uniqueness of the network equilibrium.

Lemma 7.3 (Uniqueness of the network equilibrium) *Let* **C1** *hold true and denote the eigenvalues of A by $\bar{\lambda}_j$, $j = 1, \ldots, N$. Let λ^* be the smallest real-valued eigenvalue of A. Choose $\bar{\sigma} \in (0, \infty]$ as the largest number for which the matrix*

---

[3] As $CB > 0$ the system (7.1) has relative degree one.

$$J(\xi) - \sigma(1 - \lambda^*)BC$$

is nonsingular for all $\xi \in \mathbb{R}^n$ *and all* $\sigma \in [0, \bar{\sigma})$. *Then the network equilibrium solution* $X_0 = \mathbf{1}_N \otimes x_0$ *is the unique equilibrium solution of* (7.5) *for* $\sigma \in [0, \bar{\sigma})$.

The proof of the lemma is provided in the Appendix.

7.4.3 Oscillations in Networks of Inert Systems

Lemmata 7.1, 7.2, and 7.3 provide conditions for the coupled systems to have bounded solutions, the network equilibrium to be unique, and the existence of a time-delay $\tau > 0$ for which this equilibrium is unstable. The following theorem summarizes these results.

Theorem 7.1 (Conditions for oscillation) *Consider the coupled system (7.1), (7.2) and suppose that* **C1** *and* **C2** *hold true. Suppose in addition that*

- *the systems (7.1) are strictly* $\mathscr{C}^1$*-semipassive with a storage function that satisfies the conditions of Lemma 7.1;*
- *the matrix*
$$J(\xi) - \sigma(1 - \lambda^*)BC$$
is nonsingular for all $\xi \in \mathbb{R}^n$ *and all* $\sigma \in [0, \bar{\sigma})$, *where* λ^* *is the smallest real-valued eigenvalue of* A*;*
- $\eta = \inf_{\omega > 0} \Re\left(\mathscr{H}(i\omega)\right) < 0$ *and*
$$\frac{-1}{2\eta} < \bar{\sigma}.$$

Then for each $\sigma \in \left[\frac{-1}{2\eta}, \bar{\sigma}\right)$ *there exists a* $\tau > 0$ *for which the coupled system is oscillatory.*

Using the second and third condition of the theorem (see also Lemmas 7.2 and 7.3) one easily determines a (range of) coupling strength(s) for which there exist τ such that oscillations emerge. In particular, for any $\sigma \in \left[\frac{-1}{2\eta}, \bar{\sigma}\right)$ one can use bifurcation software such as DDE-Biftool [7] for finding the values of τ for which the characteristic equation (7.7) has a root in $\mathbb{C}_+$. A viable strategy for computing the bifurcation diagram in the (σ, τ)-parameter space is to start with some $\sigma_H \in \left[\frac{-1}{2\eta}, \bar{\sigma}\right)$ and $\tau = 0$. Then increase τ until at $\tau = \tau_H$ a Hopf bifurcation is detected. (We remark that the bifurcation that causes instability of the network equilibrium is necessarily a Hopf bifurcation because otherwise the condition of Lemma 7.3 would be violated.) A curve of Hopf bifurcation points can then be computed using a continuation algorithm starting from (σ_H, τ_H). See, for instance, [12] for an example.

Theorem 7.1 also provides almost immediately a necessary condition on the dimension n of the systems (7.1) for oscillations to emerge.

Corollary 7.1 *If* **C1** *and* **C2** *hold true, then a necessary condition for a network of inert systems (7.1) that interact via coupling functions (7.2) to be oscillatory is that* $n \geq 2$.

Proof An example with $n = 2$ is provided in the next section and examples of systems of order larger than two can be easily constructed. We complete the proof by showing that the equilibrium of a network of coupled inert systems with $n = 1$ is always stable. Let U be a nonsingular matrix such that

$$U^{-1}AU = \bar{\Lambda}$$

with $\bar{\Lambda}$ the *Jordan normal form* of A. Denote by $\bar{\lambda}_j$, $j = 1, \ldots, N$ the eigenvalues of A. After pre-multiplication of (7.7) by $\det(U^{-1} \otimes I_n)$ and post-multiplication of (7.7) by $\det(U \otimes I_n)$ it is straightforward to see that the characteristic equation (7.7) can have a root $\lambda \in \mathbb{C}_+$ only if there is a $j \in \{1, 2, \ldots, N\}$ such that

$$\lambda - J_0 + \sigma[1 - \bar{\lambda}_j \exp(-\lambda\tau)] = 0 \tag{7.8}$$

for some $\tau > 0$. (See also the proof of Lemma 7.2 in the Appendix.) However, as J_0 is a negative constant by **C1** and $|\bar{\lambda}_j| \leq 1$ for all j by **C2** there exists no $\lambda \in \mathbb{C}_+$ that solves (7.8). □

7.5 Example

We shall illustrate our results in networks of inert FitzHugh–Nagumo (FHN) model neurons [8]. The dynamics of this model neuron are given by the following equations:

$$\begin{cases} \dot{x}_1^j(t) = 0.08(x_2^j(t) - 0.8x_1^j(t)) \\ \dot{x}_2^j(t) = x_2^j(t) - \frac{1}{3}(x_2^j(t))^3 - x_1^j(t) - 0.559 + u^j(t) \end{cases}$$

with output $y^j(t) = x_2^j(t)$. One can easily verify that the isolated FHN model neuron has a locally exponentially stable equilibrium at $x_0 = \begin{pmatrix} -1.225 & -0.980 \end{pmatrix}^\top$. Moreover, it is shown in [21] that the FHN model neuron is strictly $\mathscr{C}^\infty$-semipassive with a quadratic storage function. Hence, we conclude that the solutions of any network of FHN model neurons are uniformly (ultimately) bounded for any nonnegative σ and τ. To check whether we can have oscillations in a network of FHN model neurons, we determine the transfer function $\mathscr{H}(s)$:

$$\mathscr{H}(s) = \begin{pmatrix} 0 & 1 \end{pmatrix} \left(sI - \begin{pmatrix} -0.064 & 0.08 \\ -1 & 1 - (-0.980)^2 \end{pmatrix} \right)^{-1} \begin{pmatrix} 0 \\ 1 \end{pmatrix}.$$

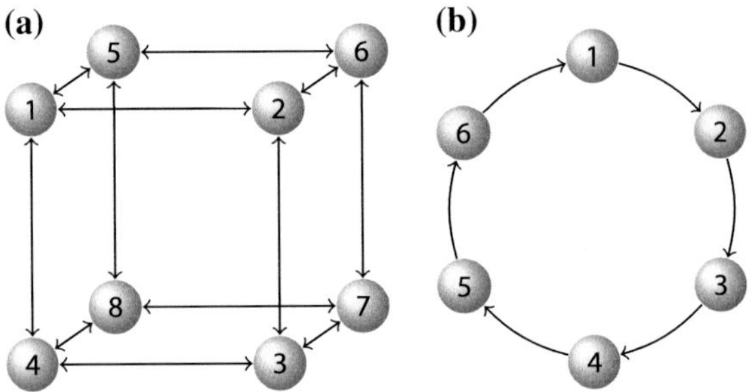

Fig. 7.1 **a** The cube network with uniform bidirectional interactions, and **b** the ring network with uniform unidirectional interactions

We find that $\eta = \inf_{\omega>0} \Re(\mathscr{H}(i\omega)) = -0.205$, which is attained at $\omega = \omega^* = 0.417$. Thus for coupling strengths

$$\sigma > \underline{\sigma} = \frac{0.5}{0.205},$$

there exist $\tau > 0$ for which the zero solution of the linearized system is unstable, hence the network equilibrium is unstable. In addition, because

$$\det(J(\xi) - \sigma(1 - \bar{\lambda}_j)BC) = (\xi_2)^2 + \tfrac{1}{4} + \sigma(1 - \bar{\lambda}_j)$$

is positive for any $\xi = \begin{pmatrix}\xi_1 & \xi_2\end{pmatrix}^\top \in \mathbb{R}^2$, any $\sigma \geq 0$ and any real-valued $\bar{\lambda}_j \in [-1, 1]$, we conclude that the network equilibrium $X_0 = \mathbf{1}_k \otimes x_0$ is unique. Thus if $\sigma > \underline{\sigma}$ there exist values of τ for which the coupled FHN model neurons are oscillatory.

We have performed a numerical analyses with a cube network and a ring network, which are shown in Fig. 7.1a and b, respectively. The example with the cube network has been taken from [22]. For both networks, we have determined the regions of instability in the (σ, τ)-parameter space with $0 \leq \sigma \leq 8$ and $0 \leq \tau \leq 20$. These regions, which are computed with DDE-Biftool [7] using the strategy explained in the previous section, are shown in Fig. 7.2a for the cube network and Fig. 7.2b for the ring network. In these plots the areas shown in gray correspond to the regions of hyperbolic instability of the network equilibrium. The thick black curves are the stability crossing curves; At a stability crossing curve the characteristic equation (7.7) has a purely imaginary root.

In addition, we present the results of a number of numerical simulations, which are performed with Matlab using the DDE23 solver. For each simulation we have used constant initial data on the interval $[-\tau, 0]$. This initial data is chosen to be a normally distributed perturbation of the network equilibrium, with mean and variance of the perturbation being set to 0 and 0.05, respectively.

Figure 7.3 show the results of numerical simulation for the cube network with $\sigma = 3$ and either $\tau = 8$ or $\tau = 15$. The left plots show the birth of oscillatory activity from the network equilibrium. The eight smaller plots show 100 time units of steady state oscillatory activity. In case of $\sigma = 3$ and $\tau = 8$ we observe an oscillation where neurons 1, 3, 6 and 8 oscillate synchronously, neurons 2, 4, 5 and 7 oscillate

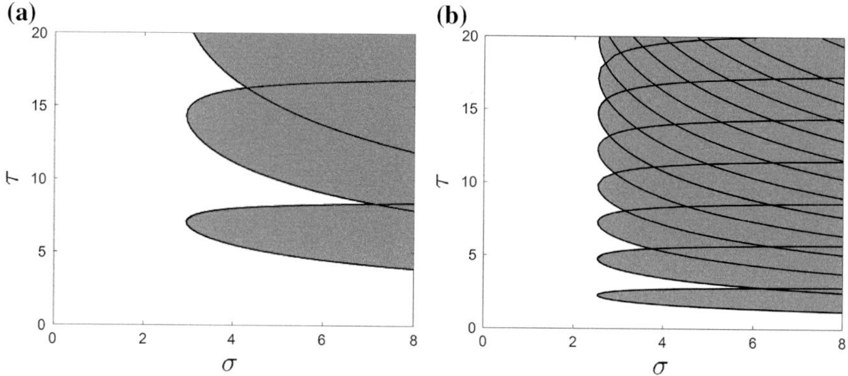

Fig. 7.2 The regions of stability (*white*) and instability (*grey*) of the network equilibrium in the (σ, τ)-parameter space for **a** the cube network, and **b** the ring network

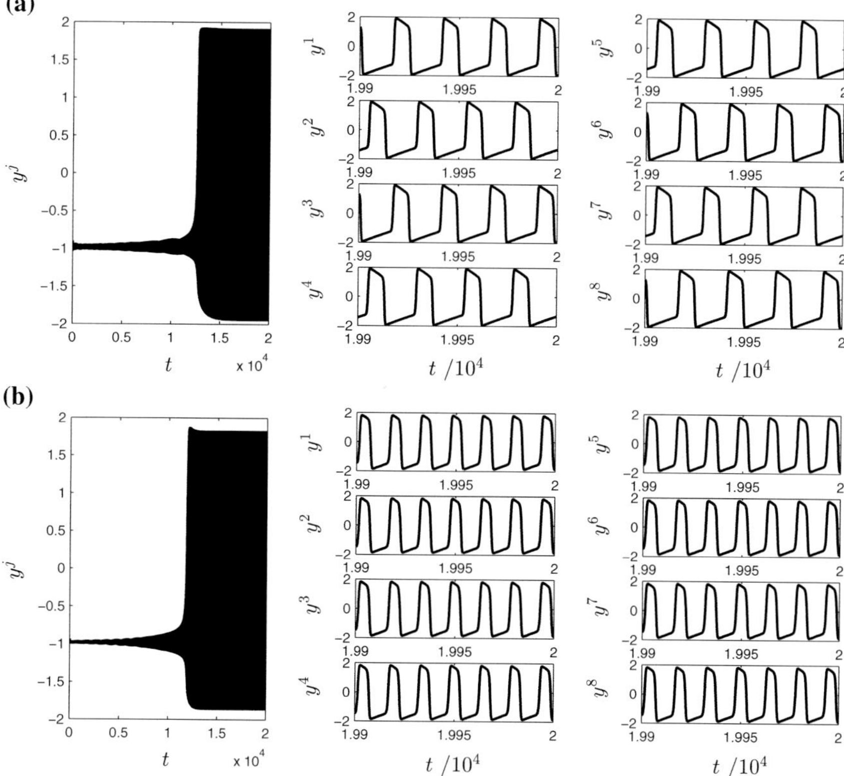

Fig. 7.3 Results of numerical simulation with the cube network for **a** $\sigma = 3$ and $\tau = 8$, and **b** $\sigma = 3$ and $\tau = 15$

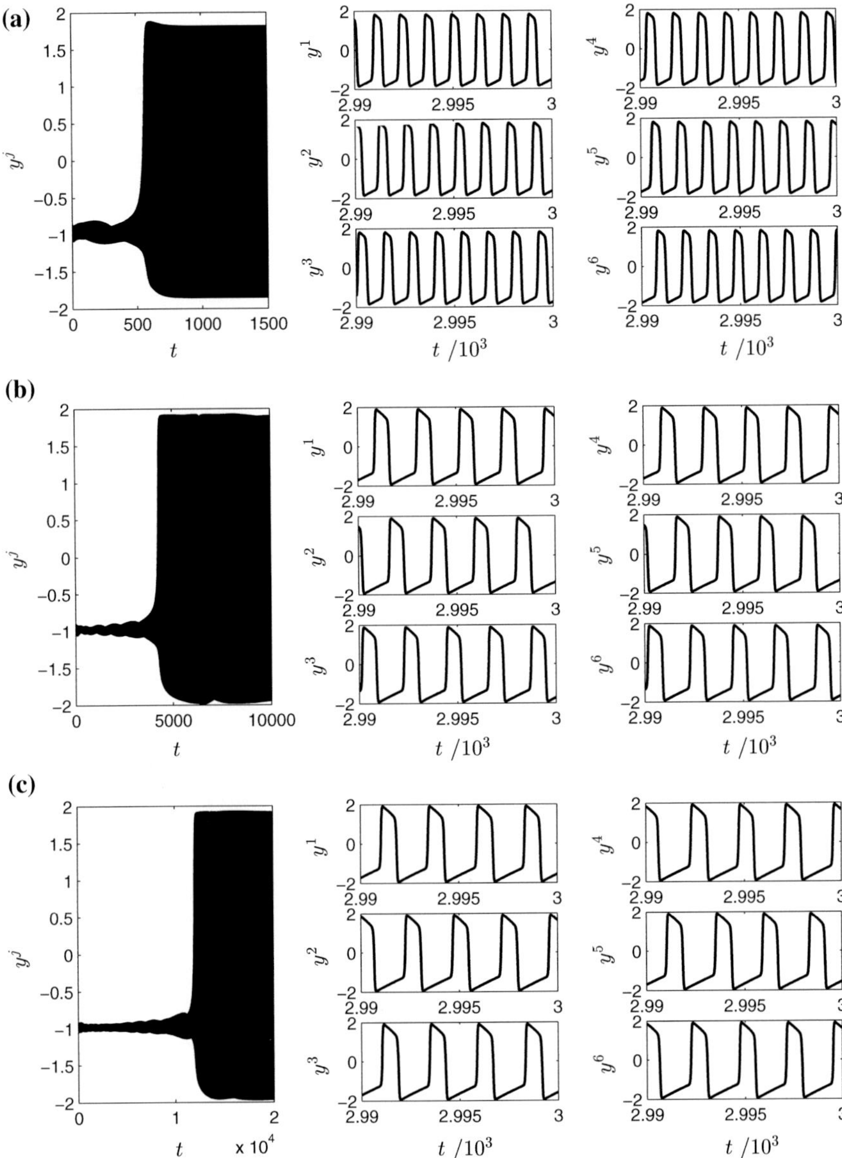

Fig. 7.4 Results of numerical simulation with the ring network for **a** $\sigma = 5$ and $\tau = 2$, **b** $\sigma = 5$ and $\tau = 7$, and **c**: $\sigma = 5$ and $\tau = 12$

synchronously, but the oscillations of the two synchronized clusters alternate. An increase of the time-delay to $\tau = 15$ results in completely synchronous oscillatory activity. For more details about the (prediction of) resulting oscillatory activity in this network we refer to [22].

Figure 7.4 show the results of numerical simulation for the ring network with $\sigma = 5$ and either $\tau = 2$, $\tau = 7$ or $\tau = 12$. Again the left plots show the onset of oscillation and the other plots show 100 time units of steady state oscillatory activity. In all cases, we observe oscillatory activity in the form of persistent propagating waves. In case of $\sigma = 5$ and $\tau = 2$ stable *rotating wave* oscillations have emerged. Indeed, the steady state oscillations are periodic and there is a constant time-shift between the oscillations of any two adjacent neurons. For time-delay $\tau = 12$ we observe the emergence of stable *standing wave* steady state activity, which is characterized by the synchronous activity of neurons 1, 3 and 5 that alternates with synchronous activity of neurons 2, 4 and 6. A somewhat intermediate oscillatory behavior is found for $\tau = 7$. In this case, neurons 1 and 4 oscillate synchronously, neurons 2 and 5 are synchronized, and the steady state oscillations of neurons 3 and 6 are completely identical. However, the oscillations of neurons 1, 2 and 3, hence those of neurons 4, 5 and 6, take the form of a rotating wave. The emerged oscillatory activity in the ring network can be analyzed and predicted using the theory presented in [12].

7.6 Conclusions

We have considered the problem of emergence of oscillatory activity in networks of identical inert systems that interact via linear time-delay coupling. We have presented conditions for

- the solutions of the coupled systems to be uniformly (ultimately) bounded;
- the network equilibrium, which is exponentially stable in absence of coupling, to become unstable in the presence of coupling;
- the network equilibrium to be unique.

If all three points are satisfied the network of time-delay coupled system will be oscillatory. Our conditions for the first two points above to hold true are expressed at the level of the systems. In particular, a strict semipassivity property of the systems ensures that the whole network has bounded solutions, and conditions for instability of the network equilibrium can be verified by evaluating the transfer function (from u^j to y^j) of the uncoupled system in equilibrium. As a corollary to these results, we have shown that a network of inert systems (7.1) with time-delay coupling (7.2) can be oscillatory only if the systems are at least of second order.

We have illustrated our results with two networks, a cube and a ring, with FHN model neurons as systems. For both networks we have determined the values of the coupling strength σ and time-delay τ for which oscillations emerge. Trajectories of the coupled systems are obtained for several values of the coupling strength and time-delay by numerical integration of the governing equations. It is shown that interesting patterns of oscillatory activity may emerge from a network equilibrium.

Afterword

This book chapter is written for the occasion of the 60th birthday of Henk Nijmeijer. Both authors have shared many ideas, thoughts and papers with Henk on the collective behavior of coupled dynamical systems. We are certain to continue working together with Henk on this fascinating topic for many more years.

Appendix

Proof of Lemma 7.2

As mentioned in [4], the characteristic equation $\Delta(\lambda; \sigma, \tau)$ can have a root in $\mathbb{C}_+$ (for some $\tau > 0$) only if (at least) one of the following conditions is violated:

- $I_N \otimes (J_0 - \sigma BC)$ is a stable matrix;
- $I_N \otimes (J_0 - \sigma BC) + \sigma A \otimes BC$ is a stable matrix;
- the spectral radius ρ[4] of the frequency dependent matrix

$$[I_N \otimes (J_0 - \sigma BC)]^{-1} [\sigma A \otimes BC]$$

is strictly smaller than one for all frequencies:

$$\rho \left([I_N \otimes (J_0 - \sigma BC)]^{-1} [\sigma A \otimes BC] \right) < 1 \quad \forall \omega > 0.$$

As already remarked in the introduction, we restrict ourselves to the case where the network equilibrium is stable in case of zero time-delay. This implies that the first two conditions are satisfied such that instability of the network equilibrium in presence of time-delay requires the third condition to be violated. We show that the condition of Lemma 7.2 implies this to be the case.

Using some elementary properties of the Kronecker product, cf. [2], we obtain that

$$\rho\left([i\omega I_{nm} - I_N \otimes (J_0 - \sigma BC)]^{-1} [\sigma (A \otimes BC)]\right) = \rho \left(\sigma A \otimes [i\omega I_n - J_0 + \sigma BC)]^{-1} BC \right).$$

Condition **C2** implies that A has an eigenvalue equal to 1 (with right eigenvector in span $\{\mathbf{1}_N\}$). Then employing the fact that the eigenvalues of $\sigma A \otimes [i\omega I_n - J_0 + \sigma BC)]^{-1} BC$ are the product of all eigenvalues of σA and all eigenvalues of $[i\omega I_n - J_0 + \sigma BC)]^{-1} BC$, cf. [2], we find that

$$\rho \left(\sigma [i\omega I_n - J_0 + \sigma BC)]^{-1} BC \right) > 1$$
$$\Rightarrow \quad \rho\left([i\omega I_{nm} - I_N \otimes (J_0 - \sigma BC)]^{-1} [\sigma (A \otimes BC)]\right) > 1.$$

[4] The spectral radius of a square (complex) matrix is the largest eigenvalue in absolute value of that matrix.

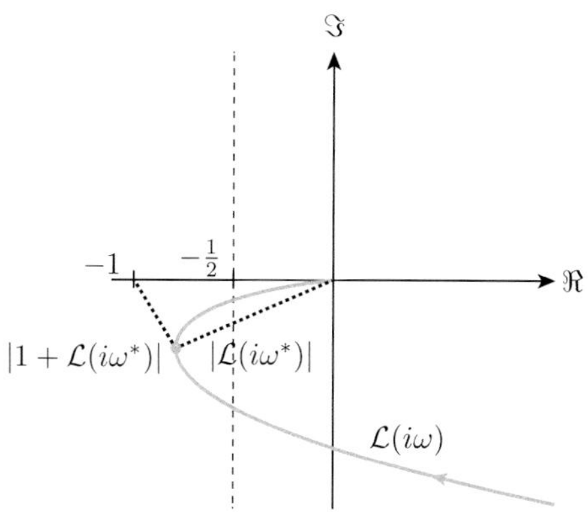

Fig. 7.5 If the Nyquist plot of $\mathscr{L}(i\omega) = \sigma\mathscr{H}(i\omega))$ intersects with $\left\{s \in \mathbb{C} \mid \Re(s) < -\frac{1}{2}\right\}$, then $|\mathscr{L}(i\omega^*)| > |1 + \mathscr{L}(i\omega^*)|$, where $\omega^* = \arg\min_{\omega>0} \Re\left(\mathscr{L}(i\omega)\right)$

Some straightforward manipulations (that involve some theory about the inverse of the sum of two matrices, cf. [13]) show that

$$\begin{aligned}\bar{\rho}(\omega;\sigma) &:= \rho\left(\sigma[i\omega I_n - J_0 + \sigma BC)]^{-1}BC\right)\\ &= \frac{|\sigma p(i\omega)|}{|q(i\omega) + \sigma p(i\omega)|} = \left|\frac{\sigma\mathscr{H}(i\omega)}{1+\sigma\mathscr{H}(i\omega)}\right|.\end{aligned}$$

It follows from Fig. 7.5 that if the Nyquist plot of $\mathscr{L}(i\omega) = \sigma\mathscr{H}(i\omega)$ intersects with $\left\{s \in \mathbb{C} \mid \Re(s) < \frac{1}{2}\right\}$, then there exists $\omega^* = \arg\min_{\omega>0} \Re\left(\mathscr{L}(i\omega)\right) > 0$ such that

$$|\mathscr{L}(i\omega^*)| > |1 + \mathscr{L}(i\omega^*)| \Rightarrow \bar{\rho}(\omega^*;\sigma) > 1.$$

In other words, if $\eta = \inf_{\omega>0} \Re\left(\mathscr{H}(i\omega)\right) = \Re\left(\mathscr{H}(i\omega^*)\right) < 0$, then for each $\sigma \geq \frac{-1}{2\eta}$,

$$\bar{\rho}(\omega^*;\sigma) > 1 \quad \Rightarrow \quad \rho\left(\left[i\omega^* I_{nm} - I_N \otimes (J_0 - \sigma BC)\right]^{-1}[\sigma(A \otimes BC)]\right) > 1.$$

Fix $\sigma^* \geq \frac{-1}{2\eta}$. We now show that $\bar{\rho}(\omega^*;\sigma^*) > 1$ implies (7.7) to have a root in $\mathbb{C}_+$. Define

$$\beta(\lambda;\sigma^*,\tau) = 1 - \alpha(\lambda;\sigma^*)\exp(-\lambda\tau)$$

with

$$\alpha(\lambda;\sigma^*) = \frac{\sigma^*\mathscr{H}(\lambda)}{1+\sigma^*\mathscr{H}(\lambda)}.$$

Note that $\bar{\rho}(\omega;\sigma^*) = |\alpha(i\omega;\sigma^*)|$. Consider the function $\kappa : \mathbb{R}_+ \to \mathbb{R}$, $\kappa(\omega) = 1 - |\alpha(i\omega;\sigma^*)|^2 = 1 - \bar{\rho}^2(\omega;\sigma^*)$. Note that $\lim_{\omega\to\infty}\kappa(\omega) = 1$ as $\lim_{\omega\to\infty}\bar{\rho}(\omega;\sigma^*) = 0$. Because $\bar{\rho}(\omega^*;\sigma^*) > 1$ there exist a $\omega_0 > 0$ such that $\bar{\rho}(\omega_0;\sigma^*) = 1$. Let us choose, without loss of generality, this ω_0 such that for any small number $\delta > 0$ we have $\bar{\rho}(\omega_0 - \delta;\sigma^*) > 1$ and $\bar{\rho}(\omega_0 + \delta;\sigma^*) < 1$, i.e., $\kappa' = \frac{d\kappa}{d\omega} > 0$ at $\omega = \omega_0$. In addition, there is a $\tau_0 > 0$ for which $\beta(i\omega_0; \sigma^*, \tau_0) = 0$. Following [11], pp. 95, if we differentiate $\beta(\lambda; \sigma, \tau) = 0$ at $\lambda = i\omega_0$ with respect to τ, we find

$$\Re\left(\frac{\mathrm{d}\lambda}{\mathrm{d}\tau}\right)^{-1} = \frac{1}{2\omega_0}\kappa'(\omega_0) < 0.$$

This implies the existence of some $\tau^* < \tau_0$ for which $\beta(\lambda; \sigma^*, \tau^*)$ has a root in $\mathbb{C}_+$.

Let U be a nonsingular matrix such that

$$U^{-1}AU = \bar{\Lambda}$$

with $\bar{\Lambda}$ the *Jordan normal form* of A and let $\bar{\lambda}_j$, $j = 1, 2, \ldots, N$, be the eigenvalues of A. After pre-multiplication of (7.7) by $\det(U^{-1} \otimes I_n)$ and post-multiplication of (7.7) by $\det(U \otimes I_n)$, we conclude that the roots of (7.7) (of course, for $\sigma = \sigma^*$) are identical to the roots of

$$\prod_{j=1}^{N} \Delta_j(\lambda; \sigma^*, \tau)$$

with

$$\Delta_j(\lambda; \sigma^*, \tau) = \det\left(\lambda I_n - (J_0 - \sigma^* BC) - \sigma^* \bar{\lambda}_j BC \exp(-\lambda\tau)\right).$$

By **C2** there is always an eigenvalue of A equal to 1. Without loss of generality, we let $\bar{\lambda}_1 = 1$ such that

$$\Delta_1(\lambda; \sigma^*, \tau) = \det\left(\lambda I_n - (J_0 - \sigma^* BC) - \sigma^* BC \exp(-\lambda\tau)\right).$$

It is straightforward to verify that $\Delta_1(\lambda; \sigma^*, \tau)$ and $\beta(\lambda; \sigma^*, \tau)$ have the same roots. Thus $\Delta_1(\lambda; \sigma^*, \tau)$ has a root in $\mathbb{C}_+$ for some $\tau^* < \tau_0$, which implies that (7.7) has a root in $\mathbb{C}_+$ for $\sigma = \sigma^*$ and $\tau = \tau^*$. □

Proof of Lemma 7.3

The proof follows from arguments given first in [15]. First we show that the conditions of the lemma imply that the Jacobian matrix of

$$F(x) - \sigma[(I_N - A) \otimes BC)]x \tag{7.9}$$

is nonsingular at all $x \in \mathbb{R}^{Nn}$. Let again U be a nonsingular matrix such that

$$U^{-1}AU = \bar{\Lambda}$$

with $\bar{\Lambda}$ the Jordan normal form of A. Then the Jacobian matrix of (7.9) is nonsingular if and only if

$$\begin{pmatrix} J(\tilde{x}^1) & & \\ & \ddots & \\ & & J(\tilde{x}^N) \end{pmatrix} - \sigma(I_N - \bar{\Lambda}) \otimes BC$$

is nonsingular for all $\tilde{x}^j \in \mathbb{R}^n$, $j = 1, \ldots, N$. Due to the triangular structure of $\bar{\Lambda}$ the above matrix is singular if and only if (at least) one of the matrices

$$J(\xi) - \sigma(1 - \bar{\lambda}_j)BC, \quad j = 1, \ldots, N, \quad \xi \in \mathbb{R}^n$$

is singular. By construction the matrix BC has one positive diagonal entry and all other entries equal zero. Thus $J(\xi) - \sigma(1 - \lambda_j)BC$ can only be singular if $\bar{\lambda}_j$ is real valued. It follows that the conditions of the lemma imply that the Jacobian matrix is nonsingular for all $\sigma \in [0, \bar{\sigma})$.

Now we consider an auxiliary coupled system (7.5) with σ replaced by $\epsilon\sigma$ with parameter $\epsilon \in [0, 1]$. By the conditions of the lemma the Jacobian matrix of this auxiliary coupled system is, like the original coupled system, nonsingular at each point in $\mathbb{R}^{Nn}$. Now suppose that for this auxiliary coupled system there is some $\epsilon = \epsilon^* \in (0, 1)$ for which there exists an equilibrium X_0^* other than the network equilibrium X_0. Then, due to the implicit function theorem, this additional equilibrium point is determined by an equation of the form $X_0^* = \mathscr{F}(\epsilon^*)$. Decreasing ϵ from ϵ^* to zero implies the existence of an equilibrium other than X_0 for $\epsilon = 0$. This contradicts **C1**, which states that the isolated system has a globally asymptotically stable (hence unique) equilibrium. □

References

1. Alexander, J.C.: Spontaneous oscillations in two 2-component cells coupled by diffusion. J. Math. Biol. **23**, 205–219 (1986)
2. Brewer, J.W.: Kronecker products and matrix calculus in system theory. IEEE Trans. Circuits Syst. **25**(9), 772–781 (1978)
3. Burton, T.A.: Stability and Periodic Solutions of Ordinary and Functional Differential Equations. Academic, New York (1985)
4. Chen, J., Latchman, H.A.: Frequency sweeping tests for stability independent of delay. IEEE Trans. Autom. Control **40**, 1640–1645 (1995)
5. Churchill, R.V., Brown, J.W.: Complex Variables and Applications, 5th edn. McGraw-Hill, Inc., New York (1990)
6. Dhamala, M., Jirsa, V.K., Ding, M.: Enhancement of neural synchrony by time delay. Phys. Rev. Lett. **92**, 074104 (2004)
7. Engelborghs, K., Luzyanina, T., Samaey, G.: DDE-BIFTOOL v. 2.00: a Matlab package for bifurcation analysis of delay differential equations. Tw report 330, Dept. Computer Science, KU Leuven (2001)
8. FitzHugh, R.: Impulses and physiological states in theoretic models of nerve membrane. Biophys. J. **1**, 445–466 (1961)

9. Hale, J.K., Verduyn Lunel, S.M.: Introduction to Functional Differential Equations. Applied Mathematical Sciences, vol. 99. Springer, Heidelberg (1993)
10. Horn, R.A., Johnson, C.R.: Matrix Analysis, 6th edn. Cambridge University Press, Cambridge (1999)
11. Michiels, W., Niculescu, S.-I.: Stability and Stabilization of Time-Delay Systems. Advances in Design and Control. SIAM (2007)
12. Michiels, W., Nijmeijer, H.: Synchronization of delay-coupled nonlinear oscillators: an approach based on the stability analysis of synchronized equilibria. Chaos **19**, 033110 (2009)
13. Miller, K.S.: On the inverse of the sum of matrices. Math. Mag. **5**(2), 67–72 (1981)
14. Pogromsky, A., Glad, T., Nijmeijer, H.: On diffusion driven oscillations in coupled dynamical systems. Int. J. Bifurc. Chaos **9**(4), 629–644 (1999)
15. Pogromsky, A., Kuznetsov, N., Leonov, G.: Pattern generation in diffusive networks; How do those brainless centipedes walk? In: 50th IEEE Conference on Decision and Control, pp. 7849–7854 (2011)
16. Ramana Reddy, D.V., Sen, A., Johnston, G.L.: Experimental evidence of time-delay-induced death in coupled limit-cycle oscillators. Phys. Rev. Lett. **85**(16), 3381–3384 (2000)
17. Sipahi, R., Niculescu, S., Abdallah, C.T., Michiels, W., Gu, K.: Stability and stabilization of systems with time delay. IEEE Control Syst. **31**(1), 38–65 (2011)
18. Smale, S.: A mathematical model of two cells via Turing's equation. The Hopf Bifurcation and Its Applications, vol. 19. Springer, New York (1976)
19. Steur, E., Nijmeijer, H.: Synchronization in networks of diffusively time-delay coupled (semi-)passive systems. IEEE Trans. Circuits Syst. I **58**(6), 1358–1371 (2011)
20. Steur, E., Oguchi, T., van Leeuwen, C., Nijmeijer, H.: Partial synchronization in diffusively time-delay coupled oscillator networks. Chaos **22**, 043144 (2012)
21. Steur, E., Tyukin, I., Nijmeijer, H.: Semi-passivity and synchronization of diffusively coupled neuronal oscillators. Physica D **238**, 2119–2128 (2009)
22. Steur, E., van Leeuwen, C., Pogromsky, A.Y.: Synchronous oscillations in networks of time-delay coupled inert systems. In: Proceedings of the fourth IFAC Conference on Analysis and Control of Chaotic Systems (IFAC CHAOS 2015), Tokyo, Japan (2015)
23. Tomberg, E.A., Yakubovich, V.A.: Conditions for auto-oscillations in nonlinear systems. Sibirsk. Mat. Zh. **30**(4), 641–653 (1989)
24. Tomberg, E.A., Yakubovich, V.A.: On one problem of Smale. Sibirsk. Mat. Zh. **41**(4), 926–928 (2000)
25. Turing, A.: The chemical basis of morphogenesis. Phylos. Trans. Royal Soc. **B.237**, 37–72 (1952)

Part III
Control of Nonlinear Mechanical Systems

Chapter 8
Leader–Follower Synchronisation for a Class of Underactuated Systems

Dennis J.W. Belleter and Kristin Y. Pettersen

Abstract In this work, leader–follower synchronisation is considered for underactuated followers in an inhomogeneous multi-agent system. The goal is to synchronise the motion of a leader and an underactuated follower. Measurements of the leader's position, velocity, acceleration and jerk are available, while the dynamics of the leader is unknown. The leader velocities are used as input for a constant bearing guidance algorithm to assure that the follower synchronises its motion to the leader. It is also shown that the proposed leader–follower scheme can be applied to multi-agent systems that are subjected to unknown environmental disturbances. Furthermore, the trajectory of the leader does not need to be known. The closed-loop dynamics are analysed and it is shown that under certain conditions all solutions remain bounded and the synchronisation error kinematics are shown to be integral input-to-state stable with respect to changes in the unactuated sway velocity. For straight-line motions, i.e. where the desired yaw rate and sway velocity go to zero, synchronisation is achieved. Simulation results are presented to validate the proposed control strategy.

8.1 Introduction

This work considers leader–follower synchronisation for inhomogeneous multi-agent systems with underactuated agents. In particular, we consider synchronisation of underactuated marine vessels on straight-line trajectories and curved paths. Leader–follower synchronisation has several applications concerning both

---

This work was partly supported by the Research Council of Norway through its Centres of Excellence funding scheme, project No. 223254 AMOS.

D.J.W. Belleter (✉) · K.Y. Pettersen
Centre for Autonomous Marine Operations and Systems (AMOS), Department of Engineering Cybernetics, Norwegian University of Science and Technology (NTNU), NO7491 Trondheim, Norway
e-mail: dennis.belleter@itk.ntnu.no

K.Y. Pettersen
e-mail: kristin.y.pettersen@itk.ntnu.no

N. van de Wouw et al. (eds.), *Nonlinear Systems*, Lecture Notes in Control and Information Sciences 470, DOI 10.1007/978-3-319-30357-4_8

autonomous and non-autonomous vehicles. Leader–follower synchronisation has, for instance, been applied to underway replenishment operations, robot manipulator master–slave synchronisation and formation control tasks.

In the marine systems literature, leader–follower synchronisation has played an important part in research on underway replenishment of ships, see for instance [15, 19, 24]. For these operations, the supply ship is usually responsible for synchronising its motion with the ship it is supplying. In [19] the case of a fully actuated follower that synchronises its output with a leader with unknown dynamics is investigated. An observer–controller scheme is utilised to achieve synchronisation where the observers are used to estimate the unknown velocities of the leader and follower. The observer–controller scheme utilised in [19] is based on theory for master–slave synchronisation of robotic manipulators investigated in [21]. In [24] the focus is on interaction forces between two vessels during underway replenishment operations. For control purposes, the constant bearing guidance algorithm from [6] is used to synchronise the ships along a straight-line path. The vessels are underactuated, but no analysis of the underactuated internal dynamics is given. In [15] underway replenishment between fully actuated vessels is investigated and adaptive backstepping controllers are designed to reject exogenous disturbances.

Leader–follower synchronisation is also widely applied for other coordinated control applications. Applications include master–slave synchronisation of robot manipulators in [21], leader–follower synchronisation for control of mobile robots in [2, 10, 11, 23] and for formation control of marine vessels in [7]. For these applications, the models are either fully actuated or formulated only at the kinematic level. However, most commercial systems are underactuated or become underactuated at higher speed, e.g. vessels with a tunnel thruster to apply a sideways force are fully actuated for low speeds but the tunnel thruster becomes inefficient at cruising speed, see [9] and the references therein. Also cars and most mobile robots are underactuated (nonholonomic) systems. Furthermore, many marine vehicles and autonomous aerial vehicles are second-order holonomic systems with internal dynamics that are not asymptotically stable, i.e. are non-minimum phase systems. A formation control strategy that can be applied to underactuated multi-agent systems is considered in [16] where hybrid control techniques are used. The approach is based on consensus rather than leader–follower synchronisation and does not take into account disturbance rejection. In [22] formation control of underactuated vessels under the influence of constant disturbances is considered using neural network adaptive dynamic surface control to track pre-defined paths.

In [4] formation control of underactuated systems is considered, and straight-line path following in formation is achieved for underactuated marine vessels under the influence of constant ocean currents. Straight-line target tracking for underactuated unmanned surface vessels is investigated in [8]. In [8] constant bearing guidance is used to track a target moving in a straight line, experimental results are presented but closed-loop stability is not proven.

The case considered in this work is leader–follower synchronisation for an underactuated follower in an inhomogeneous multi-agent system. The multi-agent system can thus consist of a leader with arbitrary dynamics as long as it moves in the same space as the follower(s). The follower can be any type of vehicle described by the nonlinear manoeuvring model that is introduced in the next section. For formation control purposes, each follower can again be the leader of other followers, or all followers can have the same leader. Examples of possible configurations are autonomous surface vessels (ASV) following an autonomous underwater vehicle (AUV) as communication nodes during AUV search and survey operations, or a fleet of ASVs manoeuvring by following a leader. Since we consider an underactuated system, we need to take into account the full dynamic model in the control design and analysis. In particular, since the system is underactuated it is not possible to consider a purely kinematic model since then the internal sway dynamics cannot be analysed. Moreover, for the case considered here it is not possible to perform feedback linearisation of the full dynamics. The leader dynamics and the leader trajectories are assumed to be unknown. The leader is free to move as it wants independently of the follower, but the follower has access to measurements of the leader's position and velocity in the inertial frame for use in the guidance law. If the follower uses controllers with acceleration feedforward, the leader's acceleration and jerk also need to be measured. This includes cases where there is communication between the leader and follower, but also when the follower reads AIS measurements of the leader [18].

It should be noted that the leader–follower synchronisation scheme in this work has its dual problem in trajectory tracking. Hence, the input signal of the leader could easily be replaced by a virtual leader. This is true for most, if not all, leader–follower type synchronisation schemes since the leader can always be represented as a virtual vehicle with known trajectory and properties. However, when performing trajectory tracking in most cases it is preferable to use information about the dynamics of the vehicle since then perfect tracking can be achieved for all types of motions. When the leader dynamics and desired trajectory are not known a priori, the followers' internal dynamics might be perturbed by the chosen leader motion. Moreover, when the strategy is applied in a chained form, i.e. followers become leaders to other vehicles, the duality is lost. The stability properties derived in this work will still hold with respect to each leader. However, string stability is not considered in this work and should be investigated to analyse the error propagation along the chain of vehicles.

Preliminary results for this problem have been presented in [5], where the followers' yaw rate was used as a parameter to limit the motion of the follower to reduce the synchronisation error. However, in this work, the effect of the internal dynamics was not considered in the analysis of the guidance. In this chapter, we generalise the results of [5] by analysing the complete closed-loop system including the fully actuated closed-loop dynamics, the underactuated sway dynamics in addition to the synchronisation error kinematics. We discuss the conditions to achieve synchronisation and the physical meaning of these conditions. In particular, we show that the synchronisation error kinematics become integral input-to-state stable (iISS) with respect to changes in the velocity when coupled with the underactuated dynamics, i.e. perfect synchronisation is not possible on trajectories that excite the underactuated dynamics. Moreover, we also prove that the constant bearing guidance gives uniformly semiglobally exponentially stable (USGES) synchronisation error kinematics, rather than simply uniformly globally asymptotically stable (UGAS) and uniformly locally exponentially stable (ULES) as proved in previous work.

The work is organised as follows. In Sect. 8.2, the dynamic model for the follower and the constant bearing guidance algorithm are introduced. The closed-loop behaviour is investigated in Sect. 8.3. Section 8.4 presents simulations considering different scenarios. Finally, Sect. 8.5 gives the conclusions of the work.

8.2 The Follower

This section presents the model for the follower and the guidance law for the follower that is used to synchronise its motion to that of the leader. The leader–follower synchronisation scheme is developed for a class of systems described by a 3-DOF manoeuvring model. This class of systems includes underactuated autonomous surface vessels (ASV) and autonomous underwater vehicles (AUV) moving in the horizontal plane. However, it should be noted that the leader–follower scheme and analysis can be extended to different classes of systems with similar properties such as unmanned aerial vehicles by considering the appropriate dynamic model, control/guidance scheme and appropriate disturbances.

8.2.1 The Vessel Model

We consider an ASV or AUV moving in the horizontal plane. The motion of the vessel is described by the position and orientation of the vessel w.r.t. the earth-fixed reference frame, i.e. $\eta \triangleq [x, y, \psi]^T$. For marine craft, the earth-fixed north-east-down (NED) frame is usually used as an inertial frame [12]. The vector of linear and angular velocities is given in the body-fixed reference frame by $\nu \triangleq [u, v, r]^T$, containing the surge velocity u, sway velocity v and yaw rate r. The vessel is disturbed by an ocean current expressed in the inertial frame n, i.e. the earth-fixed frame. The current is denoted by $\boldsymbol{V_c}$ and satisfies the following assumption.

Assumption 8.1 The ocean current is assumed to be constant and irrotational w.r.t. n, i.e. $\boldsymbol{V_c} \triangleq [V_x, V_y, 0]^T$. Furthermore, it is bounded by $V_{\max} > 0$ such that $\|V_c\| = \sqrt{V_x^2 + V_y^2} \leq V_{\max}$.

The ocean current velocity is expressed in the body-fixed frame b and is denoted by $\nu_c \triangleq [u_c, v_c, 0]^T$. It can be obtained by $\nu_c = \boldsymbol{R}(\psi)^T \boldsymbol{V}_c$ where $\boldsymbol{R}(\psi)$ is the rotation matrix from the body to inertial frame defined as

$$\boldsymbol{R}(\psi) \triangleq \begin{bmatrix} \cos(\psi) & -\sin(\psi) & 0 \\ \sin(\psi) & \cos(\psi) & 0 \\ 0 & 0 & 1 \end{bmatrix}. \tag{8.1}$$

The vessel model is expressed in terms of the relative velocity defined as $\nu_r \triangleq \nu - \nu_c = [u_r, v_r, r]^T$ expressed in b. Since the ocean current is constant and irrotational, the vessel can be described by the 3-DOF manoeuvring model [12]

$$\dot{\eta} = \boldsymbol{R}(\psi)\nu_r + \boldsymbol{V}_c, \tag{8.2a}$$

$$\boldsymbol{M}\dot{\nu}_r + \boldsymbol{C}(\nu_r)\nu_r + \boldsymbol{D}\nu_r = \boldsymbol{B}\boldsymbol{f}. \tag{8.2b}$$

The vector $\boldsymbol{f} \triangleq [T_u, T_r]^T$ is the control input vector, containing the surge thrust T_u and the rudder angle T_r. The matrix $\boldsymbol{M} = \boldsymbol{M}^T > 0$ is the system inertia matrix including added mass, $\boldsymbol{C}$ is the Coriolis and centripetal matrix, $\boldsymbol{D} > 0$ is the hydrodynamic damping matrix and $\boldsymbol{B}$ is the actuator configuration matrix.

Remark 8.1 By expressing the model in relative velocities, the environmental disturbances can be incorporated in the model more easily and controlled more straightforwardly.

Assumption 8.2 We assume port–starboard symmetry.

Remark 8.2 Assumption 8.2 is to the authors' best knowledge satisfied for all commercially available surface and underwater vessels.

The matrices $\boldsymbol{M}$, $\boldsymbol{D}$ and $\boldsymbol{B}$ are constant and are defined as

$$\boldsymbol{M} \triangleq \begin{bmatrix} m_{11} & 0 & 0 \\ 0 & m_{22} & m_{23} \\ 0 & m_{23} & m_{33} \end{bmatrix}, \quad \boldsymbol{D} \triangleq \begin{bmatrix} d_{11} & 0 & 0 \\ 0 & d_{22} & d_{23} \\ 0 & d_{32} & d_{33} \end{bmatrix}, \quad \boldsymbol{B} \triangleq \begin{bmatrix} b_{11} & 0 \\ 0 & b_{22} \\ 0 & b_{32} \end{bmatrix}.$$

The non-constant matrix $\boldsymbol{C}(\nu_r)$ can be derived from $\boldsymbol{M}$ (see [12]).

Assumption 8.3 It is assumed that the position of the body-fixed frame is chosen such that $\boldsymbol{M}^{-1}\boldsymbol{B}\boldsymbol{f} = [\tau_u, 0, \tau_r]^T$.

Remark 8.3 This is possible as long as the centre of mass is located along the centreline of the vessel. Coordinate transformations for this translation can be found in [14].

The model can be written in component form as

$$\dot{x} = u_r \cos(\psi) - v_r \sin(\psi) + V_x, \tag{8.3a}$$

$$\dot{y} = u_r \sin(\psi) + v_r \cos(\psi) + V_y, \tag{8.3b}$$

$$\dot{\psi} = r, \tag{8.3c}$$

$$\dot{u}_r = F_{u_r}(v_r, r) + \tau_u, \tag{8.3d}$$

$$\dot{v}_r = X(u_r)r + Y(u_r)v_r, \tag{8.3e}$$

$$\dot{r} = F_r(u_r, v_r, r) + \tau_r, \tag{8.3f}$$

which is clearly underactuated in sway. Therefore, tracking has to be achieved by a suitable velocity and heading assignment that takes into account the underactuation. For this purpose, constant bearing guidance is used. The definitions of F_{u_r}, $X(u_r)$, $Y(u_r)$ and F_r are given by

$$F_{u_r} \triangleq \frac{1}{m_{11}}(m_{22}v_r + m_{23}r)r - \frac{d_{11}}{m_{11}}u_r, \tag{8.4}$$

$$X(u_r) \triangleq \frac{m_{23}^2 - m_{11}m_{33}}{m_{22}m_{33} - m_{23}^2}u_r + \frac{d_{33}m_{23} - d_{23}m_{33}}{m_{22}m_{33} - m_{23}^2}, \tag{8.5}$$

$$Y(u_r) \triangleq \frac{(m_{22} - m_{11})m_{23}}{m_{22}m_{33} - m_{23}^2}u_r - \frac{d_{22}m_{33} - d_{32}m_{23}}{m_{22}m_{33} - m_{23}^2}, \tag{8.6}$$

$$\begin{aligned} F_r(u_r, v_r, r) \triangleq & \frac{m_{23}d_{22} - m_{22}(d_{32} + (m_{22} - m_{11})u_r)}{m_{22}m_{33} - m_{23}^2}v_r \\ & + \frac{m_{23}(d_{23} + m_{11}u_r) - m_{22}(d_{33} + m_{23}u_r)}{m_{22}m_{33} - m_{23}^2}r. \end{aligned} \tag{8.7}$$

Note that $X(u_r)$ and $Y(u_r)$ are bounded for bounded arguments and $Y(u_r)$ satisfies the following assumption.

Assumption 8.4 It is assumed that $Y(u_r)$ satisfies

$$Y(u_r) \leq -Y^{\min} < 0, \ \forall u_r \in [-V_{\max}, U_{\max}].$$

with $U_{\max}$ the maximal surge speed of the follower.

Remark 8.4 This assumption is satisfied for commercial vessels by design, since $Y(u_r) \geq 0$ would imply an undamped or nominally unstable vessel in sway direction.

8.2.2 Constant Bearing Guidance

This subsection briefly describes constant bearing guidance (CB) as presented in [6, 12]. CB guidance assigns a desired velocity based on two different components expressed in the earth-fixed frame. The first component is the velocity of the leader $\boldsymbol{v}_l^n = [\dot{x}_l, \dot{y}_l]^T$ which needs to be matched. The second component is the follower–leader approach velocity $\boldsymbol{v}_a^n$ which is proportional, but upper-bounded by a maximum, to the relative position in the earth-fixed frame between the follower and the leader $\tilde{\boldsymbol{p}}^n = [\tilde{x}^n, \tilde{y}^n]^T$ and is aligned along the line-of-sight (LOS) vector. The superscript n denotes that the variable is expressed in the earth-fixed frame. An illustration of the constant bearing guidance can be seen in Fig. 8.1. The desired velocity assignment for constant bearing guidance is given by

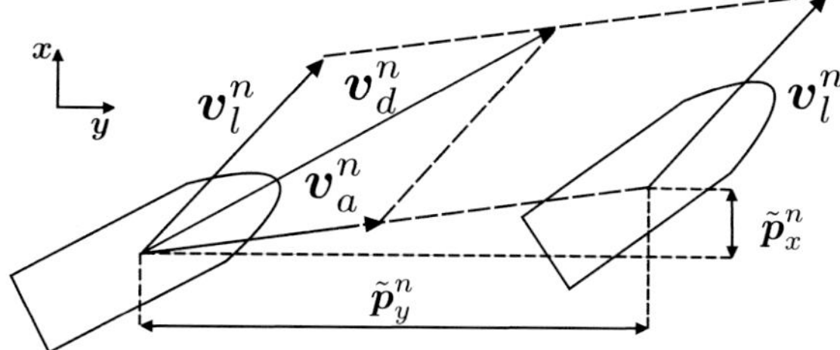

Fig. 8.1 Constant bearing guidance velocity assignments and position error

$$v_d^n = v_l^n + v_a^n, \tag{8.8}$$

$$v_a^n = -\kappa \frac{\tilde{p}^n}{\|\tilde{p}^n\|}, \tag{8.9}$$

with v_l^n the leader velocity, v_a^n the approach velocity and

$$\tilde{p}^n \triangleq p^n - p_l^n, \tag{8.10}$$

is the LOS vector between the follower and the leader, where $\|\tilde{p}^n\| \geq 0$ is the Euclidean length of this vector and

$$\kappa = U_a^{\max} \frac{\|\tilde{p}^n\|}{\sqrt{(\tilde{p}^n)^T \tilde{p}^n + \Delta_{\tilde{p}}^2}}, \tag{8.11}$$

with $U_a^{\max}$ the maximum approach speed and $\Delta_{\tilde{p}}$ a tuning parameter to affect the transient leader–follower rendezvous behaviour, which results in the synchronisation error kinematics

$$\dot{\tilde{p}}^n = v_d^n - v_l^n = -U_a^{\max} \frac{\tilde{p}^n}{\sqrt{(\tilde{p}^n)^T \tilde{p}^n + \Delta_{\tilde{p}}^2}}. \tag{8.12}$$

From (8.9) and (8.11) it can be seen that as $\tilde{p}^n \to 0$ the approach speed goes to zero and the velocity of the follower approaches the leader velocity. Conversely, when $\tilde{p}^n \to \infty$ the approach velocity approaches $U_a^{\max}$ and the guidance commands the maximum allowed velocity to close the gap.

Assumption 8.5 To assure that the problem is feasible, we assume that the sum of the magnitude of the leader velocity, the maximum approach speed and the ocean current is smaller than the maximum feasible surge velocity of the follower U_{feas}, i.e.

$$\|v_l^n\| + U_a^{\max} + \|V_c\| \leq U_{\text{feas}} \tag{8.13}$$

for all $t > 0$. Moreover, the desired speed is required to be positive, and we therefore need to assume that

$$\|\boldsymbol{v}_l^n\| - U_a^{\max} - \|\boldsymbol{V}_c\| > 0 \tag{8.14}$$

for all $t > 0$.

Remark 8.5 Note that in order to converge to a point that is at a desired off-set w.r.t the leader $\boldsymbol{p}_r$, the position of the leader should be included in (8.10) as $\boldsymbol{p}_l^n \triangleq \boldsymbol{p}_{l,\text{true}}^n + \boldsymbol{R}(\psi_l)\boldsymbol{p}_r$ where $\boldsymbol{R}(\psi_l)$ is a rotation matrix describing the orientation of the leader. For curved paths, the velocity $\boldsymbol{v}_l^n$ should then also be calculated in the off-set point to track the curvature with minimal error which requires the leader's yaw rate.

As shown in [12] the stability and convergence of the CB guidance scheme, i.e. (8.8)–(8.9) and (8.11), can be investigated using the positive definite, radially unbounded Lyapunov function candidate (LFC)

$$V = \frac{1}{2}(\tilde{\boldsymbol{p}}^n)^T\tilde{\boldsymbol{p}}^n. \tag{8.15}$$

Time differentiation of (8.15) along the trajectories of $\tilde{\boldsymbol{p}}^n$ gives

$$\dot{V} = (\tilde{\boldsymbol{p}}^n)^T(\boldsymbol{v}_d^n - \boldsymbol{v}_l^n) = -\kappa\frac{(\tilde{\boldsymbol{p}}^n)^T\tilde{\boldsymbol{p}}^n}{\|\tilde{\boldsymbol{p}}^n\|} \tag{8.16a}$$

$$= -U_{a,\max}\frac{(\tilde{\boldsymbol{p}}^n)^T\tilde{\boldsymbol{p}}^n}{\sqrt{(\tilde{\boldsymbol{p}}^n)^T\tilde{\boldsymbol{p}}^n + \Delta_{\tilde{p}}^2}} < 0, \quad \forall\,\tilde{\boldsymbol{p}}^n \neq 0 \tag{8.16b}$$

with $\boldsymbol{v}_d^n - \boldsymbol{v}_l^n = \boldsymbol{v}_a^n$ by definition. Hence, the origin $\tilde{\boldsymbol{p}}^n = 0$ is uniformly globally asymptotically stable (UGAS), which is the result given in [12].

Note however that by defining

$$\phi^*(t,\tilde{\boldsymbol{p}}^n) \triangleq \frac{U_a^{\max}}{\sqrt{(\tilde{\boldsymbol{p}}^n)^T\tilde{\boldsymbol{p}}^n + \Delta_{\tilde{p}}^2}} \tag{8.17}$$

which for each $r > 0$ and $|\tilde{\boldsymbol{p}}^n(t)| \leq r$ gives

$$\phi^*(t,\tilde{\boldsymbol{p}}^n) \leq \frac{U_a^{\max}}{\sqrt{r^2 + \Delta_{\tilde{p}}^2}} \triangleq c^*(r) \tag{8.18}$$

which substituted in (8.16) gives

$$\dot{V} \leq -2c^*(r)V(t,\tilde{\boldsymbol{p}}^n) \tag{8.19}$$

for all $|\tilde{\boldsymbol{p}}^n(t_0)| \leq r$ and any $r > 0$. The solutions of a linear system of the form $\dot{x} = -2c^*(r)x$ are given by

$$x(t) = e^{-2c^*(r)(t-t_0)}x(t_0) \tag{8.20}$$

so by the comparison lemma [17, Lemma 3.4] we have

$$V(t, \tilde{\boldsymbol{p}}^n) \leq e^{-2c^*(r)(t-t_0)}V(t_0, \tilde{\boldsymbol{p}}^n(t_0)) \tag{8.21}$$

and consequently

$$\|\tilde{\boldsymbol{p}}^n(t)\| \leq \|\tilde{\boldsymbol{p}}^n(t_0)\| e^{-c^*(r)(t-t_0)} \tag{8.22}$$

for all $t > t_0$, $|\tilde{\boldsymbol{p}}^n(t_0)| \leq r$, and any $r > 0$. Therefore, we can conclude that (8.12) is a uniformly semi-globally exponentially stable (USGES) system according to [20, Definition 2.7], a result which has not previously been shown in [12] nor [5].

Theorem 8.1 *Using the constant bearing guidance scheme, i.e. (8.8)–(8.9) and (8.11), the origin of the synchronisation error kinematics (8.12) is uniformly semi-globally exponentially stable (USGES).*

The desired heading ψ_d and its derivative, the desired yaw rate r_d, are calculated by extracting heading information from the inner and outer products of the desired velocity $\boldsymbol{v}_d^n$ and the actual velocity $\boldsymbol{v}^n$ [8]. This assures that $\boldsymbol{v}^n$ is aligned with $\boldsymbol{v}_d^n$. Moreover, since it provides us with the course, and equivalently heading, information it allows for compensation of the environmental disturbance. More details about constant bearing guidance can be found in [12] and the references therein.

8.2.3 The Controller

The control goals are

$$\lim_{t\to\infty} \tilde{\boldsymbol{p}}^n = \boldsymbol{0}, \tag{8.23}$$

$$\lim_{t\to\infty} \tilde{\boldsymbol{v}}^n \triangleq \boldsymbol{v}^n - \boldsymbol{v}_d^n = 0, \tag{8.24}$$

which correspond to synchronisation with the leader, i.e. that the follower vessel follows the leader, with a constant desired relative position and the same inertial frame velocity. Note that the body frame velocity may be different due to differences in actuation topology etc. In this section, we present feedback linearising controllers using the desired velocity and heading angle from Sect. 8.2.2, in order to achieve these control goals. In the following section, it will be shown that the feasibility of these goals depends on the type of motion the leader executes.

Since the follower is underactuated, we cannot directly control the velocity in the earth-fixed coordinates, but rather the forward velocity and yaw rate in body-fixed coordinates. Therefore, we transform the velocity error in the earth-fixed frame to an error in the body-fixed frame using the coordinate transformation

$$\begin{bmatrix} \tilde{\psi} \\ \tilde{u}_r \\ \tilde{v}_r \end{bmatrix} = \begin{bmatrix} 1 & 0 & 0 \\ 0 & \cos(\tilde{\psi} + \psi_d) & \sin(\tilde{\psi} + \psi_d) \\ 0 & -\sin(\tilde{\psi} + \psi_d) & \cos(\tilde{\psi} + \psi_d) \end{bmatrix} \begin{bmatrix} \tilde{\psi} \\ \tilde{\boldsymbol{v}}^n \end{bmatrix}. \tag{8.25}$$

It is straightforward to show that the Jacobian of this transformation is given by

$$\frac{\partial T}{\partial(\tilde{\psi}, \tilde{\boldsymbol{v}}^n)} = \begin{bmatrix} 1 & 0 & 0 \\ -\tilde{v}_x^n \mathrm{s}(\cdot) + \tilde{v}_y^n \mathrm{c}(\cdot) & \mathrm{c}(\cdot) & \mathrm{s}(\cdot) \\ -\tilde{v}_x^n \mathrm{c}(\cdot) - \tilde{v}_y^n \mathrm{s}(\cdot) & -\mathrm{s}(\cdot) & \mathrm{c}(\cdot) \end{bmatrix} \tag{8.26}$$

with $\mathrm{s}(\cdot) = \sin(\tilde{\psi} + \psi_d)$ and $\mathrm{c}(\cdot) = \cos(\tilde{\psi} + \psi_d)$. The Jacobian (8.26) can easily be verified to be non-singular. Consequently, T is a global diffeomorphism. A physical interpretation of this is that when $\tilde{\psi}$ is driven to zero, i.e. $\boldsymbol{v}^n$ is aligned with $\boldsymbol{v}_d^n$ by the CB guidance algorithm, the relative surge velocity error can be used to control $\boldsymbol{v}^n$ to $\boldsymbol{v}_d^n$. Note that perturbation of the underactuated sway motion will disturb this balance which will be shown in the analysis of the next section.

Remark 8.6 For the underactuated model considered here only $\tilde{u}_r = u_r - u_d$ can be used for control purposes, while for the fully actuated case $\tilde{v}_r = v_r - v_d$ could be used to control the sway velocity and the perturbation problem does not exist. For the underactuated case, the heading controller needs to assure that $\boldsymbol{v}^n$ is aligned with $\boldsymbol{v}_d^n$ and the control action can be prescribed solely by the surge actuator, something which prevents the magnitude from being matched on curved trajectories and in the presence of accelerations.

Remark 8.7 Note that the coupling between the heading and velocity control is what allows for disturbance rejection. Since if a larger (or smaller) velocity is needed to compensate for the effect of the current, the heading controller will assure that the vessel is rotated such that $\boldsymbol{v}^n$ and $\boldsymbol{v}_d^n$ are aligned and hence the vessel keeps the correct course.

We will use the following feedback linearising P controller for the surge velocity:

$$\tau_u = -F_{u_r}(v_r, r) + \dot{u}_d - k_{u_r}(u_r - u_d), \tag{8.27}$$

with $k_{u_r} > 0$ a constant controller gain.

Using (8.27), we can control u_r towards u_d provided that we have the acceleration of the leader available to calculate $\dot{u}_d$, but we cannot directly control v_r. Along the lines of [8], we aim to control v_r indirectly by using a proper yaw rate controller. Following [8], we have for $\tilde{\chi} = \chi - \chi_d$:

$$\sin(\tilde{\chi}) = \frac{\boldsymbol{v}_d^n \times \boldsymbol{v}^n}{\|\boldsymbol{v}_d^n\|\|\boldsymbol{v}^n\|} = \frac{\dot{y}v_{d,x}^n - \dot{x}v_{d,y}^n}{\sqrt{\left((v_{d,x}^n)^2 + (v_{d,y}^n)^2\right)(\dot{x}^2 + \dot{y}^2)}} \tag{8.28a}$$

$$\cos(\tilde{\chi}) = \frac{(\boldsymbol{v}_d^n)^T \boldsymbol{v}^n}{\|\boldsymbol{v}_d^n\|\|\boldsymbol{v}^n\|} = \frac{\dot{x}v_{d,x}^n + \dot{y}v_{d,y}^n}{\sqrt{\left((v_{d,x}^n)^2 + (v_{d,y}^n)^2\right)(\dot{x}^2 + \dot{y}^2)}} \tag{8.28b}$$

$$\tan(\tilde{\chi}) = \frac{\boldsymbol{v}_d^n \times \boldsymbol{v}^n}{(\boldsymbol{v}_d^n)^T \boldsymbol{v}^n} = \frac{\dot{y}v_{d,x}^n - \dot{x}v_{d,y}^n}{\dot{x}v_{d,x}^n + \dot{y}v_{d,y}^n}$$

$$\Rightarrow \tilde{\chi} = -\text{atan2}(\dot{y}v_{d,x}^n - \dot{x}v_{d,y}^n, \dot{x}v_{d,x}^n + \dot{y}v_{d,y}^n) \tag{8.28c}$$

where $\tilde{\chi} \triangleq \psi - \psi_d + \beta - \beta_d \triangleq \tilde{\psi} + \tilde{\beta}$ with $\tilde{\beta}$ the difference in side-slip angle between different orientations. We can thus define

$$\psi_d - \tilde{\beta} = \psi - \text{atan2}(\dot{y}v_{d,x}^n - \dot{x}v_{d,y}^n, \dot{x}v_{d,x}^n + \dot{y}v_{d,y}^n) \tag{8.29}$$

Note that from (8.28) we also have

$$\dot{\tilde{\chi}} = \frac{\dot{x}\ddot{y} - \dot{y}\ddot{x}}{\dot{x}^2 + \dot{y}^2} + \frac{v_{d,y}^n \dot{v}_{d,x}^n - v_{d,x}^n \dot{v}_{d,y}^n}{(v_{d,x}^n)^2 + (v_{d,y}^n)^2} \tag{8.30}$$

so we can write

$$r_d - \dot{\tilde{\beta}} = r - \frac{\dot{x}\ddot{y} - \dot{y}\ddot{x}}{\dot{x}^2 + \dot{y}^2} - \frac{v_{d,y}^n \dot{v}_{d,x}^n - v_{d,x}^n \dot{v}_{d,y}^n}{(v_{d,x}^n)^2 + (v_{d,y}^n)^2} \tag{8.31a}$$

$$\triangleq r - R_1(u_r, v_r, \dot{x}, \dot{y}, \boldsymbol{v}_d^n)r - R_2(\boldsymbol{v}_d^n, \dot{\boldsymbol{v}}_d^n) - R_3(u_r, v_r, \dot{x}, \dot{y}, \boldsymbol{v}_d^n, \dot{\boldsymbol{v}}_d^n) \tag{8.31b}$$

where

$$R_1(\cdot) \triangleq \frac{u_r^2 + v_r^2 + V_x^2 + X(u_r)(u_r + v_{c,u}^b) - v_r v_{c,v}^b + u_r v_{c,u}^b - v_{d,x}^n(V_x - v_r\sin(\psi))}{u_r^2 + v_r^2 + 2(u_r v_{c,x}^b + v_r v_{c,y}^b) + V_x^2 + V_y^2}$$
$$+ \frac{-(v_{d,y}^n(v_r - v_{c,v}^b) - v_{d,x}^n v_{c,u}^b)\cos(\psi) + \cos^2(\psi)(V_y^2 - V_x^2)}{u_r^2 + v_r^2 + 2(u_r v_{c,x}^b + v_r v_{c,y}^b) + V_x^2 + V_y^2} \leq C_{R_1}^{\max}$$

$$R_2(\cdot) \triangleq \frac{v_{d,y}^n \dot{v}_{d,x}^n - v_{d,x}^n \dot{v}_{d,y}^n}{(v_{d,x}^n)^2 + (v_{d,y}^n)^2}$$

$$R_3(\cdot) \triangleq \frac{Y(u_r)v_r(u_r + v_{c,u}^b) + k_{u_r}(u_r - u_d)(v_r - v_{c,v}^b)}{u_r^2 + v_r^2 + 2(u_r v_{c,x}^b + v_r v_{c,y}^b) + V_x^2 + V_y^2}$$
$$+ \frac{\dot{v}_{d,x}^n(v_{c,u}^b - v_r)\cos(\psi) + \dot{v}_{d,y}^n(V_x - v_r\sin(\psi) - v_{c,u}^b\cos(\psi))}{u_r^2 + v_r^2 + 2(u_r v_{c,x}^b + v_r v_{c,y}^b) + V_x^2 + V_y^2} \leq C_{R_3}$$

and $v_{c,u}^b \triangleq V_x\cos(\psi) + V_y\sin(\psi)$ and $v_{c,v}^b \triangleq -V_x\sin(\psi) + V_y\cos(\psi)$ are the components of the current expressed in the body frame axis.

Note that R_1 can be bounded by the constant $C_{R_1}^{\max}$ since R_1 has the same growth rate in v_r and u_r for the denominator and numerator while the ocean current components are bounded (in body frame) and constant (in inertial frame). The term R_3 can be bounded by the constant C_{R_3} since the denominator and numerator grow at the same rate with respect to v_r and u_r and the current is bounded. Note that the denominator of R_1, R_2 and R_3 are larger than zero for non-zero $\|\boldsymbol{v}^n\|$ and $\|\boldsymbol{v}_d^n\|$ which is verified by Assumption 8.5. Boundedness of R_2 will be considered later since its numerator grows linearly with v_r and its denominator does not grow with v_r.

Since the inertial frame velocities, i.e. $\dot{x}$ and $\dot{y}$, are measured V_x and V_y can be substituted in expression (8.31) using the model equations (8.3a) and (8.3b), respectively, for implementation purposes. Alternatively, a kinematic ocean current observer as in [1] can be used to estimate $\dot{x}$, $\dot{y}$, V_x and V_y based on measurements of the positions and relative velocities. Hence, all the variables in (8.31) are known and can thus be substituted in the yaw rate controller. A further derivative of (8.31) can be taken to obtain $\ddot{\psi}_d - \ddot{\tilde{\beta}}$ as an acceleration feedforward. Note that this will also require knowledge of the jerk of the leader motion since it contains $\dot{R}_2(\boldsymbol{v}_d^n, \dot{\boldsymbol{v}}_d^n)$ and $\dot{R}_3(u_r, v_r, \dot{x}, \dot{y}, \boldsymbol{v}_d^n, \dot{\boldsymbol{v}}_d^n)$.

To control the yaw rate, we use the following controller:

$$\begin{aligned}\tau_r &= -F_r(u_r, v_r, r) + \frac{1}{R_1(u_r, v_r, \dot{x}, \dot{y}, \boldsymbol{v}_d^n)}\Big(- \dot{R}_1(u_r, v_r, \dot{x}, \dot{y}, \boldsymbol{v}_d^n)r - \dot{R}_2(\boldsymbol{v}_d^n, \dot{\boldsymbol{v}}_d^n) \\ &\quad - \dot{R}_3(u_r, v_r, \dot{x}, \dot{y}, \boldsymbol{v}_d^n, \dot{\boldsymbol{v}}_d^n) - k_\psi(\psi - \psi_d + \tilde{\beta}) - k_r(\dot{\psi} - \dot{\psi}_d + \dot{\tilde{\beta}})\Big) \qquad (8.32a) \\ &= -F_r(u_r, v_r, r) + \frac{1}{R_1(u_r, v_r, \dot{x}, \dot{y}, \boldsymbol{v}_d^n)}\Big(- \dot{R}_1(u_r, v_r, \dot{x}, \dot{y}, \boldsymbol{v}_d^n)r - \dot{R}_2(\boldsymbol{v}_d^n, \dot{\boldsymbol{v}}_d^n) \\ &\quad - \dot{R}_3(u_r, v_r, \dot{x}, \dot{y}, \boldsymbol{v}_d^n, \dot{\boldsymbol{v}}_d^n) - k_\psi\tilde{\chi} - k_r\dot{\tilde{\chi}}\Big) \qquad (8.32b)\end{aligned}$$

with $k_\psi > 0$ and $k_r > 0$ constant controller gains. This control action is well defined if $R_1(u_r, v_r, \dot{x}, \dot{y}, \boldsymbol{v}_d^n)$ satisfies certain conditions, which is something discussed in the following when considering the boundedness of r. We introduce the vector $\boldsymbol{\xi} \triangleq [\tilde{u}_r, \tilde{\chi}, \dot{\tilde{\chi}}]^T$, with the tracking errors $\tilde{u}_r \triangleq u_r - u_d$, $\tilde{\chi} \triangleq \tilde{\psi} + \tilde{\beta}$ and $\dot{\tilde{\chi}} \triangleq \dot{\tilde{\psi}} - \dot{\tilde{\beta}}$. The dynamics of $\boldsymbol{\xi}$ can be found by applying the controllers (8.27) and (8.32) to the dynamical system (8.3) resulting in:

$$\dot{\boldsymbol{\xi}} = \begin{bmatrix} -k_{u_r} & 0 & 0 \\ 0 & 0 & 1 \\ 0 & -k_\psi & -k_r \end{bmatrix} \boldsymbol{\xi} \triangleq \Sigma\boldsymbol{\xi}. \qquad (8.33)$$

The system (8.33) is linear and time invariant and k_{u_r}, k_ψ and k_r are strictly positive. Consequently, Σ is Hurwitz and the origin of (8.33) is uniformly globally exponentially stable (UGES) and hence the controllers guarantee exponential tracking of the desired surge velocity and course.

Note that through the assignment of (8.32) we use the heading controller to perform course control, i.e. we force the direction of $\boldsymbol{v}_d^n$ and $\boldsymbol{v}^n$ to be equal. To investigate how the course controller affects r, we start by rewriting (8.31) to obtain

$$r = \frac{1}{R_1(u_r, v_r, \dot{x}, \dot{y}, \boldsymbol{v}_d^n)} \left(\dot{\tilde{\chi}} - R_2(\boldsymbol{v}_d^n, \dot{\boldsymbol{v}}_d^n) - R_3(u_r, v_r, \dot{x}, \dot{y}, \boldsymbol{v}_d^n, \dot{\boldsymbol{v}}_d^n) \right) \tag{8.34}$$

This function is well defined if the numerator of R_1 given in (8.31) is larger than zero. This condition is satisfied if u_d is sufficiently large at all time and if u_r starts sufficiently close to u_d The term $\dot{\tilde{\chi}}/R_1$ will be bounded since $\dot{\tilde{\chi}}$ is bounded and R_1 is bounded by constant C_{R_1} as shown earlier. The same holds for the term R_3/R_1. The term R_2/R_1 however grows linearly in v_r since $\dot{v}_{d,x}^n$ and $\dot{v}_{d,y}^n$ depend linearly on v_r since the derivative of the approach speed $\boldsymbol{v}_a^n$ depends on $\dot{x}$ and $\dot{y}$. When (8.34) is substituted in (8.3e), the linear growth will assure that there is no finite escape time for v_r but some conditions have to be satisfied to show boundedness. Summarising the above, we have that the course controller results in a well-defined yaw rate if the following condition is satisfied.

Condition 8.1 If the numerator of R_1 is strictly larger than zero, then the yaw rate equation (8.34) is well defined and bounded. In particular, besides being upper-bounded there also exists a lower bound for R_1 such that $0 < C_{R_1}^{\min} \leq R_1(u_r, v_r, \dot{x}, \dot{y})$.

Remark 8.8 Condition 8.1 is satisfied for a sufficiently large desired surge velocity u_d if u_r starts in a neighbourhood of u_d. Further analysis has to be performed to find the precise physical meaning of the bound, but it appears to be that inertial frame velocity vector has to have a positive magnitude for all time. This can be satisfied by keeping the surge velocity u_r sufficiently large to be able to dominate the effects of the ocean current and the sway velocity v_r. In particular, if the inertial frame velocity vector would have a zero crossing, the rotation would change instantaneously and when the magnitude of the inertial frame velocity vector is zero then the desired rotation is undefined.

Remark 8.9 Note that Condition 8.1 is a condition that plays a role in the initial behaviour when the difference between the initial orientation of the follower and the leader is large, e.g. if they point in opposite directions. In this case, u_d obtained from (8.25) needs to be saturated to a lower bound such that it stays positive and well defined. As soon as the follower is oriented in the same direction as the leader, Condition 8.1 is easily satisfied for physically sensible motions of the leader and u_d can simply be obtained from (8.25).

The term R_2 can be interpreted as dependent on the desired curvature of the motion. In particular, it can be rewritten as $R_2 = \|\boldsymbol{v}_d^n\|\kappa$ where κ denotes the curvature of the desired trajectory. This term grows linearly with the inertial frame velocities of the follower since it depends on $\dot{\boldsymbol{v}}_a^n$

$$R_2(\boldsymbol{v}_d^n, \dot{\boldsymbol{v}}_d^n) = \frac{v_{d,y}^n \dot{v}_{d,x}^n - v_{d,x}^n \dot{v}_{d,y}^n}{(v_{d,x}^n)^2 + (v_{d,y}^n)^2} = \frac{v_{d,y}^n \dot{v}_{l,x}^n - v_{d,x}^n \dot{v}_{l,y}^n}{(v_{d,x}^n)^2 + (v_{d,y}^n)^2} + \frac{v_{d,y}^n \dot{v}_{a,x}^n - v_{d,x}^n \dot{v}_{a,y}^n}{(v_{d,x}^n)^2 + (v_{d,y}^n)^2} \tag{8.35}$$

which using the transformation (8.25) can be bounded by

$$\begin{aligned} R_2 &\leq \frac{U_a^{\max}}{(v_{d,x}^n)^2 + (v_{d,y}^n)^2} \left(\frac{v_{d,y}^n + v_{d,x}^n}{\sqrt{\tilde{x}^2 + \tilde{y}^2 + \Delta_{\tilde{p}}^2}} + \frac{v_{d,y}^n (\tilde{x}^2 + \tilde{x}\tilde{y}) + v_{d,x}^n (\tilde{y}^2 + \tilde{x}\tilde{y})}{(\tilde{x}^2 + \tilde{y}^2 + \Delta_{\tilde{p}}^2)^{3/2}} \right) \tilde{v}_r + C_{R_2} \\ &\triangleq R_2' \tilde{v}_r + C_{R_2} \end{aligned} \tag{8.36}$$

where C_{R_2} is some constant which magnitude will depend on the leader's velocity and acceleration. Note that the term R_2' is uniformly bounded for desired velocities greater than zero and that it decreases as the positional error grows. Moreover, it contains two parameters that can be tuned, i.e. the maximum approach speed $U_{a,\max}$ and the interaction tuning parameter $\Delta_{\tilde{p}}$. Hence, these tuning parameters can be used to influence the interaction behaviour between r and v_r.

8.3 Closed-Loop Analysis

In this section, the closed-loop system, i.e. the fully actuated closed-loop dynamics, the underactuated sway dynamics and the synchronisation error kinematics, are investigated. In particular, the closed-loop path-following error kinematics and dynamics for (8.2) with the proposed leader–follower synchronisation scheme is given by

$$\dot{\tilde{\boldsymbol{p}}}^n = -\frac{U_a^{\max} \tilde{\boldsymbol{p}}^n}{\sqrt{(\tilde{\boldsymbol{p}}^n)^T \tilde{\boldsymbol{p}}^n + \Delta_{\tilde{p}}^2}} + \begin{bmatrix} \tilde{u}_r \cos(\tilde{\chi} - \tilde{\beta} + \psi_d) - \tilde{v}_r \sin(\tilde{\chi} - \tilde{\beta} + \psi_d) \\ \tilde{u}_r \sin(\tilde{\chi} - \tilde{\beta} + \psi_d) + \tilde{v}_r \cos(\tilde{\chi} - \tilde{\beta} + \psi_d) \end{bmatrix} \tag{8.37a}$$

$$\dot{\tilde{v}}_r = Y(u_r)\tilde{v}_r + X(u_r)r - \dot{v}_d - Y(u_r)v_d \tag{8.37b}$$

$$\dot{\boldsymbol{\xi}} = \Sigma \boldsymbol{\xi} \tag{8.37c}$$

where v_d and $\dot{v}_d$ can be verified to be given by

$$v_d = (V_x - v_{d,x}^n)\sin(\psi) - (V_y - v_{d,y}^n)\cos(\psi) \tag{8.38}$$

$$\begin{aligned} \dot{v}_d &= -\dot{v}_{d,x}^n \sin(\psi) + (V_x - v_{d,x}^n) r \cos(\psi) + \dot{v}_{d,y}^n \cos(\psi) + (V_y - v_{d,y}^n) r \sin(\psi) \\ &= -(\dot{v}_{d,x}^n + \dot{v}_{a,x}^n + (V_y - v_{d,y}^n) r) \sin(\psi) + (\dot{v}_{l,y}^n + \dot{v}_{a,y}^n + (V_x - v_{d,x}^n) r) \cos(\psi) \end{aligned} \tag{8.39}$$

with v_d bounded for a bounded leader velocity. The equation for $\dot{v}_d$ depends on $\dot{v}_{a,x}^n$, $\dot{v}_{a,y}^n$ and r which will depend on $\tilde{v}_r$. However, as in (8.36), we can derive a bound for $\dot{v}_d$

$$\dot{v}_d \leq \left(\frac{\|\boldsymbol{V}_c - \boldsymbol{v}_d^n\|^2 R_2'}{C_{R_1}^{\min}} + \frac{U_a^{\max}}{\sqrt{\tilde{x}^2 + \tilde{y}^2 + \Delta_{\tilde{p}}^2}} \left(1 + \frac{(\tilde{x} + \tilde{y})^2}{\tilde{x}^2 + \tilde{y}^2 + \Delta_{\tilde{p}}^2} \right) \right) \tilde{v}_r + C_2 \tag{8.40a}$$

$$\leq C_3 \tilde{v}_r + C_2 \tag{8.40b}$$

where C_2 is a constant which will depend on the leader's maximum velocity and acceleration and on the magnitude of the ocean current. The magnitude of the constant C_3 can again be adjusted by tuning $U_a^{\max}$ and $\Delta_{\tilde{p}}$.

Please note that the terms perturbing the CB path following error system in (8.37a) compared to (8.12) arise since we here do not only consider the kinematic model, but instead take into account the (underactuated) dynamics given in (8.37b)–(8.37c). We thus take into account that the desired inertial frame velocity may not be matched since part of the error in the inertial frame velocity error is transferred to the sway direction as seen in (8.25). Note that this coupling between the underactuated dynamics and the synchronisation error kinematics was not taken into account in [5].

In order to not violate Condition 8.1, we analyse the system (8.40) under the following assumption.

Assumption 8.6 The desired relative surge velocity is saturated to a sufficiently large lower bound $u_{d,\min}$ such that Condition 8.1 is not violated. It is assumed that there exists such a lower bound that satisfies $u_{d,\min} < \|\boldsymbol{v}_d^n\|$, i.e. that the leader velocity can be matched without violating Condition 8.1.

Since $u_r = u_d$ is a stable equilibrium point the surge velocity dynamics, for any $\delta > 0$ there exists a positively invariant neighbourhood of the equilibrium point such that all solutions originating in this neighbourhood satisfy $|u_r - u_d| < \delta$. Therefore, in the remainder, we only consider solutions starting in the neighbourhood of $u_r = u_d$ such that Condition 8.1 is not violated and there are no finite escape times.

Since substituting (8.34) in (8.37b) shows that there is no finite escape time for v_r and the tracking dynamics (8.37c) are UGES, it suffices to investigate local boundedness of v_r near the set where $u_r - u_d \leq \delta$ such that r is well defined. Therefore, we consider the system

$$\dot{\tilde{v}}_r = Y(u_r)\tilde{v}_r + X(u_r)r - \dot{v}_d - Y(u_r)v_d \tag{8.41}$$

We substitute (8.34) and we obtain

$$\dot{\tilde{v}}_r = Y(u_r)\tilde{v}_r + \frac{X(u_r)}{R_1(u_r, v_r, \dot{x}, \dot{y}, \boldsymbol{v}_d^n)} \left(\dot{\tilde{\chi}} - R_2(\boldsymbol{v}_d^n, \dot{\boldsymbol{v}}_d^n) - R_3(u_r, v_r, \dot{x}, \dot{y}, \boldsymbol{v}_d^n, \dot{\boldsymbol{v}}_d^n) \right)$$
$$- \dot{v}_d - Y(u_r)v_d \tag{8.42}$$

Using the following Lyapunov function, we show boundedness for all solutions starting in the neighbourhood of $u_r = u_d$ by considering the Lyapunov function

$$V(\tilde{v}_r) = \frac{1}{2}\tilde{v}_r^2 \tag{8.43}$$

The derivative of (8.43) along the solutions of (8.41) is given by

$$\begin{aligned}\dot{V}(\tilde{v}_r) &= Y(u_r)\tilde{v}_r^2 + \frac{X(u_r)\left(\dot{\tilde{\chi}} - R_2(\boldsymbol{v}_d^n, \dot{\boldsymbol{v}}_d^n) - R_3(u_r, v_r, \dot{x}, \dot{y}, \boldsymbol{v}_d^n, \dot{\boldsymbol{v}}_d^n)\right)}{R_1(u_r, v_r, \dot{x}, \dot{y}, \boldsymbol{v}_d^n)}\tilde{v}_r \\ &\quad + (\dot{v}_d - Y(u_r)v_d)\,\tilde{v}_r \qquad (8.44a)\\ &\leq -\left(|Y^{\min}| - \frac{|X^{\max}|R_2'}{C_{R_1}^{\min}} - C_3\right)\tilde{v}_r^2 + \frac{|X^{\max}|\left(|\dot{\tilde{\chi}}| + C_{R_2} + C_{R_3}\right)}{C_{R_1}^{\min}}\tilde{v}_r \\ &\quad + (C_2 + |Y^{\max}|v_d)\tilde{v}_r \qquad (8.44b)\end{aligned}$$

where $Y^{\min}$, $Y^{\max}$ and $X^{\max}$ are the minimum and maximum values over the interval of velocities considered and will exist for sufficiently small δ. From which we can conclude boundedness if

$$\frac{|Y^{\min}|}{|X^{\max}|} > \frac{|R_2'|}{C_{R_1}^{\min}} + \frac{C_3}{|X^{\max}|} \tag{8.45}$$

which is a bound that depends on the leader motion, the environmental disturbance and parameters $U_a^{\max}$ and $\Delta_{\tilde{p}}$. From (8.36) and (8.40), we can see that the term R_2' can be tuned using the parameters $U_a^{\max}$ and $\Delta_{\tilde{p}}$. In particular, if we increase $\Delta_{\tilde{p}}$, i.e. choose a smoother leader–follower rendezvous behaviour, then the terms R_2' and C_3 will be reduced. Hence, condition (8.45) can be guaranteed to hold by appropriate tuning of the constant bearing guidance algorithm and all solutions of (8.37b) originating in a neighbourhood of $u_r = u_d$ are uniformly bounded.

Remark 8.10 Note that increasing $\Delta_{\tilde{p}}$ has an effect on the dissipating term in (8.37a). In particular, it lowers the 'gain' of the synchronisation around the origin, i.e. the turning manoeuvre required will be less severe which has a positive effect on (8.45), but the synchronisation error increases since the follower takes a smoother trajectory.

We can now investigate the interconnection between (8.37a) and (8.37b). In particular, we show that the synchronisation error kinematics are integral input-to-state stable with respect to the output of (8.37b) and (8.37c). If we lump the perturbations into a new input $\nu(t) \triangleq [\nu_1(t), \nu_2(t)]^T$, we can rewrite (8.37a) as

$$\dot{\tilde{\boldsymbol{p}}}^n = -\frac{U_a^{\max}}{\sqrt{(\tilde{\boldsymbol{p}}^n)^T\tilde{\boldsymbol{p}}^n + \Delta_{\tilde{p}}^2}}\tilde{\boldsymbol{p}}^n + \nu(t) \tag{8.46}$$

If we consider the Lyapunov function

$$V(\tilde{\boldsymbol{p}}^n) = \frac{(\tilde{\boldsymbol{p}}^n)^T \tilde{\boldsymbol{p}}^n}{\sqrt{(\tilde{\boldsymbol{p}}^n)^T \tilde{\boldsymbol{p}}^n + \Delta^2}} \tag{8.47}$$

we obtain

$$\dot{V}(\tilde{\boldsymbol{p}}^n) = \frac{2(\tilde{\boldsymbol{p}}^n)^T \dot{\tilde{\boldsymbol{p}}}^n}{\sqrt{(\tilde{\boldsymbol{p}}^n)^T \tilde{\boldsymbol{p}}^n + \Delta_{\tilde{p}}^2}} - \frac{\left((\tilde{\boldsymbol{p}}^n)^T \dot{\tilde{\boldsymbol{p}}}^n\right)\left((\tilde{\boldsymbol{p}}^n)^T \tilde{\boldsymbol{p}}^n\right)}{2\left((\tilde{\boldsymbol{p}}^n)^T \tilde{\boldsymbol{p}}^n + \Delta_{\tilde{p}}^2\right)^{3/2}} \tag{8.48a}$$

$$\leq -\frac{2U_a^{\max}(\tilde{\boldsymbol{p}}^n)^T \tilde{\boldsymbol{p}}^n}{(\tilde{\boldsymbol{p}}^n)^T \tilde{\boldsymbol{p}}^n + \Delta_{\tilde{p}}^2} - \frac{U_a^{\max}\left((\tilde{\boldsymbol{p}}^n)^T \tilde{\boldsymbol{p}}^n\right)}{\left((\tilde{\boldsymbol{p}}^n)^T \tilde{\boldsymbol{p}}^n + \Delta_{\tilde{p}}^2\right)^2} + \frac{3}{2}\|\nu(t)\| \tag{8.48b}$$

$$\leq -\frac{2U_a^{\max}(\tilde{\boldsymbol{p}}^n)^T \tilde{\boldsymbol{p}}^n}{(\tilde{\boldsymbol{p}}^n)^T \tilde{\boldsymbol{p}}^n + \Delta_{\tilde{p}}^2} - \frac{U_a^{\max}\left((\tilde{\boldsymbol{p}}^n)^T \tilde{\boldsymbol{p}}^n\right)}{\left((\tilde{\boldsymbol{p}}^n)^T \tilde{\boldsymbol{p}}^n + \Delta_{\tilde{p}}^2\right)^2} + \frac{3\sqrt{2}}{2}\|[\tilde{u}_r, \tilde{v}_r]^T\| \tag{8.48c}$$

The first two terms are clearly negative definite and the third term is a class $\mathscr{K}$ function of the input. Consequently, (8.47) is an iISS Lyapunov function for (8.37a) [3] and the system (8.37) is iISS with respect to $\tilde{u}_r$ and $\tilde{v}_r$. The results of this section can be summarised in the following theorem.

Theorem 8.2 *Consider the system (8.37). Under Assumptions 8.1–8.6 all the solutions of (8.37) starting in a neighbourhood of $u_r = u_d$ are bounded if the CB guidance algorithm is tuned such that it holds that*

$$\frac{|Y^{\min}|}{|X^{\max}|} > \frac{|R_2'|}{C_{R_1}^{\min}} + \frac{C_3}{|X^{\max}|} \tag{8.49}$$

for the given leader motion. Moreover, the synchronisation error kinematics (8.37a) are integral input-to-state stable with respect to the output of (8.37b)–(8.37c).

Corollary 8.1 *If the leader trajectory is a straight-line with constant velocity then, under the conditions of Theorem 8.2, the synchronisation error converges to zero.*

Proof In this case, the course of the leader and its inertial frame velocity are constant. Therefore, as the follower synchronises with the leader its course will converge to the leader's course. Since $\tilde{v}_r$ is not directly controllable, the only stable configuration the follower can be regulated to, to keep a constant course, will be when $r \to 0$ and $v_r \to 0$. Consequently, both $\tilde{v}_r$ and $\tilde{u}_r$ go to zero and we arrive at the unperturbed version of (8.37a), i.e. (8.12), which has a USGES equilibrium according to Theorem 8.1.

8.4 Simulations

In this section, two scenarios are used as case studies to validate the control strategy

1. the leader moves along a straight-line path that is at an angle with respect to the earth-fixed frame.
2. the leader moves along a sinusoidal path.

In both cases, the follower ship is affected by a constant ocean current. The leader is represented by a point moving in the horizontal plane that is to be followed. This allows for a very straightforward implementation of the desired path and illustrates that the leader dynamics are not needed for the control strategy. Some parameters for the simulations are given in Table 8.1. This includes the parameters for the controllers and guidance law, and the magnitude of the ocean current. The follower vessel in the simulation is described by the ship model from [13].

Table 8.1 Simulation parameters

Variable	Value	Unit	Variable	Value	Unit
$U_{a,\max}$	2	m/s	k_ψ	0.04	–
$\Delta_{\bar{p}}$	500	m	k_r	0.9	–
V_x	-1.1028	m/s	k_{u_r}	0.1	–
V_y	0.8854	m/s			

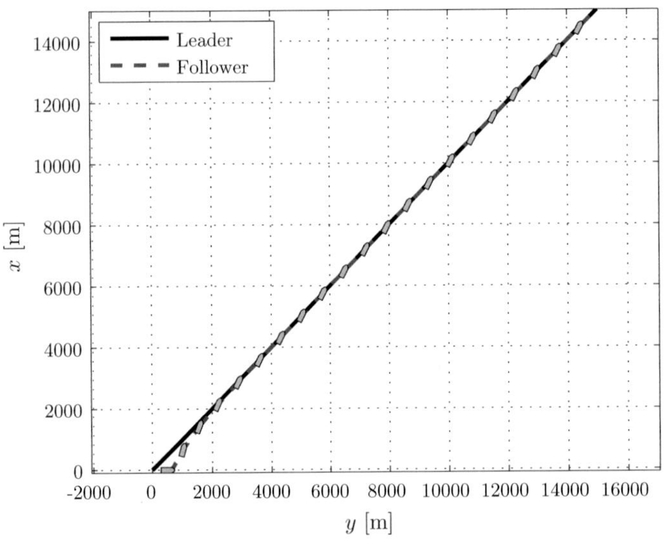

Fig. 8.2 Motion in the horizontal plane

8.4.1 Straight-Line Path Following

The motion of the leader and the follower in the horizontal plane can be seen in Fig. 8.2. From Fig. 8.2 it can be seen that the follower converges to the trajectory of the leader and compensates for the current by side-slipping to maintain the desired path. The side-slipping is a desired result of the control strategy and is necessary to remain on the straight-line path in the presence of ocean currents. In particular, since the vessel is underactuated in sway, a side-slip angle w.r.t. the path is necessary to compensate for the force pushing the vessel in the transverse direction of the path. Since the desired heading angle is calculated from the inner and outer products of the desired and actual velocity, the desired angle is the angle for which the velocity error is zero, which is the necessary side-slip angle.

The synchronisation error in x and y can be seen in Fig. 8.3. Figure 8.3 clearly shows that $\tilde{x}^n$ and $\tilde{y}^n$ converge to zero. Hence, target tracking or leader–follower

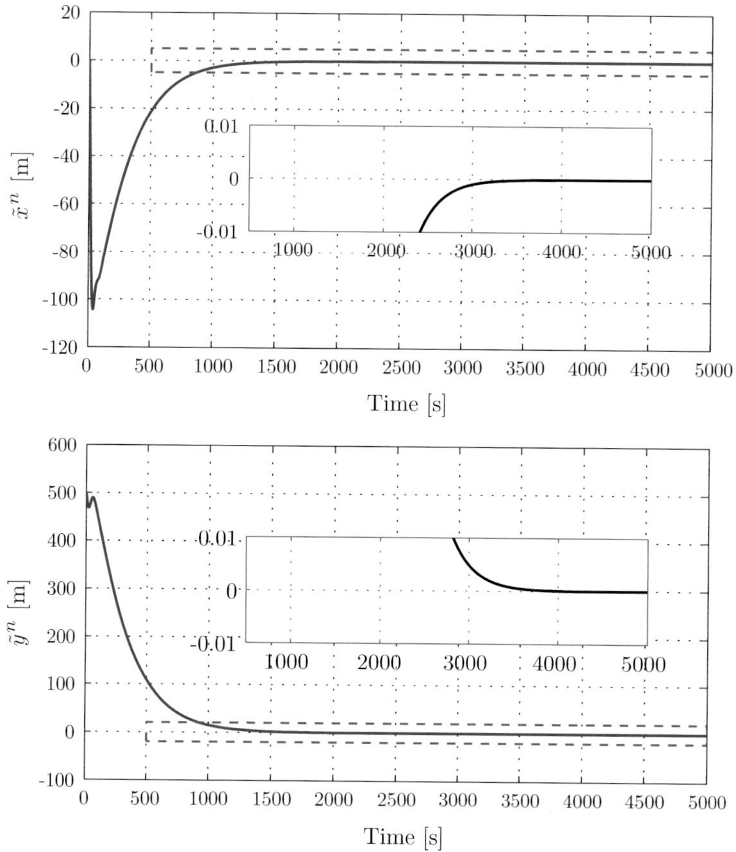

Fig. 8.3 x (*top*) and y (*bottom*) synchronisation error

synchronisation with zero synchronisation error is attained for straight-line motions with $r_d \to 0$ which is in-line with our analysis of Sect. 8.3.

8.4.2 Sinusoidal Path Following

In the second case study, the leader generates a sinusoidal reference for the follower which demands a constantly changing desired yaw rate. Hence, the synchronisation error kinematics are perturbed.

The trajectory of the leader and the follower for tracking of a sinusoidal path can be seen in Fig. 8.4. From Fig. 8.4 it can be seen that the follower gets close to the trajectory of the leader and compensates for the current to maintain the desired path. Figure 8.5 shows the position synchronisation error in the x and y direction. From Fig. 8.5 it can be seen that the synchronisation error in x decreases below an amplitude of about 1.5 m, while the error in y direction, which is the direction transversal to the propagation of the sinusoid and most prone to drift, decreases to below 2.5 m. Note that the error plots are asymmetric due to the vessel changing its direction with respect to the current which causes different behaviour.

The behaviour in the test-case is in-line with the analysis of Sect. 8.3 since we have convergence from large initial errors, the follower converges towards the trajectory of the leader. When the follower is close to the leader, the follower exhibits integral input-to-state stable behaviour and stays in a neighbourhood of the leader dependent on the size of the desired yaw rate to track this motion.

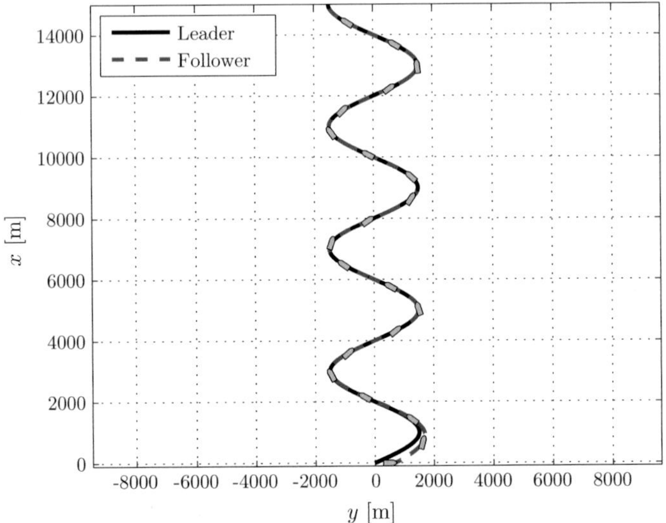

Fig. 8.4 Motion in the horizontal plane

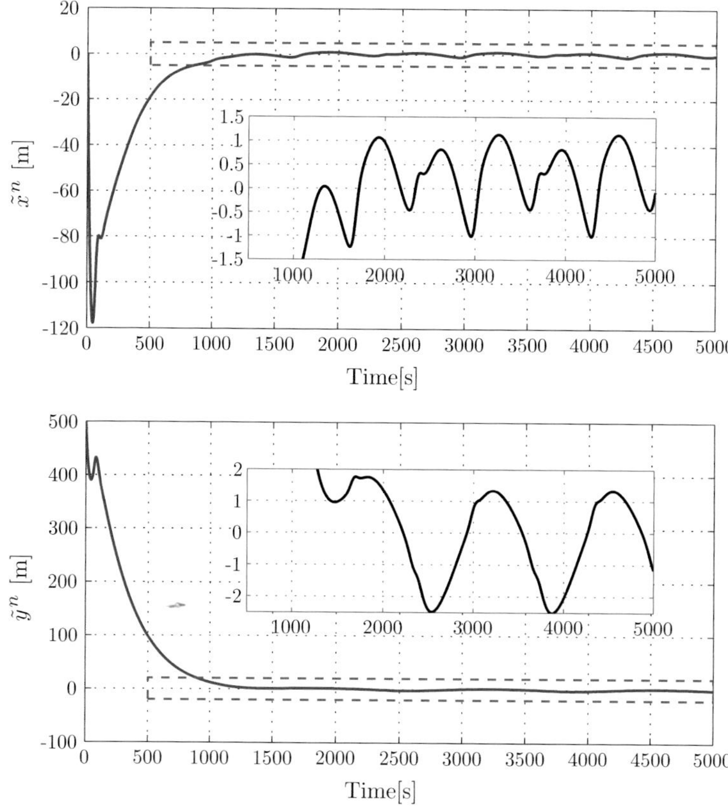

Fig. 8.5 x (*top*) and y (*bottom*) synchronisation error

8.5 Conclusions

This chapter has presented and analysed a control scheme for leader–follower synchronisation for inhomogeneous multi-agent systems consisting of an underactuated follower and a leader vessel with unknown dynamics. The developed leader–follower scheme can be applied to multi-agent systems with underactuated follower agents that are subjected to environmental disturbances. The dynamics of the leader is unknown, and the leader may be fully actuated or underactuated. Position and velocity measurements of the leader are available to the follower for use in the guidance law. If the follower uses controllers with acceleration feedforward, acceleration and jerk measurements of the leader also need to be available to the follower. The leader is free to move as it wants independently of the follower(s), and can for instance be manually controlled. The follower thus has no information about the future motion of the leader. The follower uses a constant bearing guidance algorithm to track the leader. The constant bearing guidance algorithm is shown to provide USGES synchronisation error kinematics. The constant bearing guidance algorithm is then coupled to controllers designed for the underactuated follower vehicle. This results in a closed-loop system consisting of the fully actuated controlled dynamics, underactuated dynamics and synchronisation error kinematics. The solutions of the underactuated and the fully actuated dynamics, have been shown to be bounded under certain conditions. Furthermore, the synchronisation error kinematics has been shown to be integral input-to-state stable with respect to changes in the unactuated sway velocity. Moreover, it has been shown that synchronisation can be achieved when the leader moves along a straight-line since in this case the perturbation of the underactuated dynamics to the synchronisation error kinematics vanishes. The validity of the control scheme has been shown in a case study.

Acknowledgments The authors would like to thank Erjen Lefeber for the valuable discussions on and inputs to the material presented in this chapter.

References

1. Aguiar, A.P., Pascoal, A.M.: Dynamic positioning and way-point tracking of underactuated auvs in the presence of ocean currents. Int. J. Control **80**(7), 1092–1108 (2007)
2. Aguirre, A.A: Remote Control and Motion Coordination of Mobile Robots. Ph.D. thesis, Technische Universiteit Eindhoven (2011)
3. Angeli, D., Sontag, E.D., Wang, Y.: A characterization of integral input-to-state stability. IEEE Trans. Autom. Control **45**(6), 1082–1097 (2000)
4. Belleter, D.J.W., Pettersen, K.Y.: Path following for formations of underactuated marine vessels under influence of constant ocean currents. In: Proceedings of the 53th IEEE Conference on Decision and Control, Los Angeles, USA, Dec. 15–17, pp. 4521–4528 (2014)
5. Belleter, D.J.W., Pettersen, K.Y.: Underactuated leader-follower synchronisation for multi-agent systems with rejection of unknown disturbances. In: American Control Conference (ACC), Chicago, USA, pp. 3094–3100 (2015)
6. Breivik, M., Fossen, T.I.: Guidance laws for planar motion control. In: Proceedings of the 47th IEEE Conference on Decision and Control, Cancun, Mexico, pp. 570–577 (2008)
7. Breivik, M., Hovstein, V.E., Fossen, T.I.: Ship formation control: a guided leader-follower approach. In: Proceedings of the IFAC World Congress, Seoul, Korea (2008)
8. Breivik, M., Hovstein, V.E., Fossen, T.I.: Straight-line target tracking for unmanned surface vehicles. Model. Identif. Control **29**(4), 131–149 (2008)
9. Caharija, W., Pettersen, K.Y., Gravdahl, J.T.: Path following of marine surface vessels with saturated transverse actuators. In: American Control Conference (ACC), Washington, USA, pp. 546–553 (2013)
10. Dasdemir, J., Loría, A.: Robust formation-tracking control of mobile robots via one-to-one time-varying communication. Int. J. Control, pp. 1–17 (2014)
11. Desai, J.P., Ostrowski, J.P., Kumar, V.: Modeling and control of formations of nonholonomic mobile robots. IEEE Trans. Robot. Autom. **17**(6), 905–908 (2001)
12. Fossen, T.I.: Handbook of Marine Craft Hydrodynamics and Motion Control. Wiley, Chichester (2011)
13. Fredriksen, E, Pettersen, K.Y.: Global κ-exponential way-point manoeuvering of ships. In: 43rd IEEE Conference on Decision and Control, 2004. CDC, vol. 5, pp. 5360–5367. IEEE (2004)
14. Fredriksen, E., Pettersen, K.Y.: Global κ-exponential way-point maneuvering of ships: theory and experiments. Automatica **42**(4), 677–687 (2006)

15. Fu, S.H., Haddad, W.M.: Nonlinear adaptive tracking of surface vessels with exogenous disturbances. Asian J. Control **5**(1), 88–103 (2003)
16. Haddad, W.M., Nersesov, S.G., Hui, Q., Ghasemi, M.: Flocking and rendezvous control protocols for nonlinear dynamical systems via hybrid stabilization of sets. In: IEEE Proceedings of 52nd Conference on Decision and Control (CDC), 2013
17. Khalil, H.K.: Nonlinear Systems. Prentice Hall Inc, Upper Saddle River (2002)
18. Kyrkjebø, E.: Motion Coordination of Mechanical Systems. Ph.D. thesis, Norwegian University of Science and Technology (2007)
19. Kyrkjebø, E., Pettersen, K.Y., Wondergem, M., Nijmeijer, H.: Output synchronization control of ship replenishment operations: Theory and experiments. Control Eng. Prac. **15**(6), 741–755 (2007)
20. Loría, A., Panteley, E.: Cascaded nonlinear time-varying systems: analysis and design. In: Advanced Topics in Control Systems Theory, pp. 23–64. Springer, New York (2005)
21. Nijmeijer, H., Rodriguez-Angeles, A.: Synchronization of Mechanical Systems. World Scientific, London (2003)
22. Peng, Z., Wang, D., Chen, Z., Hu, X., Lan, W.: Adaptive dynamic surface control for formations of autonomous surface vehicles with uncertain dynamics. IEEE Trans. Control Syst. Technol. **21**(2), 513–520 (2013)
23. Poonawala, H.A., Satici, A.C., Spong, M.W.: Leader-follower formation control of nonholonomic wheeled mobile robots using only position measurements, istanbul, turkey. In: 9th Asian Control Conference (ASCC), pp. 1–6 (2013)
24. Skejic, R., Breivik, M., Fossen, T.I., Faltinsen, O.M.: Modeling and control of underway replenishment operations in calm water. In: 8th Manoeuvring and Control of Marine Craft, Guarujá, Brazil, pp. 78–85 (2009)

Chapter 9
Position Control via Force Feedback in the Port-Hamiltonian Framework

Mauricio Muñoz-Arias, Jacquelien M.A. Scherpen and Daniel A. Dirksz

Abstract In this chapter, position control strategies via force feedback are presented for standard mechanical systems in the port-Hamiltonian framework. The presented control strategies require a set of coordinate transformations, since force feedback in the port-Hamiltonian framework is not straightforward. With the coordinate transformations force feedback can be realized while preserving the port-Hamiltonian structure. The port-Hamiltonian formalism offers a modeling framework with a clear physical structure and other properties that can often be exploited for control design purposes, which is why we believe it is important to preserve the structure. The proposed control strategies offer an alternative solution to position control with more tuning freedom and exploit knowledge of the system dynamics.

9.1 Introduction

We are honored to write this chapter at the occasion of the 60th birthday of Henk Nijmeijer. I (the second author) know Henk as an influential and stimulating teacher during my study Applied Mathematics at the University of Twente more than 25 years ago. When I performed my traineeship and master thesis project under his supervision, he raised my interest in doing a Ph.D. in the field of systems and control. The corresponding research environment in Twente was open, international and inspiring, and Henk contributed significantly to that, being one of the leaders in the field of nonlinear control systems. In recent years, we started to collaborate and

M. Muñoz-Arias (✉) · J.M.A. Scherpen
Faculty of Mathematics and Natural Sciences, University of Groningen,
Nijenborgh 4, 9747 AG Groningen, The Netherlands
e-mail: m.munoz.arias@rug.nl

J.M.A. Scherpen
e-mail: j.m.a.scherpen@rug.nl

D.A. Dirksz
Irmato Group, Oliemolenstraat 5, 9203 Zn Drachten, The Netherlands
e-mail: ddirksz@irmato.com

N. van de Wouw et al. (eds.), *Nonlinear Systems*, Lecture Notes
in Control and Information Sciences 470, DOI 10.1007/978-3-319-30357-4_9

publish together. I wish for new collaborations and exchange of ideas in the future. Congratulations Henk!

The current technological advances continuously increase the demand for robots and intelligent systems that are fast, accurate, and able to perform tasks under different circumstances. Sensing and using force measurements are examples of how reliability and performance of such robotic systems can be improved for almost all tasks in which a manipulator comes in contact with external objects [3, 10, 24]. Position control with force feedback for robotic systems has been thoroughly discussed in [3, 10, 18, 19, 25] and the references therein for the Euler–Langrange (EL) framework. In the EL framework, control design is based on selecting a suitable storage function that ensures position control. However, the desired storage function under the EL framework does not qualify as an energy function in any physically meaningful sense as stated in [3, 19].

In this chapter, we present position control strategies via force feedback for standard mechanical systems in the PH framework. The port-Hamiltonian (PH) modeling framework of [13, 26] has received a considerable amount of interest in the last decade due to its insightful physical structure. Moreover, it is well known that a larger class of (nonlinear) physical systems can be described in the PH framework. The popularity of PH systems can be largely accredited to its application for analysis and control design of physical systems, as shown in [6–8, 20, 21, 26] and many others. Control laws in the PH framework are derived with a clear physical interpretation via direct shaping of the closed-loop energy, interconnection, and dissipation structure, see [6, 26]. In this chapter, we apply the PH modeling framework, since it allows extensions on the system coordinates, which facilitates the incorporation of force feedback in the input of the systems. Lastly, the presented control strategy preserves the PH structure, thus granting the aforementioned advantages to the closed-loop system.

The results presented in this chapter are based on [15], and extend the results presented in [16] and [17]. In [16] a class of standard mechanical systems in the PH framework with force feedback and zero external forces has been introduced, for mechanical systems with a constant mass-inertia matrix. However, applying the results from [16] to systems with a nonconstant mass-inertia matrix is not trivial. In [17] preliminary results are presented for the more general class of mechanical systems with a nonconstant mass-inertia matrix. In this chapter, we combine these previous results into a PH framework for position control with force feedback for standard mechanical systems.

The main contribution of this chapter is the introduction of an alternative position control strategy for mechanical systems that includes force feedback, in the PH framework. We present a control approach based on the *modeled* internal forces of a standard mechanical system; for this approach the system is extended with the internal forces into a PH system, which is then asymptotically stabilized. Furthermore, we analyze the disturbance attenuation properties to external forces, i.e., when the external forces are constant we show that the system has a constant steady-state error, and we apply an integral type control to compensate for position errors caused by these constant forces. We reformulate the stability analysis and analyze the robustness against external forces of the control strategy. The resulting controller has nicely tunable properties and interpretations, outperforming most of the existing force feedback control strategies. In addition, we develop a strategy assuming that we have force sensors that give measurements of the (real) total forces in the system, i.e., the internal plus external forces. Those measurements can be used to realize rejection of the external forces in the system.

The chapter is organized as follows. In Sect. 9.2.1, we provide a general background in the PH framework [6]. In Sect. 9.2.2, we apply the results of [27] to equivalently describe the original PH system in a PH form which has a constant mass-inertia matrix in the Hamiltonian via a change of coordinates. This coordinate transformation simplifies the extension of the results in [16] to systems with a nonconstant mass-inertia matrix. A PH model of a robot manipulator of two-DOF is introduced in order to show a mass-inertia decomposition case. Furthermore, in Sect. 9.2.3 we briefly recall the Hamilton–Jacobi inequality related to $\mathscr{L}_2$ analysis. In Sect. 9.2.4, we recap the constructive procedure of [14] to modify the Hamiltonian function of a forced PH system in order to generate Lyapunov functions for nonzero equilibria, i.e., a system in the presence of nonzero constant external forces. In Sect. 9.3, we realize a dynamic extension in order to include the modeled internal forces, while preserving the PH structure. In Sect. 9.4, we present the position control which uses feedback of the modeled forces. We also look at the disturbance attenuation properties when there are external forces, and we apply a type of integral control when the external forces are constant. For constant forces the system converges to a constant position different than the desired one, justifying the application of integral control. In Sect. 9.5, we assume that we have measurements of the total forces in the system, and use these measurements for control. Consequently, we show that we can realize rejection of the total forces in the system while preserving the PH structure. Finally, simulations are given in Sect. 9.6 to motivate our results for position control, and concluding remarks are provided in Sect. 9.7.

9.2 Preliminaries

This section provides the background for the main contributions presented in this chapter. We deal here with the analysis of physical systems described in the PH framework, canonical transformations, and stability analysis in the presence of a disturbance, and a constant force in the input of system.

9.2.1 Port-Hamiltonian Systems

We briefly recap the definition, properties and advantages of modeling and control with the PH formalism.

The PH framework is based on the description of systems in terms of energy variables, their interconnection structure, and power ports. PH systems include a large family of physical nonlinear systems. The transfer of energy between the physical system and the environment is given through energy elements, dissipation elements, and power preserving ports [6, 13], based on the study of Dirac structures.

A class of PH system, introduced in [13], is described by

$$\Sigma = \begin{cases} \dot{x} = [J(x) - R(x)] \dfrac{\partial H(x)}{\partial x} + g(x) w \\ \\ y = g(x)^{\top} \dfrac{\partial H(x)}{\partial x} \end{cases} \tag{9.1}$$

with $x \in \mathbb{R}^{\mathcal{N}}$ the states of the system, the skew-symmetric interconnection matrix $J(x) \in \mathbb{R}^{\mathcal{N} \times \mathcal{N}}$, the positive-semidefinite damping matrix $R(x) \in \mathbb{R}^{\mathcal{N} \times \mathcal{N}}$, and the Hamiltonian $H(x) \in \mathbb{R}$. The matrix $g(x) \in \mathbb{R}^{\mathcal{N} \times \mathcal{M}}$ weights the action of the control inputs $w \in \mathbb{R}^{\mathcal{M}}$ on the system, and w, $y \in \mathbb{R}^{\mathcal{M}}$ with $\mathcal{M} \leq \mathcal{N}$, form a power port pair. We now restrict the description to a class of standard mechanical systems.

Consider a class of standard mechanical systems of n-DOF as in (9.1), e.g., an n-DOF rigid robot manipulator. Consider furthermore the addition of an external force vector. The resulting system is then given by

$$\begin{bmatrix} \dot{q} \\ \dot{p} \end{bmatrix} = \begin{bmatrix} 0_{n \times n} & I_{n \times n} \\ -I_{n \times n} & -D(q, p) \end{bmatrix} \begin{bmatrix} \dfrac{\partial H(q, p)}{\partial q} \\ \dfrac{\partial H(q, p)}{\partial p} \end{bmatrix} + \begin{bmatrix} 0_{n \times n} \\ G(q) \end{bmatrix} u + \begin{bmatrix} 0_{n \times n} \\ B(q) \end{bmatrix} f_e \tag{9.2}$$

$$y = G(q)^{\top} \frac{\partial H(q, p)}{\partial p}, \tag{9.3}$$

with the vector of generalized configuration coordinates $q \in \mathbb{R}^n$, the vector of generalized momenta $p \in \mathbb{R}^n$, the identity matrix $I_{n \times n}$, the damping matrix $D(q, p) \in \mathbb{R}^{n \times n}$, $D(q, p) = D(q, p)^{\top} \geq 0$, $y \in \mathbb{R}^n$ the output vector, $u \in \mathbb{R}^n$ the input vector, $f_e \in \mathbb{R}^n$ the vector of external forces, $\mathcal{N} = 2n$, matrix $B(q) \in \mathbb{R}^{n \times n}$, and the input matrix $G(q) \in \mathbb{R}^{n \times n}$ everywhere invertible, i.e., the PH system is *fully actuated*. The Hamiltonian of the system is equal to the sum of kinetic and potential energy,

$$H(q, p) = \frac{1}{2} p^{\top} M^{-1}(q) p + V(q), \tag{9.4}$$

where $M(q) = M^{\top}(q) > 0$ is the $n \times n$ inertia (generalized mass) matrix and $V(q)$ is the potential energy.

We consider the PH system (9.2) as a class of standard mechanical systems with external forces.

Remark 9.1 The robot dynamics is given in joint space in (9.2), and here the *external forces* $f_e \in \mathbb{R}^n$ are introduced. The geometric Jacobian maps the external forces in the *work space*, F_e, to the (generalized) external forces in the joint space, f_e, [25]. In this chapter the following holds,

$$f_e = \mathscr{J}(q)^\top F_e, \qquad F_e \in \mathbb{R}^N, \tag{9.5}$$

and the geometric Jacobian is given by

$$\mathscr{J}(q) = \begin{bmatrix} \mathscr{J}_v(q) \\ \mathscr{J}_\omega(q) \end{bmatrix} \in \mathbb{R}^{6\times n}, \tag{9.6}$$

where $\mathscr{J}_v(q) \in \mathbb{R}^{3\times n}$, and $\mathscr{J}_\omega(q) \in \mathbb{R}^{3\times n}$ are the linear, and angular geometric Jacobians, respectively, and $N = \{3, 6\}$. If the Jacobian is full rank, we can always find $f_e \in \mathbb{R}^n$ that corresponds to F_e. Then, it is not a limitation to suppose $B(q) = I_n$. This separation between joint and work spaces is important here, because we control the robot by acting on the generalized coordinates q, i.e., in the joint space, but we grasp objects with the end-effector in the work space.

Example 9.1 Consider the system given by the two-DOF shoulder of the PERA, [23]. A picture of the PERA is shown in Fig. 9.1. A *Denavit–Hartenberg* representation of the PERA, see [25], is given in Fig. 9.2. The shoulder consists of a link actuated by two motors. The model of the shoulder consists of a mass m_s, a link length l_s, and a linear damping $d_s > 0$. The states of the system are $x = (q, p)^\top$, where $(q, p) \in \mathbb{R}^2$ are the generalized coordinates q_1, and q_2, and p_1, p_2 are the generalized momenta of the system. The system is described in the PH form by

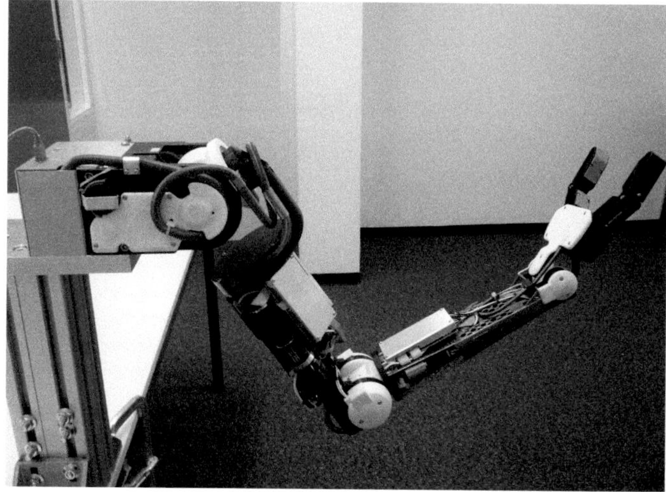

Fig. 9.1 PERA at the University of Groningen

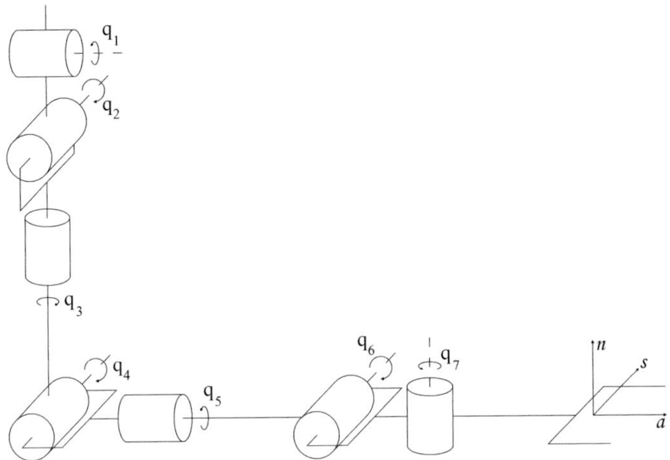

Fig. 9.2 *Denavit–Hartenberg* representation of the PERA [12]

$$\begin{bmatrix} \dot{q} \\ \dot{p} \end{bmatrix} = \begin{bmatrix} 0_{2\times 2} & I_{2\times 2} \\ -I_{2\times 2} & D(q,p) \end{bmatrix} \begin{bmatrix} \frac{\partial V(q)}{\partial q} \\ M(q)^{-1} p \end{bmatrix} + \begin{bmatrix} 0 \\ G \end{bmatrix} u_s + \begin{bmatrix} 0 \\ B \end{bmatrix} f_e \tag{9.7}$$

$$y_s = G^{\top} M(q)^{-1} p \tag{9.8}$$

with an input matrix $G = I_{2\times 2}$ (fully actuated), a vector of external forces $f_e \in \mathbb{R}^2$, an input–output port pair (u_s, y_s), Hamiltonian of the form

$$H(q,p) = \frac{1}{2} p^{\top} M(q)^{-1} p + V(q) \tag{9.9}$$

with $V(q)$ the potential energy, and a mass-inertia matrix $M(q) \in \mathbb{R}^{2\times 2}$, s.t., $M(q) = diag(a, b)$ where

$$a = m_s l_s^2 \cos(q_2)^2 + \mathscr{I}_1 + \mathscr{I}_2 \tag{9.10}$$

$$b = m_s l_s^2 + \mathscr{I}_2 \tag{9.11}$$

and with $\mathscr{I}_1$, and $\mathscr{I}_2$ the inertias of the joints. Furthermore, the gravity vector is

$$\frac{\partial V(q)}{\partial q} = \begin{bmatrix} g m_s l_s \cos(q_2) \sin(q_1) \\ g m_s l_s \sin(q_2) \cos(q_1) \end{bmatrix} \tag{9.12}$$

with g the gravitational acceleration. The shoulder is experiencing Coulomb friction that we have determined, and validated experimentally, [2, 12]. The dissipation matrix has the form

$$D(q, p) = D(\dot{q}) = diag\left(d_{s_1}(\dot{q}_1), d_{s_2}(\dot{q}_2)\right), \tag{9.13}$$

where $\dot{q} = M^{-1}(q)\,p$, and with

$$d_{s_i} = \left(F_{c_i} + \left(F_{s_i} - F_{c_i}\right)e^{|\dot{q}_i|\dot{q}_{s_i}^{-1}|}\right)\left(\alpha_{f_i} + \dot{q}_i^2\right)^{-0.5} + F_{v_i}\dot{q}_i, \tag{9.14}$$

where F_{c_i}, F_{s_i}, and F_{v_i} are the Coulomb, static, and viscous friction coefficients, respectively, and the Coulomb friction force is approximated as in [9] with positive (small) constants α_i, $\dot{q}_{s_i}$ is the constant due to the Stribeck velocity [1], and $i = 1, 2$. □

9.2.2 Canonical Transformations of Port-Hamiltonian Systems

We recap here the results of [7, 8] in terms of generalized coordinate transformations for PH systems, and we apply the results of [27] to equivalently describe the original PH system in a PH form which has a constant mass-inertia matrix in the Hamiltonian.

A generalized canonical transformation of [7] is applied in (9.1) via a set of transformations

$$\bar{x} = \Phi(x) \tag{9.15}$$

$$\bar{H}(\bar{x}) = H(x) + U(x) \tag{9.16}$$

$$\bar{y} = y + \alpha(x) \tag{9.17}$$

$$\bar{u} = u + \beta(x) \tag{9.18}$$

that changes the coordinates x into $\bar{x}$, the Hamiltonian H into $\bar{H}$, the output y into $\bar{y}$, and the input u into $\bar{u}$. It is said to be a generalized canonical transformation for PH systems if it transforms a PH system (9.1) into another one.

The class of generalized canonical transformations are characterized by the following theorems.

Theorem 9.1 *([8]) Consider the PH system (9.1). For any smooth scalar function $U(x) \in \mathbb{R}$, and any smooth vector function $\beta(x) \in \mathbb{R}^{\mathscr{M}}$, there exists a pair of smooth functions $\Phi(x) \in \mathbb{R}^{\mathscr{N}}$ and $\alpha(x) \in \mathbb{R}^{\mathscr{M}}$ such that the set of equations (9.15)–(9.18) yields a generalized canonical transformation. The function $\Phi(x)$ yields a generalized canonical transformation with $U(x)$ and $\beta(x)$ if and only if the partial differential equation (PDE)*

$$\frac{\partial \Phi}{\partial (x, t)} \left((J - R) \frac{\partial U}{\partial x}^{\top} + (K - S) \frac{\partial (H + U)}{\partial x}^{\top} + g\beta \right) = 0 \qquad (9.19)$$
$$-1$$

holds with a skew-symmetric matrix $K(x)$, *and a symmetric matrix* $S(x)$ *satisfying* $R(x) + S(x) \geq 0$. *We have left out the arguments of* $\Phi(x)$, $H(x)$, $J(x)$, $R(x)$, $S(x)$, $K(x)$, $U(x)$, $g(x)$, *and* $\beta(x)$, *for notational simplicity. Furthermore, the change of output* $\alpha(x)$, *and the matrices* $\bar{J}(\bar{x})$, $\bar{R}(\bar{x})$, *and* $\bar{g}(\bar{x})$, *are given by*

$$\alpha(x) = g(x)^{\top} \frac{\partial U(x)}{\partial x} \qquad (9.20)$$

$$\bar{J}(\bar{x}) = \frac{\partial \Phi(x)}{\partial x} (J(x) + K(x)) \frac{\partial \Phi(x)}{\partial x}^{\top} \qquad (9.21)$$

$$\bar{g}(\bar{x}) = \frac{\partial \Phi(x)}{\partial x} g(x) \qquad (9.22)$$

$$\bar{R}(\bar{x}) = \frac{\partial \Phi}{\partial x} (R(x) + S(x)) \frac{\partial \Phi(x)}{\partial x}^{\top}. \qquad (9.23)$$

Theorem 9.2 (*[8]*) *Consider the PH system described by (9.1) and transform it by the generalized canonical transformation with* $U(x)$ *and* $\beta(x)$ *such that* $H(x) + U(x) \geq 0$. *Then, the new input-output mapping* $\bar{u} \to \bar{y}$ *is passive with storage function* $\bar{H}(\bar{x})$ *if and only if*

$$\frac{\partial (H + U)}{\partial (x)}^{\top} \left((J - R) \frac{\partial U}{\partial x}^{\top} - S \frac{\partial (H + U)}{\partial x}^{\top} + g\beta \right) \geq 0. \qquad (9.24)$$
$$-1$$

Suppose that (9.19) *holds, that* $H(x) + U(x)$ *is positive-definite and that the system is zero-state detectable. Then, the feedback* $u = -\beta(x) - \mathcal{C}(x)(y + \alpha(x))$ *with* $\mathcal{C}(x) \geq \epsilon I > 0$ *renders the system asymptotically stable. Suppose moreover that* $H + U$ *is decrescent and that the transformed system is periodic, then, the feedback renders the system uniformly asymptotically stable.*

Consider a class of standard mechanical systems (9.2) in the PH framework with a nonconstant mass-inertia matrix $M(q)$. The aim of this section is to transform the original system (9.2) into a PH formulation with a constant mass-inertia matrix via a generalized canonical transformation [7]. The presented change of variables to deal with a nonconstant mass- inertia matrix has first been proposed in [27].

Consider the system (9.1) with nonconstant $M(q)$, and a coordinate transformation as

$$\bar{x} = \Phi(x) = \Phi(q, p) \triangleq \begin{pmatrix} \bar{q} \\ \bar{p} \end{pmatrix} = \begin{pmatrix} q - q_d \\ T(q)^{-1} p \end{pmatrix} = \begin{pmatrix} q - q_d \\ T(q)^{\top} \dot{q} \end{pmatrix} \qquad (9.25)$$

with a constant desired position $q_d \in \mathbb{R}^n$, and where $T(q)$ is a lower triangular matrix such that

$$T(q) = T\left(\Phi^{-1}(q, p)\right) = \bar{T}(\bar{q}) \tag{9.26}$$

and

$$M(q) = T(q)\,T(q)^\top = \bar{T}(\bar{q})\,\bar{T}(\bar{q})^\top . \tag{9.27}$$

Consider now the Hamiltonian $H(q, p)$ as in (9.4), and using (9.25), we realize $\bar{H}(\bar{x}) = H\left(\Phi^{-1}(\bar{x})\right)$ and $\bar{V}(\bar{q}) = V\left(\Phi^{-1}(\bar{q})\right)$ as

$$\bar{H}(\bar{x}) = \frac{1}{2}\bar{p}^\top \bar{p} + \bar{V}(\bar{q}) . \tag{9.28}$$

The new form of the interconnection and damping matrices of the PH system are realized via the coordinate transformation (9.25), the mass-inertia matrix decomposition (9.27), and the new Hamiltonian (9.28), [26]. The resulting PH system is then given by

$$\begin{bmatrix} \dot{\bar{q}} \\ \dot{\bar{p}} \end{bmatrix} = \begin{bmatrix} 0_{n\times n} & \bar{T}(\bar{q})^{-\top} \\ -\bar{T}(\bar{q})^{-1} & \bar{J}_2(\bar{q}, \bar{p}) - \bar{D}(\bar{q}, \bar{p}) \end{bmatrix} \begin{bmatrix} \dfrac{\partial \bar{H}(\bar{q}, \bar{p})}{\partial \bar{q}} \\ \dfrac{\partial \bar{H}(\bar{q}, \bar{p})}{\partial \bar{p}} \end{bmatrix} + \begin{bmatrix} 0_{n\times n} \\ \bar{G}(\bar{q}) \end{bmatrix} v + \begin{bmatrix} 0_{n\times n} \\ \bar{B}(\bar{q}) \end{bmatrix} f_e \tag{9.29}$$

$$\bar{y} = \bar{G}(\bar{q})^\top \frac{\partial \bar{H}(\bar{q}, \bar{p})}{\partial \bar{p}} \tag{9.30}$$

with a new input $u = v \in \mathbb{R}^n$, and where the skew-symmetric matrix $\bar{J}_2(\bar{q}, \bar{p})$ takes the form

$$\bar{J}_2(\bar{q}, \bar{p}) = \frac{\partial\left(\bar{T}(\bar{q})^{-1}\bar{p}\right)}{\partial \bar{q}} \bar{T}(\bar{q})^{-\top} - \bar{T}(\bar{q})^{-1} \frac{\partial\left(\bar{T}(\bar{q})^{-1}\bar{p}\right)^\top}{\partial \bar{q}} \tag{9.31}$$

with

$$(q, p) = \Phi^{-1}(\bar{q}, \bar{p}) \tag{9.32}$$

together with the matrix $\bar{D}(\bar{q}, \bar{p})$, and the input matrices $\bar{G}(\bar{q})$, and $\bar{B}(\bar{q})$, are described by

$$\bar{D}(\bar{q},\bar{p}) = \bar{T}(\bar{q})^{-1} D\left(\Phi^{-1}(\bar{q},\bar{p})\right) \bar{T}(\bar{q})^{-\top}, \tag{9.33}$$

$$\bar{G}(\bar{q}) = \bar{T}(\bar{q})^{-1} G(\bar{q}), \tag{9.34}$$

$$\bar{B}(\bar{q}) = \bar{T}(\bar{q})^{-1} B(\bar{q}), \tag{9.35}$$

respectively. Via the transformation (9.25), we then obtain a class of mechanical systems with a constant (identity) mass-inertia matrix in the Hamiltonian function as in (9.28), which equivalently describes the original system (9.2) with nonconstant mass-inertia matrix.

Example 9.2 Consider the robot manipulator of Example 9.1. Given the mass-inertia matrix $M(q) = diag(a, b)$ with a, and b as in (9.10) and (9.11), respectively, we compute a $T(q)$ as in (9.27), s.t.,

$$T(q) = \begin{bmatrix} \sqrt{a} & 0 \\ 0 & \sqrt{b} \end{bmatrix} = \begin{bmatrix} \sqrt{m_s l_s^2 \cos(q_2)^2 + \mathscr{I}_1 + \mathscr{I}_2} & 0 \\ 0 & \sqrt{m_s l_s^2 + \mathscr{I}_2} \end{bmatrix} \tag{9.36}$$

with $\mathscr{I}_1$, and $\mathscr{I}_2$ the inertias of the joints, and m_s, l_s as the mass and the length of the shoulder of the robot, respectively. Based on $T(q)$, we can compute the matrices $\bar{J}_2(\bar{q},\bar{p})$, $\bar{D}(\bar{q},\bar{p})$, $\bar{G}(\bar{q})$, and $\bar{B}(\bar{q})$, as in (9.31), (9.33)–(9.35).

The coordinate transformation of this section is used in the rest of this chapter in order to deal with nonconstant mass-inertia matrices.

9.2.3 Hamilton–Jacobi Inequality

In order to show the usefulness of some results on position control with force feedback presented later, we apply the Hamilton–Jacobi inequality useful for $\mathscr{L}_2$ gain analysis of nonlinear systems [26]. Toward this end we analyze the $\mathscr{L}_2$-gain of a closed-loop system w.r.t. an $\mathscr{L}_2$ disturbance δ.

Consider the time-invariant nonlinear system

$$\begin{aligned} \dot{\hat{x}} &= \mathscr{F}\left(\hat{x}\right) + \tilde{G}\left(\hat{x}\right)\delta \\ \hat{y} &= h\left(\hat{x}\right) \end{aligned} \tag{9.37}$$

with states $\hat{x}$, input disturbance δ, output $\hat{y}$ and continuously differentiable vector functions $\mathscr{F}\left(\hat{x}\right)$, $\tilde{G}\left(\hat{x}\right)$ and $h\left(\hat{x}\right)$. Let γ be a positive constant, then the $\mathscr{L}_2$-gain bound is found if for a γ there exists a continuously differentiable, positive-semidefinite function $\mathscr{W}\left(\hat{x}\right)$ that satisfies the Hamilton–Jacobi inequality (HJI)

$$\left(\frac{\partial \mathscr{W}(\hat{x})}{\partial \hat{x}}\right)^{\top} \mathscr{F}(\hat{x}) + \frac{1}{2}\frac{1}{\gamma^2}\left(\frac{\partial \mathscr{W}(\hat{x})}{\partial \hat{x}}\right)^{\top} \tilde{G}(\hat{x})\,\tilde{G}(\hat{x})^{\top} \frac{\partial \mathscr{W}(\hat{x})}{\partial \hat{x}} + \frac{1}{2}h(\hat{x})^{\top} h(\hat{x}) \leq 0 \tag{9.38}$$

for $\hat{x} \in \mathbb{R}^{\mathscr{N}}$. The system (9.37) is then finite-gain $\mathscr{L}_2$ stable and its gain is less than or equal to γ.

9.2.4 Stability Analysis for Constant External Forces

Consider a class of PH systems as described by (9.1). We now briefly recall the procedure of [14], i.e., we analyze the stability of the system (9.1) for a constant, and nonzero, input $w = \bar{u} \in \mathbb{R}^{\mathscr{M}}$, leading to a forced equilibrium $\breve{x} \in \mathbb{R}^{\mathscr{N}}$. The forced equilibria $\breve{x}$ are solutions of

$$\left[J(\breve{x}) - R(\breve{x})\right] \frac{\partial H}{\partial x}(\breve{x}) + g(\breve{x})\,\bar{u} = 0 \tag{9.39}$$

and if $[J(x) - R(x)]$ is invertible for every $x \in \mathbb{R}^{\mathscr{N}}$, the unique solution of (9.39) is $\dfrac{\partial H}{\partial x}(x) = \mathscr{K}(x)\,\bar{u}$ where

$$\mathscr{K}(x) = -\left[J(x) - R(x)\right]^{-1} g(x). \tag{9.40}$$

Based on (9.40), we define the matrices

$$J_s(x) \triangleq \mathscr{K}^{\top}(x)\, J(x)\, \mathscr{K}(x) \tag{9.41}$$

and

$$R_s(x) \triangleq \mathscr{K}^{\top}(x)\, R(x)\, \mathscr{K}(x) \tag{9.42}$$

which we use below to find the embedded Hamiltonian system. Clearly, $J_s(x)$ and $R_s(x)$ satisfy $J_s(x) = -J_s^{\top}(x)$, and $R_s(x) = R_s^{\top}(x) \geq 0$, respectively. Let us now consider the following PH system

$$\begin{bmatrix} \dot{x} \\ \dot{\zeta} \end{bmatrix} = \left[J_a(x) - R_a(x)\right] \begin{bmatrix} \dfrac{\partial H_a(x)}{\partial x} \\ \dfrac{\partial H_a(x)}{\partial \zeta} \end{bmatrix} \tag{9.43}$$

on the augmented state space $(x, \zeta) \in \mathbb{R}^{\mathscr{N}} \times \mathbb{R}^{\mathscr{M}}$, endowed with the structure matrices

$$J_a(x) = \begin{bmatrix} J(x) & J(x)\mathscr{K}(x) \\ -(J(x)\mathscr{K}(x))^\top & J_s(x) \end{bmatrix} \tag{9.44}$$

$$R_a(x) = \begin{bmatrix} R(x) & R(x)\mathscr{K}(x) \\ (R(x)\mathscr{K}(x))^\top & R_s(x) \end{bmatrix} \tag{9.45}$$

with $\mathscr{K}(x)$, $J_s(x)$, and $R_s(x)$ as in (9.40)–(9.42), respectively, and with an augmented Hamiltonian

$$H_a(x, \zeta) \triangleq H(x) + H_s(\zeta), \quad H_s(\zeta) \triangleq -\bar{u}^\top \zeta. \tag{9.46}$$

Theorem 9.3 ([14]) *Consider, a class of PH systems (9.1) with a constant input $w = \bar{u}$, and the matrix $[J(x) - R(x)]$ invertible for every $x \in \mathbb{R}^{\mathscr{N}}$. Define $\mathscr{K}(x)$ by (9.40), and assume the functions $\mathscr{K}_{ij}$ to satisfy*

$$\frac{\partial \mathscr{K}_{ij}}{\partial x_k} = \frac{\partial \mathscr{K}_{kj}}{\partial x_i}, \; i, k \in \bar{n} \triangleq \{1, \ldots, \mathscr{N}\}, \; j \in \bar{m} \triangleq \{1, \ldots, \mathscr{M}\}. \tag{9.47}$$

Also, assume that there exist locally smooth functions $\mathscr{C}_j : \mathbb{R}^{\mathscr{N}} \to \mathbb{R}$, called Casimirs [14], satisfying

$$\mathscr{K}_{ij}(x) = \frac{\partial \mathscr{C}_j}{\partial x_i}(x), \quad j \in \bar{m}, \quad i \in \bar{n} \tag{9.48}$$

and $\zeta_j = \mathscr{C}_j(x) + c_j$, where $c_1, \ldots, c_{\mathscr{M}}$ depend on the initial conditions of $\zeta(t)$ in (9.43). Then, the dynamics of (9.1) with input $u = \bar{u}$ is asymptotically stable at the equilibrium point $\check{x}$ fulfilling (9.39), and it can be alternatively represented by

$$\dot{x} = [J(x) - R(x)] \frac{\partial H_r}{\partial x}(x) \tag{9.49}$$

where

$$H_r(x) \triangleq H(x) - \sum_{j=1}^{\mathscr{M}} \bar{u}_j \zeta_j \tag{9.50}$$

and H_r qualifies as a Lyapunov function for the forced dynamics (9.49).

Remark 9.2 The $\mathscr{L}_2$-gain analysis of Sect. 9.2.3 gives a bound on the relation between an input δ and a output $\hat{y}$ as in (9.37) of a proposed closed-loop system for a $\mathscr{L}_2$-input disturbance δ. The $\mathscr{L}_2$-gain analysis differs from Theorem 9.3 in the sense that the $\mathscr{L}_2$-gain analysis is related to the output $\hat{y}$ while the analysis in this section is for the case where the system is asymptotically stable, i.e., the system (9.29) has a new equilibrium point caused by a constant f_e.

9.3 Force Feedback via Dynamic Extension

In this section, a force feedback strategy is introduced for a mechanical system in the PH framework. The force feedback is included to bring robustness and better tunable properties in the position control strategy. In comparison with force feedback in the EL framework [3, 24], the force feedback in the PH framework has nicely interpretable control strategies, as well as cleaner tuning opportunities that grant a better performance. The force feedback is achieved via a dynamic extension and a change of variables that introduces a new state for the PH system (9.29). The dynamics of the new state is realized such that it depends on the internal forces of the mechanical system. The internal forces are given by a set of kinetic, potential, and energy dissipating elements. The dynamic extension is realized such that the extended system also has a PH structure. The present work is inspired by the results of [4, 5, 22], which treat position feedback.

Denote the internal forces on the system (9.2) by $f_{in}(q, p)$, i.e.,

$$f_{in}(q, p) = -\frac{\partial H(q, p)}{\partial q} - D(q, p)\frac{\partial H(q, p)}{\partial p} \tag{9.51}$$

with $H(q, p)$ as in (9.4). Define a new state $z \in \mathbb{R}^n$ with dynamics depending on the internal forces $f_{in}(q, p)$, such that,

$$\dot{z} = Y^\top T(q)^{-1} f_{in}(q, p) \tag{9.52}$$

with Y a constant matrix, to be defined later on. Consider now the coordinate transformation

$$\hat{p} = \bar{p} - Az \tag{9.53}$$

with $\bar{p}$ defined in (9.25), and with A a constant matrix that we use later to tune our controller. Furthermore, we can define for system (9.29) the control input

$$v = \bar{G}(\bar{q})^{-1} A\dot{z} + \bar{v} \tag{9.54}$$

where $\bar{v}$ is a new input, which realizes an extended PH system with states $\hat{p}$ and z, i.e.,

$$\begin{bmatrix} \dot{\bar{q}} \\ \dot{\hat{p}} \\ \dot{z} \end{bmatrix} = \underbrace{\begin{bmatrix} 0_{n\times n} & \bar{T}^{-\top} & \bar{T}^{-\top}Y \\ -\bar{T}^{-1} & \bar{J}_2 - \bar{D} & \left(\bar{J}_2 - \bar{D}\right)Y \\ -Y^\top\bar{T}^{-1} & -Y^\top\left(\bar{J}_2^\top + \bar{D}\right) & -Y^\top\left(\bar{J}_2^\top + \bar{D}\right)Y \end{bmatrix}}_{\hat{J}(\bar{q},\hat{p},z)-\hat{R}(\bar{q},\hat{p},z)} \begin{bmatrix} \dfrac{\partial \hat{H}(\bar{q}, \hat{p}, z)}{\partial \bar{q}} \\ \dfrac{\partial \hat{H}(\bar{q}, \hat{p}, z)}{\partial \hat{p}} \\ \dfrac{\partial \hat{H}(\bar{q}, \hat{p}, z)}{\partial z} \end{bmatrix}$$

$$+\begin{bmatrix} 0_{n\times n} \\ \bar{G}(\bar{q}) \\ 0_{n\times n} \end{bmatrix} \bar{v} + \begin{bmatrix} 0_{n\times n} \\ \bar{B}(\bar{q}) \\ 0_{n\times n} \end{bmatrix} f_e \tag{9.55}$$

$$\hat{y} = \bar{G}(\bar{q})^\top \frac{\partial \hat{H}(\bar{q}, \hat{p}, z)}{\partial \hat{p}} \tag{9.56}$$

with Hamiltonian

$$\hat{H}(\bar{q}, \hat{p}, z) = \frac{1}{2}\hat{p}^\top \hat{p} + \frac{1}{2} z^\top K_z^{-1} z + \bar{V}(\bar{q}) \tag{9.57}$$

where $K_z > 0$, and $Y = AK_z$. In (9.55) the arguments of $T(\bar{q})$, $\bar{J}_2(\bar{q}, \hat{p})$, and $\bar{D}(\bar{q}, \hat{p})$, are left out for notational simplicity.

Remark 9.3 Although in (9.55) the $\dot{z}$ dynamics are described in terms of $\bar{J}_2(\bar{q}, \hat{p})$, $\bar{D}(\bar{q}, \hat{p})$, and $\hat{H}(\bar{q}, \hat{p}, z)$, they are still the same as described by (9.52) with (9.51), in the new coordinates (9.25).

It can be verified that system (9.55) is PH, since

$$\hat{J}(\bar{q}, \hat{p}) = \begin{bmatrix} 0_{n\times n} & \bar{T}(\bar{q})^{-\top} & \bar{T}(\bar{q})^{-\top} Y \\ -\bar{T}(\bar{q})^{-1} & \bar{J}_2(\bar{q}, \hat{p}) & \bar{J}_2(\bar{q}, \hat{p}) Y \\ -Y^\top \bar{T}(\bar{q})^{-1} & -Y^\top \bar{J}_2(\bar{q}, \hat{p})^\top & -Y^\top \bar{J}_2(\bar{q}, \hat{p})^\top Y \end{bmatrix} \tag{9.58}$$

is skew-symmetric, while

$$\hat{R}(\bar{q}, \hat{p}) = \begin{bmatrix} 0_{n\times n} & 0_{n\times n} & 0_{n\times n} \\ 0_{n\times n} & \bar{D}(\bar{q}, \hat{p}) & \bar{D}(\bar{q}, \hat{p}) Y \\ 0_{n\times n} & Y^\top \bar{D}(\bar{q}, \hat{p}) & Y^\top \bar{D}(\bar{q}, \hat{p}) Y \end{bmatrix} \tag{9.59}$$

can be shown to be positive-semidefinite via the Schur complement. Notice that by extending the dynamics of (9.29) with the internal forces $\dot{z}$ in the input (9.54), we include force feedback and preserve the PH structure.

Remark 9.4 In [16] we present results for the case when the mass-inertia matrix is constant. The case for a constant M does not require the coordinate transformation (9.25), and system (9.55) is then described by $T = I$, $\bar{J}_2 = 0$, $\bar{D} = D$, $\bar{G} = G$, $\bar{B} = B$, $Y = M^{-1}AK_z$ and Hamiltonian

$$\hat{H}_c = \frac{1}{2}\hat{p}^\top M^{-1} \hat{p} + \frac{1}{2} z^\top K_z z + \bar{V}(\bar{q}) \tag{9.60}$$

instead of (9.57).

In this section, we have realized an extended mechanical system that includes force feedback and preserves the PH structure. In the next section, we deal in more detail with position control and stability analysis.

9.4 Position Control with Modeled Internal Forces

In this section, a position control strategy with force feedback is introduced. We feed back the modeled internal forces, and the resulting system preserves the PH structure. The control laws here presented are better tunable and more insightful solutions in comparison with the solutions given in the EL framework [3, 19].

9.4.1 Position Control with Zero External Forces

In this section, energy-shaping [11, 19, 26] and damping injection are combined with force feedback (of modeled forces) to realize position control.

Theorem 9.4 *Consider system (9.55) and assume* $f_e = 0$*. Then, the control input*

$$v = \bar{G}(\bar{q})^{-1}\left(\frac{\partial \bar{V}(\bar{q})}{\partial \bar{q}} - K_p(\bar{q} - q_d)\right) - C\hat{y} \tag{9.61}$$

with $K_p > 0$*,* $C > 0$*, and* q_d *being the desired constant position, asymptotically stabilizes the extended system* (9.55) *at* $(\bar{q}, \hat{p}, z) = (q_d, 0, 0)$.

Proof This is a well-known result, see [26], but we repeat the proof here for notational reasons and for ease of reading. The control input (9.61) applied to system (9.55) with $f_e = 0$ realizes the closed-loop system described by

$$\begin{bmatrix} \dot{\bar{q}} \\ \dot{\hat{p}} \\ \dot{z} \end{bmatrix} = \underbrace{\begin{bmatrix} 0_{n\times n} & \bar{T}^{-\top} & \bar{T}^{-\top}Y \\ -\bar{T}^{-1} & \bar{J}_2 - \bar{D} - \bar{G}C\bar{G}^\top & \left(\bar{J}_2 - \bar{D}\right)Y \\ -Y^\top\bar{T}^{-1} & -Y^\top\left(\bar{J}_2^\top + \bar{D}\right) & -Y^\top\left(\bar{J}_2^\top + \bar{D}\right)Y \end{bmatrix}}_{\mathscr{F}(\hat{x})} \begin{bmatrix} \dfrac{\partial \hat{H}_d}{\partial \bar{q}} \\ \dfrac{\partial \hat{H}_d}{\partial \hat{p}} \\ \dfrac{\partial \hat{H}_d}{\partial z} \end{bmatrix} \tag{9.62}$$

$$\hat{y} = \bar{G}^\top \frac{\partial \hat{H}_d}{\partial \hat{p}} \tag{9.63}$$

with Hamiltonian

$$\hat{H}_d = \frac{1}{2}\hat{p}^\top\hat{p} + \frac{1}{2}(\bar{q} - q_d)^\top K_p(\bar{q} - q_d) + \frac{1}{2}z^\top K_z^{-1}z, \tag{9.64}$$

where the arguments of $\hat{H}_d(\bar{q}, \hat{p}, z)$, $T(\bar{q})$, $\bar{J}_2(\bar{q}, \hat{p})$, $\bar{D}(\bar{q}, \hat{p})$, $\bar{G}(\bar{q})$, and $\bar{B}(\bar{q})$ are left out for simplicity. Take (9.64) as candidate Lyapunov function, which then gives

$$\dot{\hat{H}}_d = -\begin{bmatrix} \frac{\partial \hat{H}_d}{\partial \hat{p}} \\ \frac{\partial \hat{H}_d}{\partial z} \end{bmatrix}^\top \underbrace{\begin{bmatrix} \bar{D} + \bar{G}C\bar{G}^\top & -\bar{D}Y \\ -Y^\top \bar{D} & Y^\top \bar{D} Y \end{bmatrix}}_{K} \begin{bmatrix} \frac{\partial \hat{H}_d}{\partial \hat{p}} \\ \frac{\partial \hat{H}_d}{\partial z} \end{bmatrix}. \tag{9.65}$$

Since $\bar{G}(\bar{q})$ is full rank and $C > 0$, via the Schur complement it can be shown that matrix K in (9.65) is positive definite. Subsequently, via LaSalle's invariance principle we can prove that that the closed-loop system (9.62) is asymptotically stable in $\bar{q} = q_d$. □

Substituting v in (9.54) by (9.61) then gives the total control input u for the original system (9.2), which in terms of the original coordinates (q, p) becomes

$$u = G(q)^{-1}T(q)\left(A\dot{z} + \frac{\partial V(q)}{\partial q} - K_p(q - q_d)\right) - CG(q)^\top \left(M(q)^{-1}p - T(q)^{-\top}Az\right) \tag{9.66}$$

with $\dot{z}$ as in (9.51). The above results correspond to the case when the external forces on the system are zero, i.e., $f_e = 0$. In the next subsection we look more in detail at the case when $f_e \neq 0$.

9.4.2 Disturbance Attenuation Properties

We now show the advantages of the proposed extended system with force feedback for disturbance attenuation to unknown external forces. The closed-loop PH system (9.62) with force feedback is asymptotically stable in the desired position q_d when it has zero forces exerted from the environment, i.e., $f_e = 0$. To look at the effect of f_e being different from zero, we analyze the $\mathscr{L}_2$-gain w.r.t. an $\mathscr{L}_2$ disturbance f_e, [26]. It follows that

Theorem 9.5 *Consider a closed-loop system (9.62), an $\mathscr{L}_2$ disturbance f_e, and a constant matrix C with $\lambda_c \in \mathbb{R}^n$ being its set of eigenvalues. We then obtain a disturbance attenuation of f_e when the following conditions hold:*

$$\Gamma_1(q, p) = -D(q, p) + G(q)^\top \left(-C + \frac{1}{2}I_{n\times n}\right) G(q) + \frac{1}{2}\frac{1}{\gamma^2} B(q)\, B(q)^\top \frac{1}{2} \leq 0 \tag{9.67}$$

$$\Gamma_2(q) = AT(q)^{-\top} G(q)^\top \left(-C + \frac{1}{2}I_{n\times n}\right) G(q)^\top T(q)^{-1} A + \frac{1}{2}\frac{1}{\gamma^2} AT(q)^{-1} B(q)\, B(q)^\top T(q)^{-\top} A \leq 0 \tag{9.68}$$

$$\Gamma_3(q) = \frac{1}{2}\frac{1}{\gamma^2} AT(q)^{-1} B(q)\, B(q)^\top \geq 0 \tag{9.69}$$

$$\lambda_c \geq \frac{1}{2} \tag{9.70}$$

with γ being a positive constant.

Proof Consider the closed-loop system (9.62), but with $f_e \neq 0$, i.e.,

$$\begin{bmatrix} \dot{\bar{q}} \\ \dot{\hat{p}} \\ \dot{z} \end{bmatrix} = \begin{bmatrix} 0_{n\times n} & \bar{T}^{-\top} & \bar{T}^{-\top} Y \\ -\bar{T}^{-1} & \bar{J}_2 - \bar{D} - \bar{G} C \bar{G}^\top & \left(\bar{J}_2 - \bar{D}\right) Y \\ -Y^\top \bar{T}^{-1} & -Y^\top \left(\bar{J}_2^\top + \bar{D}\right) & -Y^\top \left(\bar{J}_2^\top + \bar{D}\right) Y \end{bmatrix} \begin{bmatrix} \frac{\partial \hat{H}_d}{\partial \bar{q}} \\ \frac{\partial \hat{H}_d}{\partial \hat{p}} \\ \frac{\partial \hat{H}_d}{\partial z} \end{bmatrix} + \begin{bmatrix} 0 \\ \bar{B} \\ 0 \end{bmatrix} f_e \tag{9.71}$$

$$\hat{y} = \bar{G}^\top \frac{\partial \hat{H}_d}{\partial \hat{p}}, \tag{9.72}$$

where the arguments of $\hat{H}_d(\bar{q}, \hat{p}, z)$, $T(\bar{q})$, $\bar{J}_2(\bar{q}, \hat{p})$, $\bar{D}(\bar{q}, \hat{p})$, $\bar{G}(\bar{q})$, and $\bar{B}(\bar{q})$ are left out for notational simplicity. We analyze the HJI (9.38) first for system (9.71) with $\mathscr{W}\left(\hat{x}\right) = \hat{H}_d\left(\hat{x}\right)$ to determine if this could be a solution. Given $\delta = f_e$ we obtain

$$-\left(\frac{\partial \hat{H}_d}{\partial \hat{x}}\right)^\top \tilde{R} \frac{\partial \hat{H}_d}{\partial \hat{x}} + \frac{1}{2}\frac{1}{\gamma^2}\left(\frac{\partial \hat{H}_d}{\partial \hat{p}}\right)^\top \bar{B}\bar{B}^\top \frac{\partial \hat{H}_d}{\partial \hat{p}} + \frac{1}{2}\hat{y}^\top \hat{y} \leq 0 \tag{9.73}$$

with $\hat{x} = (\bar{q}, \hat{p}, z)$, and

$$\tilde{R}\left(\hat{x}\right) = \begin{bmatrix} 0_{n\times n} & 0_{n\times n} & 0_{n\times n} \\ 0_{n\times n} & \bar{D}\left(\bar{q}, \hat{p}\right) + \bar{G}\left(\bar{q}\right) C \bar{G}\left(\bar{q}\right)^\top & \bar{D}\left(\bar{q}, \hat{p}\right) Y \\ 0_{n\times n} & Y^\top \bar{D}\left(\bar{q}, \hat{p}\right) & Y^\top \bar{D}\left(\bar{q}, \hat{p}\right) Y \end{bmatrix}. \tag{9.74}$$

We compute the left-hand side term of the Hamilton–Jacobi inequality (9.38) based on the function $W\left(\hat{x}\right) = \hat{H}_d\left(\hat{x}\right)$ with $\hat{H}_d\left(\hat{x}\right)$ as in (9.64), and on the function $\mathscr{F}\left(\hat{x}\right)$ of the closed-loop (9.62). Consequently, we obtain

$$\frac{\partial \mathscr{W}}{\partial \hat{x}}^\top \mathscr{F} = \begin{bmatrix} \frac{\partial \hat{H}_d}{\partial \bar{q}} \\ \frac{\partial \hat{H}_d}{\partial \hat{p}} \\ \frac{\partial \hat{H}_d}{\partial z} \end{bmatrix}^\top \begin{bmatrix} 0 & \bar{T}^{-\top} & \bar{T}^{-\top} Y \\ -\bar{T}^{-1} & \bar{J}_2 - \bar{D} - \bar{G} C \bar{G}^\top & \left(\bar{J}_2 - \bar{D}\right) Y \\ -Y^\top \bar{T}^{-1} & -Y^\top \left(\bar{J}_2^\top + \bar{D}\right) & -Y^\top \left(\bar{J}_2^\top + \bar{D}\right) Y \end{bmatrix} \begin{bmatrix} \frac{\partial \hat{H}_d}{\partial \bar{q}} \\ \frac{\partial \hat{H}_d}{\partial \hat{p}} \\ \frac{\partial \hat{H}_d}{\partial z} \end{bmatrix}$$

$$= -\left(\frac{\partial \hat{H}_d}{\partial \hat{p}} + Y\frac{\partial \hat{H}_d}{\partial z}\right)^\top \bar{D}\left(\frac{\partial \hat{H}_d}{\partial \hat{p}} + Y\frac{\partial \hat{H}_d}{\partial z}\right) - \frac{\partial \hat{H}_d}{\partial \hat{p}}^\top \bar{G}C\bar{G}^\top \frac{\partial \hat{H}_d}{\partial \hat{p}}, \tag{9.75}$$

where we have left out the arguments of $\mathscr{W}(\hat{x})$, $\mathscr{F}(\hat{x})$, $\bar{G}(\bar{q})$, $\bar{T}(\bar{q})$, $\hat{H}_d(\bar{q}, \hat{p}, z)$, and $\bar{D}(\bar{q}, \hat{p})$, for notational simplicity. From $\hat{y}$ as in (9.63), $\hat{p}$ as in (9.53), $\bar{D}(\bar{q}, \hat{p})$ as in (9.33), $\hat{x} = (\bar{q}, \hat{p}, z)$, and $Y = AK_z$, we rewrite (9.75) as

$$\begin{aligned}\frac{\partial \mathscr{W}(\hat{x})}{\partial \hat{x}}^\top \mathscr{F}(\hat{x}) &= -\left(\hat{p} + YK_z^{-1}z\right)^\top \bar{D}(\bar{q}, \hat{p})\left(\hat{p} + YK_z^{-1}z\right) - \hat{y}^\top C\hat{y} \\ &= -\frac{\partial H(q,p)}{\partial p}^\top D(q,p)\frac{\partial H(q,p)}{\partial p} - \hat{y}^\top C\hat{y}.\end{aligned} \tag{9.76}$$

Based on a input matrix $\tilde{G}(\bar{q})$ defined as

$$\tilde{G}(\hat{x}) = \begin{bmatrix} 0_{n\times n} \\ \bar{B}(\bar{q}) \\ 0_{n\times n}\end{bmatrix} \tag{9.77}$$

with $\bar{B}(\bar{q})$ as in (9.35), we compute the second term of the left-hand side of the Hamilton–Jacobi inequality (9.38) as

$$\begin{aligned}\tilde{Z}(\hat{x}) &= \frac{1}{2}\frac{1}{\gamma^2}\left(\frac{\partial \mathscr{W}(\hat{x})}{\partial \hat{x}}\right)^\top \tilde{G}(\hat{x})\tilde{G}^\top(\hat{x})\frac{\partial \mathscr{W}(\hat{x})}{\partial \hat{x}} \\ &= \frac{1}{2}\frac{1}{\gamma^2}\hat{p}^\top \bar{B}(\bar{q})\bar{B}(\bar{q})^\top \hat{p}\end{aligned} \tag{9.78}$$

and we now substitute $\hat{p}$ as in (9.53) in (9.78). Hence, we obtain

$$\begin{aligned}\tilde{Z}(\hat{x}) &= \frac{1}{2}\frac{1}{\gamma^2}(\bar{p} - Az)^\top \bar{B}(\bar{q})\bar{B}(\bar{q})^\top(\bar{p} - Az) \\ &= \frac{1}{2}\frac{1}{\gamma^2}\left(\Upsilon(q,p)^\top \Upsilon(q,p) - \Upsilon(q,p)^\top Z - Z^\top \Upsilon(q,p) + Z^\top Z\right)\end{aligned} \tag{9.79}$$

where

$$Z(\hat{x}) = B(q)^\top T(q)^{-\top} Az \tag{9.80}$$

$$\Upsilon(q,p) = B(q)^\top \frac{\partial H(q,p)}{\partial p}. \tag{9.81}$$

Finally, based on the output $\hat{y} = h(\hat{x})$, and the results (9.76), and (9.79), the Hamilton–Jacobi inequality (9.38) is rewritten as

$$-\frac{\partial H(q,p)}{\partial p}^{\top} D(q,p)\frac{\partial H(q,p)}{\partial p} - \hat{y}^{\top}C\hat{y} + \tilde{Z} + \frac{1}{2}\hat{y}^{\top}\hat{y} \le 0 \tag{9.82}$$

with $\tilde{Z}(\hat{x})$ as in (9.79). We now rewrite $\hat{y}$ as

$$\hat{y} = \bar{G}(\bar{q})^{\top}\frac{\partial \hat{H}(\hat{x})}{\partial \hat{p}} = \bar{G}(\bar{q})^{\top}\hat{p} = G(q)^{\top}\frac{\partial H(q,p)}{\partial p} - G(q)^{\top}T(q)^{-1}A\hat{z} \tag{9.83}$$

and we replace (9.83) in (9.82). Lastly, we have that

$$\begin{bmatrix} \frac{\partial H(q,p)}{\partial p} \\ z \end{bmatrix}^{\top} \underbrace{\begin{bmatrix} \Gamma_1(q,p) & -\Gamma_3(q)^{\top} \\ -\Gamma_3(q) & \Gamma_2(q) \end{bmatrix}}_{P_{HJi}} \begin{bmatrix} \frac{\partial H(q,p)}{\partial p} \\ z \end{bmatrix} \le 0 \tag{9.84}$$

The inequality (9.84) is satisfied when matrix $P_{HJi} \le 0$, which is the case if matrix C of the control law (9.61) is designed such that the inequalities (9.67)–(9.70) hold, with $\lambda_c \in \mathbb{R}^n$ being the set of eigenvalues of C. □

Remark 9.5 The Hamilton–Jacobi inequality (9.84) based on the closed-loop system (9.71) holds when the set of eigenvalues of the matrix C are chosen such that the conditions for $\Gamma_1(q,p)$, $\Gamma_2(q)$, $\Gamma_3(q)$, and λ_c are satisfied. It follows that increasing the eigenvalues of C allows for a smaller γ, and thus, a smaller $\mathcal{L}_2$-gain bound. Increasing the eigenvalues of C corresponds to increasing the damping injection.

In the next subsection we look at the special case when f_e is unknown, but constant.

9.4.3 Stability Analysis for Constant External Forces

Here, we propose an equivalent description of the system (9.62), with a different Hamiltonian function which can be used as a Lyapunov function for constant nonzero external forces, i.e., $f_e \in \mathbb{R}^n / \{0\}$. We embed the extended system into a larger PH system for which a series of Casimir functions are constructed. The analysis is based on the results of [14].

We proceed to apply the results in Sect. 9.2.4 to the closed-loop system (9.62) with constant nonzero external forces as input, i.e., $\bar{u} = f_e$. We compute matrix $\mathcal{K}(\hat{x})$ as in (9.40), and obtain

$$\mathcal{K}(\hat{x}) = -\begin{bmatrix} \bar{T}\left(-\bar{J}_2 + \bar{D}\right)\bar{T}^{\top} & 0_{n\times n} & \bar{T}Y^{-\top} \\ 0_{n\times n} & \left(\bar{G}C\bar{G}^{\top}\right)^{-1} & -\left(\bar{G}C\bar{G}^{\top}\right)^{-1}Y^{-\top} \\ -Y^{-1}\bar{T}^{\top} & -Y^{-1}\left(\bar{G}C\bar{G}^{\top}\right)^{-1} & Y^{-1}\left(\bar{G}C\bar{G}^{\top}\right)^{-1}Y^{-\top} \end{bmatrix} \begin{bmatrix} 0_{n\times n} \\ \bar{B}(\bar{q}) \\ 0_{n\times n} \end{bmatrix}. \tag{9.85}$$

Here, we left out the arguments of $\bar{T}(\bar{q})$, $\bar{G}(\bar{q})$, $\bar{J}_2(\bar{q},\hat{p})$, and $\bar{D}(\bar{q},\hat{p})$ for notational simplicity. If $\hat{G}(\bar{q}) = \left(\bar{G}(\bar{q})\, C\bar{G}(\bar{q})^\top\right)^{-1}$, then (9.85) leads to

$$\mathscr{K}\left(\hat{x}\right) = \begin{bmatrix} 0_{n\times n} \\ -\hat{G}(\bar{q})\,\bar{B}(\bar{q}) \\ Y^{-1}\hat{G}(\bar{q})\,\bar{B}(\bar{q}) \end{bmatrix}. \tag{9.86}$$

Following, the results of Theorem 9.3, we assume that the local smooth functions $\mathscr{C}_j(x)$, $j \in n$, satisfy the integrability condition (9.47). It follows that the dynamics of (9.71) can be alternatively represented by (9.49) where $H_r\left(\hat{x}\right)$ is

$$\begin{aligned} H_r\left(\hat{x}\right) &= \hat{H}_d\left(\hat{x}\right) - \sum_{j=1}^{n} f_{e_j}\mathscr{C}_j(x) \\ &= \hat{H}_d\left(\hat{x}\right) + f_e^\top \hat{G}(\bar{q})\,\hat{p} - f_e^\top Y^{-1}\hat{G}(\bar{q})\,z + f_e^\top c, \end{aligned} \tag{9.87}$$

where $\hat{x} = \left(\bar{q},\hat{p},z\right)$, and $\hat{H}_d\left(\hat{x}\right)$ as in (9.64). If we choose, the constant $c = -K_f f_e \in \mathbb{R}^n$, with $K_f > 0$. Then, we can rewrite (9.87) as

$$H_r\left(\hat{x}\right) = \frac{1}{2}\begin{bmatrix} \bar{q}-q_d \\ \hat{p} \\ z \\ f_e \end{bmatrix}^\top \underbrace{\begin{bmatrix} K_p & 0_{n\times n} & 0_{n\times n} & 0_{n\times n} \\ 0_{n\times n} & I_{n\times n} & 0_{n\times n} & \bar{B}^\top\hat{G}^\top \\ 0_{n\times n} & 0_{n\times n} & K_z^{-1} & -\bar{B}^\top\hat{G}^\top Y^{-\top} \\ 0_{n\times n} & \hat{G}\bar{B} & -Y^{-1}\hat{G}\bar{B} & K_f \end{bmatrix}}_{\hat{P}(\bar{q})} \begin{bmatrix} \bar{q}-q_d \\ \hat{p} \\ z \\ f_e \end{bmatrix} > 0 \tag{9.88}$$

where we have left out the arguments of $\bar{G}(\bar{q})$ and $\bar{B}(\bar{q})$ for notational simplicity. Since $\bar{G}(\bar{q})$ and $\bar{B}(\bar{q})$ are full rank, and $C > 0$, via the Schur complement it can be shown that matrix $\hat{P}(\bar{q})$ in (9.88) is positive definite, and then the inequality (9.88) holds. Furthermore, via Theorem 9.3, we have that

$$\dot{H}_r\left(\hat{x}\right) = -\frac{\partial H_r\left(\hat{x}\right)}{\partial \hat{x}}^\top \tilde{R}\left(\hat{x}\right) \frac{\partial H_r\left(\hat{x}\right)}{\partial \hat{x}} \leq 0 \tag{9.89}$$

and thus $H_r\left(\hat{x}\right)$ qualifies as a Lyapunov function for the forced dynamics (9.49). Then, $\dot{H}_r\left(\hat{x}\right) \leq 0$, and given that $\dfrac{\partial \hat{H}_d\left(\hat{x}\right)}{\partial \hat{p}} = \hat{p}$, and $\dfrac{\partial \hat{H}_d\left(\hat{x}\right)}{\partial z} = K_z^{-1}z$, we know that $\hat{p}, z \to 0$ as $t \to \infty$. Given the dynamics of system (9.49), $\dot{\hat{p}} = \dot{z} = 0$, it can be verified that the largest invariant set for $\dot{H}_r\left(\hat{x}\right) = 0$ equals $\left(\bar{q} - q_d - K_p^{-1}\bar{B}(\bar{q})\, f_e, \hat{p}, z\right) = (0, 0, 0)$. LaSalle's invariance then implies that the system is asymptotically stable in

$$\bar{q} = q_d + K_p^{-1}\bar{B}(\bar{q})\, f_e. \tag{9.90}$$

Remark 9.6 The $\mathscr{L}_2$-gain analysis of Sect. 9.4.2 gives a bound on the relation between input $\delta = f_e$ and the output $\hat{y}$ of the closed-loop system (9.71) for an $\mathscr{L}_2$-input disturbance δ. The $\mathscr{L}_2$-gain analysis differs from the results of Sect. 9.4.3 in the sense that the $\mathscr{L}_2$-gain analysis evaluates a bound on the output $\hat{y}$ in relation to the size of the input δ, while the analysis in Sect. 9.4.3 is for the case where the system is asymptotically stable, i.e., $\hat{y} \to 0$, with a new equilibrium point caused by a constant f_e, i.e., $\bar{q}$, which is different from the desired equilibrium point q_d. Notice that the $\mathscr{L}_2$-gain bound is related to the amount of damping injected, while the new equilibrium point (steady-state position) is related to the stiffness parameter K_p.

9.4.4 Integral Position Control

The analysis in the previous section shows that, under the assumption that f_e is constant, we can expect a constant steady-state error in the position of system (9.71). Furthermore, the analysis also justifies the application of integral control, since integral control compensates for constant steady-state errors. The main contribution of this section is to realize a type of integral position control for a class of standard mechanical systems with dissipation in the PH framework. For the extended system (9.55), with f_e constant, we propose a coordinate transformation to include the position error in the new output. By having the position error in the passive output, we can interconnect the closed-loop with an integrator in a passivity-preserving way, i.e., preserving the PH structure. The results of this section are inspired by the works of [4, 5, 22].

Theorem 9.6 *Consider system (9.55) and assume $f_e \neq 0$ and constant. Define the integrator state ξ with dynamics*

$$\dot{\xi} = -\bar{B}(\bar{q})^\top (\hat{p} + K_i(\bar{q} - q_d)) \tag{9.91}$$

q_d the desired constant position and $K_i > 0$ is a constant matrix. Then, the control input

$$v = \bar{G}(\bar{q})^{-1} \left(\frac{\partial \bar{V}(\bar{q})}{\partial \bar{q}} - K_p(\bar{q} - q_d) - K_i \dot{\bar{q}} - \bar{B}(\bar{q})\xi \right) - C\bar{G}(\bar{q})^\top \left(\hat{p} + K_i(\bar{q} - q_d)\right) \tag{9.92}$$

with $K_p > 0$, and $C > 0$, asymptotically stabilizes the extended system (9.55) at $(\bar{q}, \hat{p}, z) = (q_d, 0, 0)$, i.e., zero steady-state error.

Proof We use the results of [5]. Consider first the coordinate transformation

$$\tilde{p} = \hat{p} + K_i(\bar{q} - q_d) \tag{9.93}$$

with a constant matrix $K_i > 0$, which then implies that

$$\dot{\tilde{p}} = \dot{\hat{p}} + K_i \dot{\bar{q}} \tag{9.94}$$

since q_d is constant. The control input (9.92) with integrator dynamics (9.91) then realizes the closed-loop system

$$\begin{bmatrix} \dot{\bar{q}} \\ \dot{\tilde{p}} \\ \dot{z} \\ \dot{\xi} \end{bmatrix} = \begin{bmatrix} -K_i K_p^{-1} & \bar{T}^{-\top} & \bar{T}^{-\top} Y & 0 \\ -\bar{T}^{-1} & \bar{J}_2 - \bar{D} - \bar{G} C \bar{G}^\top & \left(\bar{J}_2 - \bar{D}\right) Y & \bar{B} \\ -Y^\top \bar{T}^{-1} & -Y^\top \left(\bar{J}_2^\top + \bar{D}\right) & -Y^\top \left(\bar{J}_2^\top + \bar{D}\right) Y & 0 \\ 0 & -\bar{B}^\top & 0 & 0 \end{bmatrix} \begin{bmatrix} \frac{\partial \hat{H}_i}{\partial \bar{q}} \\ \frac{\partial \hat{H}_i}{\partial \tilde{p}} \\ \frac{\partial \hat{H}_i}{\partial z} \\ \frac{\partial \hat{H}_i}{\partial \xi} \end{bmatrix} \tag{9.95}$$

$$\tilde{y} = \bar{G}^\top \frac{\partial \hat{H}_i}{\partial \tilde{p}} \tag{9.96}$$

with Hamiltonian

$$\hat{H}_i = \frac{1}{2} \tilde{p}^\top \tilde{p} + \frac{1}{2} (\bar{q} - q_d)^\top K_p (\bar{q} - q_d) + \frac{1}{2} z^\top K_z^{-1} z + \frac{1}{2} (f_e - \xi)^\top (f_e - \xi), \tag{9.97}$$

where the arguments of $\hat{H}_i$ $(\bar{q}, \tilde{p}, z, \xi)$, $T(\bar{q})$, $\bar{J}_2(\bar{q}, \tilde{p})$, $\bar{D}(\bar{q}, \tilde{p})$, $\bar{G}(\bar{q})$, and $\bar{B}(\bar{q})$ are left out for notational simplicity. Furthermore, notice that

$$\tilde{y} = \bar{G}(\bar{q})^\top \tilde{p} = \bar{G}(\bar{q})^\top \left(\hat{p} + K_i (\bar{q} - q_d)\right). \tag{9.98}$$

Take (9.97) as candidate Lyapunov function, which then gives

$$\dot{\hat{H}}_i = - \begin{bmatrix} \frac{\partial \hat{H}_i}{\partial \bar{q}} \\ \frac{\partial \hat{H}_i}{\partial \tilde{p}} \\ \frac{\partial \hat{H}_i}{\partial z} \end{bmatrix}^\top \underbrace{\begin{bmatrix} K_i K_p^{-1} & 0 & 0 \\ 0 & \bar{D}(\bar{q}, \tilde{p}) + \bar{G}(\bar{q}) C \bar{G}(\bar{q})^\top & -\bar{D}(\bar{q}, \tilde{p}) Y \\ 0 & -Y^\top \bar{D}(\bar{q}, \tilde{p})^\top & Y^\top \bar{D}(\bar{q}, \tilde{p})^\top Y \end{bmatrix}}_{U(\bar{q}, \tilde{p})} \begin{bmatrix} \frac{\partial \hat{H}_i}{\partial \bar{q}} \\ \frac{\partial \hat{H}_i}{\partial \tilde{p}} \\ \frac{\partial \hat{H}_i}{\partial z} \end{bmatrix}. \tag{9.99}$$

Since $\bar{G}(\bar{q})$ is full rank, $\bar{D}(\bar{q}, \tilde{p}) \geq 0$, $K_i > 0$, $K_p > 0$, $C > 0$, $K_z > 0$, $Y = AK_z$ and A being a constant matrix, via the Schur complement it can be shown that matrix $U(\bar{q}, \tilde{p}) \geq 0$, and thus $\dot{\hat{H}}_i \leq 0$ holds. Define the set

$$\mathbb{O} = \left\{ (\bar{q}, \tilde{p}, z, \xi) \mid \dot{\hat{H}}_i (\bar{q}, \tilde{p}, z, \xi) = 0 \right\}. \tag{9.100}$$

Given that $\dot{\hat{H}}_i (q_d, 0, 0, \xi) = 0, \forall \xi$, we have that ξ is free. Assume $\xi - f_e = c_1 \neq 0$ constant with $c_1 \in \mathbb{R}^n$, thus $\dot{\xi} = 0$. Then, the dynamics $\dot{\tilde{p}}$ is

$$\dot{\tilde{p}} = \bar{B}(q_d)(c_1 + f_e) \neq 0. \tag{9.101}$$

Since (9.101) is constant, then $\tilde{p}$ will change over time, and hence, we have a contradiction. Thus, the largest invariant set in $\mathbb{O}$ is $\mathbb{M} = \{q_d, 0, 0, f_e\}$. Via LaSalle's invariance principle we conclude that the system (9.95) is asymptotically stable at $(\bar{q}, \tilde{p}, z, \xi) = (q_d, 0, 0, f_e)$, which means that the constant disturbance is compensated by ξ, i.e., $\xi \rightarrow f_e$. □

Substituting v in (9.54) by (9.92) then gives the total control input u for the original system (9.2), which in terms of the original coordinates q, p becomes

$$\begin{aligned} u = {} & G(q)^{-1} T(q) \left(A\dot{z} + \frac{\partial V(q)}{\partial q} - K_p (q - q_d) - K_i \dot{q} \right) - G(q)^{-1} B(q) \xi \\ & - C G(q)^{\top} \left(M(q)^{-1} p - T(q)^{-\top} A z - T(q)^{-\top} K_i (q - q_d) \right) \end{aligned} \tag{9.102}$$

with $\dot{z}$ as in (9.52).

Here, we have applied an integral type control law as in (9.92) to compensate for position errors caused by constant forces. We observe in Theorem 9.6 how our integral control strategy follows naturally from the PH structure.

9.5 Position Control with Measured Forces

In the previous section, we have presented a position control strategy that exploits feedback of the modeled internal forces. In other words, the forces used for feedback are based on the dynamical model and the measured positions and velocities. In this section, we assume we have force sensors, which provide the (real) total forces working on the system. Then, we feed back the readings of the force sensors in the input of the system (9.29). Notice that the measured total forces f in the system can be described by

$$f(q, p) = f_{in}(q, p) + B(q) f_e \tag{9.103}$$

with $f_{in}(q, p)$ as in (9.51). In the previous section, we used (9.51) to model and compute the internal forces for feedback control. We can still use (9.51) to describe the internal forces, while adding the external forces to model the total forces in the system. Let $\bar{f}(\bar{q}, \bar{p})$ be the total forces multiplied by the matrix $T(q)$ in (9.25), i.e.,

$$\bar{f}\left(\Phi^{-1}(\bar{q}, \bar{p})\right) = \bar{f}(q, p) = T(q)^{-1} f(q, p) \tag{9.104}$$

and consider now system (9.29). Notice that in terms of the coordinates $\bar{q}$, $\bar{p}$ that $\bar{f}(\bar{q}, \bar{p})$ is then described by

$$\bar{f}(\bar{q}, \bar{p}) = -\bar{T}(\bar{q})^{-1} \frac{\partial \bar{H}(\bar{q}, \bar{p})}{\partial \bar{q}} + \left(\bar{J}_2(\bar{q}, \bar{p}) - \bar{D}(\bar{q}, \bar{p})\right) \frac{\partial \bar{H}(\bar{q}, \bar{p})}{\partial \bar{p}} + \bar{B}(\bar{q}) f_e. \tag{9.105}$$

Define for system (9.29) the input

$$v = -\bar{G}(\bar{q})^{-1} \bar{f}(\bar{q}, \bar{p}) + \bar{v} \tag{9.106}$$

with $\bar{v}$ being a new input vector, which then changes (9.29) into the PH system

$$\begin{bmatrix} \dot{\bar{q}} \\ \dot{\bar{p}} \end{bmatrix} = \begin{bmatrix} 0 & \bar{T}(\bar{q})^{-\top} \\ -\bar{T}(\bar{q})^{-1} & 0 \end{bmatrix} \begin{bmatrix} \dfrac{\partial \bar{H}_\tau(\bar{q}, \bar{p})}{\partial \bar{q}} \\ \dfrac{\partial \bar{H}_\tau(\bar{q}, \bar{p})}{\partial \bar{p}} \end{bmatrix} + \begin{bmatrix} 0 \\ \bar{G}(\bar{q}) \end{bmatrix} \bar{v} \tag{9.107}$$

$$\bar{y} = \bar{G}(\bar{q})^\top \frac{\partial \bar{H}_\tau(\bar{q}, \bar{p})}{\partial \bar{p}} \tag{9.108}$$

with Hamiltonian

$$\bar{H}_\tau(\bar{q}, \bar{p}) = \frac{1}{2} \bar{p}^\top \bar{p}. \tag{9.109}$$

We then obtain (9.29), with all forces canceled. We can thus control the system without the problems described in Sect. 9.4.2. Notice that we need to describe (9.2) in the equivalent form (9.29) in order to realize force rejection and preserve the PH structure. In the original coordinates (q, p) the control input (9.106) is given by

$$u = -G(q)^{-1} f(q, p) + \bar{v} \tag{9.110}$$

with $f(q, p)$ as in (9.103). Notice that the advantage here is that we can apply control methods without having to worry about the external forces (disturbances) and internal forces (potential forces and friction). However, (9.110) implies that there is no tuning possible in the application of force feedback. In Sect. 9.4.2 the disturbances are not rejected, however, we have the possibility to tune the force feedback with the matrix A.

In the next section we illustrate, via simulation of the system (9.7), the results of Sects. 9.4 and 9.5 for obtaining asymptotic stability in a desired position.

9.6 Simulation Results: Two-DOF Shoulder System

Consider the system of Examples 9.1 and 9.2. We have determined the parameters of the two-DOF shoulder system of Fig. 9.1 as $\mathscr{I}_i = \{0.013, 1.692\}$, $F_{c_i} = \{0.005, 0.025\}$, $F_{s_i} = \{1.905, 2.257\}$, $F_{v_i} = \{4.119, 4.973\}$, and $\dot{q}_{s_i} = \{0.167, 0.170\}$.

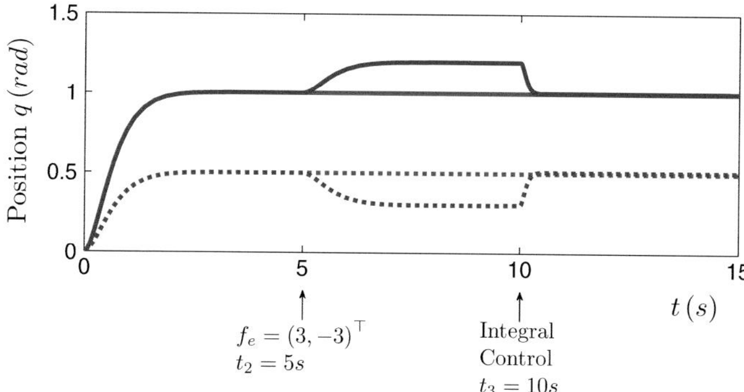

Fig. 9.3 Position control via force feedback (*blue line*) with $f_e = col\,(0, 0)^\top$ at $t \leq 5s$. Integral control (*blue line*) with $f_e = col\,(3, -3)^\top$ at $t \geq 10s$. Total force rejection (*red line*) at $t \geq 0$. Initial conditions $(q\,(0)\,, p\,(0))^\top = (0, 0, 0, 0)^\top$. *Solid line* q_1. *Dashed line* q_2

We have a link length of $l_c = 0.249m$, $m = 3.9\,kg$; matrices $A = diag\,(0.5, 0.7)$, $K_z = diag\,(2, 2)$, $K_p = diag\,(15, 15)$, and $C = diag\,(10, 10)$; an initial position $q(0) = (0, 0)$, and desired position $q_d = (1, 0.5)\,rad$. We obtain the desired position $q_d = (1, 0.5)^\top$ from an initial position $q\,(0) = (0, 0)^\top$ at $t = t_1 \geq 3s$ with the control law (9.61). Then, we apply a constant nonzero force, i.e., $f_e = (3, -3)^\top$, at $t_2 = \geq 5s$, to the closed-loop system (9.62). Results are shown in Fig. 9.3. The new position is $q = K_p^{-1} f_e + q_d = (1.2, 0.3)^\top$ (blue line) which corresponds to a different equilibrium point as in (9.90). The results presented here validate the fact that the PH system (9.62) remains stable with a constant nonzero input $\bar{u} = f_e$. Furthermore, we want to recover the desired position by applying the integral control law (9.102) to the PH system (9.62) at $t_3 \geq 10s$, with a matrix $K_i = diag\,(1, 0.5)$. We observe how the system is stabilized again at the desired position q_d at $t \geq 11s$ without a steady-state error.

Finally, we apply a constant nonzero force, i.e., $f_e = (3, -3)^\top$, to the two-DOF inputs of the to the system (9.7), and apply (9.110) at $t = t_2 \geq 5s$, which includes the measured forces of the sensors. Then, the equilibrium is achieved immediately, independent of f_e as seen in Fig. 9.3.

9.7 Concluding Remarks

We have provided a method for position control via force feedback in the PH setting. The method relies on a structure preserving extension of the system. Disturbance attenuation is studied, and robustness is obtained by extending the system once more in a structure preserving way with integral type dynamics. Finally, we present a method when forces are reconstructed and fed back directly from the sensor information.

The robotic arm example shows that performance of the method is very good. Tests with the robotic arm are under way, and show promising results, also in comparison with other control methods.

References

1. Andersson, S., Söderberg, A., Björklund, S.: Friction models for sliding dry, boundary and mixed lubricated contacts. Tribol. Intern. **40**(4), 580–587 (2007)
2. Bol, M.: Force And Position Control of the Philips Experimental Robot Arm in a Energy-Based Setting. University of Groningen, Groningen (2012)
3. Canudas de Wit, C., Siciliano, B., Bastin, G.: Theory of Robot Control. Springer, London(1996)
4. Dirksz, D.A., Scherpen, J.M.A.: Power-based control: canonical coordinate transformations. Integr. Adapt. Control Autom. **48**(6), 1046–1056 (2012)
5. Donaire, A., Junco, S.: On the addition of integral action to port-controlled Hamiltonian systems. Automatica **45**, 1910–1916 (2009)
6. Duindam, V., Macchelli, A., Stramigioli, S., Bruyninckx, H. (eds.): Modeling and Control of Complex Physical Systems: The Port-Hamiltonian Approach. Springer, Berlin (2009)
7. Fujimoto, K., Sugie, T.: Canonical transformation and stabilization of generalized hamiltonian systems. Syst. Control Lett. **42**(3), 217–227 (2001)
8. Fujimoto, K., Sakurama, K., Sugie, T.: Trajectory tracking of port-controlled Hamiltonian systems via generalized canonical transformations. Automatica **39**(12), 2059–2069 (2003)
9. Gömez-Estern, F., van der Schaft, A.J.: Physical damping in IDA-PBC controlled underactuated mechanical systems. Eur. J. Control **10**(5), 451–468 (2004)
10. Gorinevsky, D., Formalsky, A., Scheiner, A.: Force Control of Robotics Systems. CRC Press LLC, Moscow (1997)
11. Khalil, H.: Nonlinear Systems, 2nd edn. Prentice-Hall, New York (2001)
12. Koop, F.: Trajectory Tracking Control of the Philips Experimental Robot Arm in the Port-Hamiltonian Framework. University of Groningen, Groningen (2014)
13. Maschke, B.M., van der Schaft, A.J.: Port-controlled hamiltonian systems: modeling origins and system-theoretic properties. IN: Procedings of the IFAC Symposium on Nonlinear Control Systems, pp. 282–288. Bordeaux, France (1992)
14. Maschke, B.M., Ortega, R., van der Schaft, A.J.: Energy-based lyapunov functions for forced hamiltonian systems with dissipation. IEEE Trans. Autom. Control **45**(8), 1498–1502 (2000)
15. Muñoz-Arias, M., Scherpen, J.M.A., Dirksz, D.A.: Force feedback of a class of standard mechanical system in the Port-Hamiltonian framework. In: Proceedings of the 20th International Symposium on Mathematical Theory of Networks and Systems, Melbourne, Australia (2012)
16. Muñoz-Arias, M., Scherpen, J.M.A., Dirksz, D.A.: A class of standard mechanical systems with force feedback in the Port-Hamiltonian framework. In: Proceedings of the 4th IFAC Workshop on Lagrangian and Hamiltonian Methods for Nonlinear Control, pp. 90–95. Bertinoro, Italy (2012)
17. Muñoz-Arias, M., Scherpen, J.M.A., Dirksz, D.A.: Position control via force feedback for a class of standard mechanical systems in the Port-Hamiltonian framework. In: Proceedings of the 52nd IEEE Conference on Decision and Control, pp. 1622–1627. Florence, Italy (2013)
18. Murray, R., Li, Z., Sastry, S.S.: A Mathematical Introduction to Robotic Manipulation. CRC Press, Boca Raton (1994)
19. Ortega, R., Loría, A., Nicklasson, P.J., Sira-Ramirez, H.: Passivity-Based Control of Euler-Lagrange Systems. Springer, Heidelberg (1998)
20. Ortega, R., van der Schaft, A.J., Maschke, B., Escobar, G.: Interconnection and damping assignment passivity-based control of port-controlled Hamiltonian systems. Automatica **38**, 585–596 (2002)

21. Ortega, R., Spong, M., Gomez, F., Blankenstein, G.: Stabilization of underactuated mechanica systems via interconnection and damping assignment. IEEE Trans. Autom. Control **47**(8), 1218–1233 (2002)
22. Ortega, R., Romero, J.G.: Robust integral control of port-Hamiltonian systems: the case of non-passive outputs with unmatched disturbances. Syst. Control Lett. **61**, 11–17 (2011)
23. Rijs, R., Beekmans, R., Izmit, S., Bemelmans, D.: Philips Experimental Robot Arm: User Instructor Manual, Version 1.1. Koninklijke Philips Electronics N.V., Eindhoven (2010)
24. Siciliano, B., Kathib, O.: Springer Handbook of Robotics. Springer, Berlin (2008)
25. Spong, M., Hutchinson, S., Vidjasagar, M.: Robot Modeling and Control. Wiley, Hoboken (2006)
26. van der Schaft, A.J.: L_2-Gain and Passivity Techniques in Nonlinear Control: Lecture Notes in Control and Information Sciences 218. Springer, London (1999)
27. Viola, G., Ortega, R., Banavar, R., Acosta, J.A., Astolfi, A.: Total energy shaping control of mechanical systems: simplifying the matching equations via coordinate changes. IEEE Trans. Autom. Control **52**(6), 1093–1099 (2007)

Chapter 10
Endogenous Configuration Space Approach: An Intersection of Robotics and Control Theory

Krzysztof Tchoń

Abstract The endogenous configuration space approach is a control theory-oriented methodology of robotics research, dedicated to mobile manipulators. A cornerstone of the approach is a parameterized control system with output whose input–output map constitutes the mobile manipulator's kinematics. An endogenous configuration consists of the control function and of the vector of output function parameters representing the joint positions of the on-board manipulator. The mobile manipulator's Jacobian is defined as the input–output map of the linear approximation to the control system. Regular and singular endogenous configurations are introduced. The regular endogenous configuration corresponds to the local output controllability of the control system, while the singular configuration coincides with a singular optimal control-parameter pair of the control system. The inverse kinematics problem is formulated as a control problem in a driftless control system. A collection of Jacobian kinematics inverses is presented, leading to Jacobian motion planning algorithms. Performance measures of the mobile manipulator are introduced.

10.1 Introduction

It is my great honour and pleasure to make a contribution to this Festschrift in honour of the sixtieth birthday of Professor Henk Nijmeijer. I met the Jubilarian in 1983, during a control conference at the Warwick University, admiring his expertise in geometric control and enjoying his sense of humour. Our collaboration intensified over the decade after that conference. Among its fruits, there was a conference on geometric theory of control systems organized in Poland in 1984, with Henk's plenary talk [17], exchange of students and visitors, and joint research [25]. I owe to Henk my determination in applying mathematical control theory in robotics. The following text summarizes some results of such an approach to mobile manipulators.

K. Tchoń (✉)
Chair of Cybernetics and Robotics, Electronics Faculty, Wrocław University of Technology, Wrocław, Poland
e-mail: krzysztof.tchon@pwr.edu.pl

N. van de Wouw et al. (eds.), *Nonlinear Systems*, Lecture Notes in Control and Information Sciences 470, DOI 10.1007/978-3-319-30357-4_10

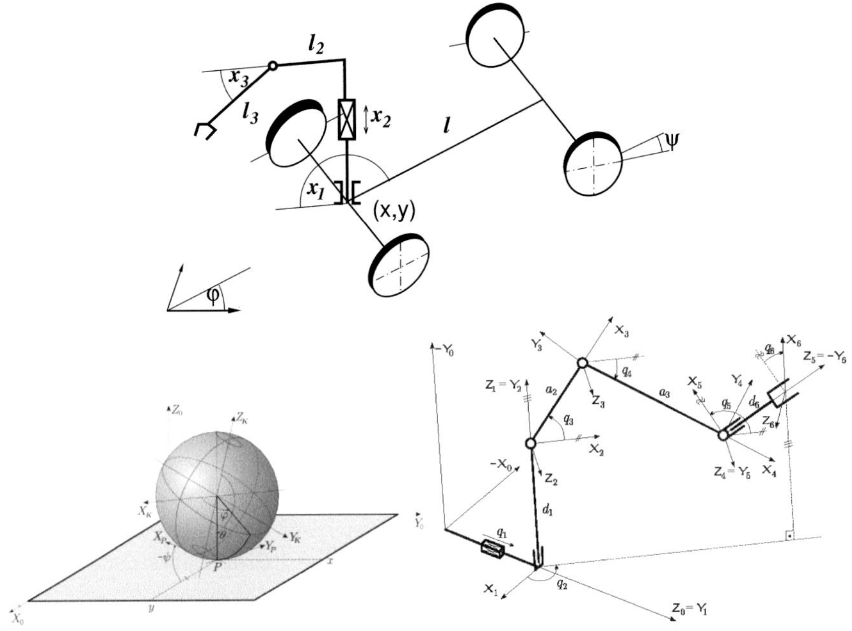

Fig. 10.1 Mobile manipulator, rolling ball, and robotic arm

A mobile manipulator is a robotic device composed of a mobile platform and a fixed base on-board manipulator. It is well known that a synergy of mobility and manipulation capabilities makes these devices of paramount usability in personal and service robotics. By their very nature, the mobile manipulators open new and challenging problems in the area of mathematical modelling, motion planning, control, and performance evaluation. For an exhaustive review of literature on this subject the reader is advised to consult the monograph [16] and the references included in [26, 28]. In this work, we present a control theoretic approach that provides a uniform theory of manipulation robots, mobile platforms and mobile manipulators. Established in [30], this approach has been then developed under the name of the endogenous configuration space approach [23, 24, 26–28, 32, 37].

In what follows, we shall concentrate on the kinematics of a mobile manipulator that consists of a nonholonomic mobile platform and a holonomic on-board manipulator. As an example of such a robot, Fig. 10.1 shows a 4-wheel mobile platform carrying a 3 degrees of freedom manipulator whose consecutive joints are rotational, prismatic and rotational. We let $q = (\mathsf{x}, \mathsf{y}, \varphi, \psi) \in R^4$ describe the position, orientation and the heading angle of the platform, while $x = (x_1, x_2, x_3) \in R^3$ denotes joint positions of the on-board manipulator. Cartesian positions of the end effector are denoted as $y = (y_1, y_2, y_3)$ and belong to the operational space R^3. The platform length equals l, the lengths of the manipulator arms are l_2 and l_3. The assumption that no side-slip of the rear and front wheels is permitted imposes on the platform motion the nonholonomic Pfaffian constraints

$$\begin{bmatrix} \sin\varphi & -\cos\varphi & 0 & 0 \\ \cos\varphi\sin\psi & \sin\varphi\sin\psi & -l\cos\psi & 0 \end{bmatrix}\dot{q} = 0$$

that lead to a representation of the kinematics of this mobile manipulator in the form of a driftless control system

$$\dot{\mathsf{x}} = lu_1\cos\varphi\cos\psi,\ \ \dot{\mathsf{y}} = lu_1\sin\varphi\cos\psi,\ \ \dot{\varphi} = u_1\sin\psi,\ \ \dot{\psi} = u_2,$$

equipped with the output function

$$y = (\mathsf{x} + (l_2 + l_3\cos x_3)\cos(\varphi + x_1), \mathsf{y} + (l_2 + l_3\cos x_3)\sin(\varphi + x_1), x_2 + l_3\sin x_3)\,.$$

The inputs driving this system consist of the longitudinal velocity and the heading angular velocity of the platform as well as of the joint positions of the on-board manipulator. The output function determines the end effector position of the on-board manipulator resulting from the configuration of its joints and the motion of the platform. The approach presented here specifies in a natural way to nonholonomic mobile platforms (the on-board manipulator is absent) and to holonomic manipulation robots (the platform is absent), like these shown in the bottom part of in Fig. 10.1.

A typical robotic problem addressed in the control system representation of a mobile manipulator is the motion planning problem consisting in defining the controls that make the output function to reach, at a given time instant, a desired value (e.g. place the end effector of the on-board manipulator at a prescribed location or move the ball to a prescribed position and orientation). An essential ingredient of the motion planning is the inversion of the kinematics. The inverse kinematics algorithms achieving this objective are often based on a concept of the mobile manipulator's Jacobian that describes an infinitesimal transformation of the controls into the motions in the operational space. The performance and efficiency of the Jacobian motion planning algorithms strongly depend on the regularity of controls. The quality of performance deteriorates in a vicinity of singular controls.

In this work, we focus on the control theoretic foundations of the endogenous configuration space approach. Conceptually, the endogenous configuration space approach is an extension to mobile manipulators of the homotopy continuation-type ideas introduced into the path planning problem of nonholonomic mobile robots by Sussmann [22] and developed further in [3, 7–9, 11, 21]. We begin with a parametrized control system with outputs whose input–output map constitutes the mobile manipulator's kinematics. Then, the endogenous configuration space is defined as a Hilbert space containing the control functions and the parameters of the control system. The input–output map of the linear approximation to the control system serves as the mobile manipulator's Jacobian. The endogenous configurations at which the Jacobian is a surjective map are referred to as regular; they correspond to the local output controllability of the control system. The singular configurations can be interpreted as singular optimal control-parameter pairs of the control system. We address the inverse kinematics problem for mobile manipulators, and derive a collection of Jacobian inverses of the kinematics. These inverses can be exploited in Jacobian motion planning algorithms, either directly or after a modification respecting constraints imposed on system's states or controls [15].

The continuation method is adopted as a guideline in this, leading to the Ważewski–Davidenko equation [10, 35]. The concept of the dynamic inverse [13] is connected with the Ważewski inequality [36], and a condition for the existence of a global inverse is provided. The dynamic Jacobian inverses constitute Jacobian motion planning algorithms. The Gram matrix of the linear approximation to the control system is used as the mobile manipulator's dexterity matrix. The eigenvalues of this matrix, on the one hand, enter into the norms of the inverse Jacobian operators, on the other, they constitute kinematic dexterity measures of mobile manipulators. In the control theoretic setting, the norm of the Jacobian and of the Jacobian pseudo-inverse operators are determined by the vector of switching functions associated with the parametric time optimal control problem.

The composition of this chapter is the following. In Sect. 10.2 we introduce the basic concepts including the kinematics, the endogenous configuration space, the Jacobian operator, regular and singular configurations and the adjoint Jacobian. Section 10.3 addresses the inverse kinematics problem, and defines several inverse Jacobian operators. In Sect. 10.4 kinematic performance issues for mobile manipulators are discussed. Section 10.5 contains conclusions.

10.2 Basic Concepts

As has been mentioned, mobile manipulators consist of a nonholonomic mobile platform and a holonomic on-board manipulator. It will be assumed that the motion of the mobile platform is characterized by n generalized coordinates $q \in R^n$, and subject to $l \leq n$ independent phase constraints in the Pfaffian form

$$A(q)\dot{q} = 0,$$

where $A(q)$ is a matrix of rank l with continuously differentiable entries. The phase constraints define a non-integrable distribution $Ker\ A(q)$ in R^n that contains admissible velocities of the platform at every q. Consequently, the platform kinematics may be represented as a driftless control system with $m = n - l$ inputs. Suppose that the platform has been equipped with an on-board manipulator with p degrees of freedom, whose joint position is denoted by $x \in R^p$, and the end effector position and orientation by $y \in R^r$. With these notations, the kinematics of the mobile manipulator can be represented as a driftless control system with outputs

$$\dot{q} = G(q)u = \sum_{i=1}^{m} g_i(q)u_i, \quad y = k(q, x) = (k_1(q, x), \ldots, k_r(q, x)). \quad (10.1)$$

10.2.1 Kinematics

The controls $(u(\cdot), x)$ driving the system (10.1) are interpreted as instantaneous platform speeds $u(t) \in R^m$, and joint positions $x \in R^p$ of the on-board manipulator. Wherever convenient, we shall use the notation $(u(\cdot), x)(t) = (u(t), x)$. The platform's control functions will be assumed Lebesgue square integrable on a time interval $[0, T]$, the joint positions are constant vectors. The control space $\mathscr{X} = L^2_m[0, T] \times R^p$ equipped with the inner product

$$\langle (u_1(\cdot), x_1), (u_2(\cdot), x_2) \rangle_{RW} = \int_0^T u_1^T(t) R(t) u_2(t) dt + x_1^T W x_2, \tag{10.2}$$

where $R(t) = R^T(t) > 0$ and $W = W^T > 0$ are symmetric, positive definite weight matrices, forms a Hilbert space called the *endogenous configuration space* of the mobile manipulator [26]. The adjective "endogenous" comes from the Greek word $\varepsilon\nu\delta o\nu = \varepsilon\nu\ \delta o\mu o\varsigma$ which means at home, inside, so the endogenous configuration means something of internal origin. Indeed, the endogenous configurations of a mobile manipulator do not manifest themselves directly, but rather through the behaviour of the mobile platform and of the end effector. Notice that on a similar etymological basis in control theory the term "endogenous feedback" introduced by M. Fliess and collaborators is used [12]. The inner product (10.2) induces in $\mathscr{X}$ a norm denoted as $||\cdot||_{RW}$. The output map of (10.1) takes its values in the operational space of the mobile manipulator, identified with R^r. It will be assumed that the operational space has been endowed with the inner product

$$\langle y_1, y_2 \rangle_Q = y_1^T Q y_2, \tag{10.3}$$

where Q denotes a symmetric, positive definite, constant matrix $Q^T = Q$, $Q > 0$. The norm induced by (10.3) will be denoted by $||\cdot||_Q$, while $||\cdot||$ will always stand for the standard Euclidean norm.

Given an initial platform posture $q_0 \in R^n$, and an endogenous configuration $(u(\cdot), x) \in \mathscr{X}$, we compute a trajectory $q(t) = \varphi_{q_0,t}(u(\cdot))$ of the platform, and an operational space trajectory $y(t) = k(q(t), x)$. It will be assumed that these trajectories are defined for every $t \in [0, T]$. The input–output map

$$K_{q_0,T} : \mathscr{X} \longrightarrow R^r,$$

of the control system (10.1), defined as

$$K_{q_0,T}(u(\cdot), x) = y(T) = k(\varphi_{q_0,T}(u(\cdot)), x), \tag{10.4}$$

will be identified with the kinematics of the mobile manipulator. With $k(q, x)$ sufficiently smooth, the continuous differentiability of the map (10.4) follows from [6, Theorem 1.1] and [20, Proposition 2.9.3].

10.2.2 Jacobian

The differential kinematics constitute the mobile manipulator's Jacobian

$$J_{q_0,T}(u(\cdot),x) : \mathscr{X} \longrightarrow R^r,$$

obtained by the differentiation

$$\begin{aligned} J_{q_0,T}(u(\cdot),x)(v(\cdot),w) &= \left.\frac{d}{d\alpha}\right|_{\alpha=0} K_{q_0,T}(u(\cdot)+\alpha v(\cdot), x+\alpha w) \\ &= C(T,x)\int_0^T \Phi(T,s)B(s)v(s)ds + D(T,x)w. \end{aligned} \tag{10.5}$$

Observe that, formally, $(v(\cdot),w)$ and $J_{q_0,T}(u(\cdot),x)(v(\cdot),w)$ belong to suitable tangent spaces, however, these spaces can be identified with $\mathscr{X}$ and R^r. The matrices determining the Jacobian result from the linear approximation of the kinematics representation (10.1) along the pair (input, trajectory) $(u(t), x, q(t) = \varphi_{q_0,t}(u(\cdot)))$, so that

$$A(t) = \frac{\partial(G(q(t))u(t))}{\partial q},\ B(t) = G(q(t)),\ C(t,x) = \frac{\partial k(q(t),x)}{\partial q},\ D(t,x) = \frac{\partial k(q(t),x)}{\partial x}, \tag{10.6}$$

and $\Phi(t,s)$ denotes the transition matrix satisfying the evolution equation

$$\frac{\partial}{\partial t}\Phi(t,s) = A(t)\Phi(t,s), \quad \Phi(s,s) = I_n.$$

Having a finite-dimensional range space, for every fixed endogenous configuration $(u(\cdot),x) \in \mathscr{X}$, the Jacobian is a compact linear operator [14]. The linear, time dependent control system

$$\dot{\xi} = A(t)\xi + B(t)v, \quad \eta = C(t,x)\xi + D(t,x)w \tag{10.7}$$

determined by the data (10.6) may be referred to as the variational system associated with (10.1). It is easily seen that the Jacobian (10.5) corresponds to the input–output map of the variational system initialized at $\xi_0 = 0$.

10.2.3 Regular and Singular Configurations

An endogenous configuration $(u(\cdot),x) \in \mathscr{X}$ of the mobile manipulator will be called regular, if the Jacobian map is surjective, otherwise the configuration is singular. The surjectivity of the Jacobian is equivalent to the output controllability of the variational system (10.7), and implies that the original system (10.1) is locally output controllable. A necessary and sufficient condition for the regularity of a configuration is that the Gram matrix

$$\mathscr{D}_{q_0,T}(u(\cdot),x) = D(T,x)W^{-1}D^T(T,x) + C(T,x)\int_0^T \Phi(T,s)B(s)R^{-1}(s)B^T(s)\Phi^T(T,s)ds\; C^T(T,x) \tag{10.8}$$

of the variational system has rank r. In the language of robotics the matrix (10.8) is named the dexterity matrix of the mobile manipulator [32]. Ultimately, the dexterity matrix characterizes the transmission of infinitesimal motions from the endogenous configuration space to the operational space of the mobile manipulator.

The following observations will provide a link between the concept of singular endogenous configuration and the singular optimal control. Consider in the system (10.1) a parametric optimal control problem, see [34, Sect. 7.5], with the manipulator joint positions $x \in R^p$ playing the role of a parameter vector, consisting in determining a control-parameter pair $(u(t), x)$ that steers the output $y(t)$ at the time instant T to a desired point $y_d \in R^r$, and simultaneously minimizes the objective function

$$\int_0^T \mathscr{L}(q(t), x, u(t))dt.$$

In order to include the parameter vector into the problem formulation, we need to complete the Eq. (10.1) with a trivial state equation $\dot{x} = 0$, obtaining the system

$$\dot{q} = G(q)u = \sum_{i=1}^m g_i(q)u_i,\ \dot{x} = 0, \quad y = k(q,x) = (k_1(q,x),\ldots,k_r(q,x)). \tag{10.9}$$

The Hamiltonian corresponding to this problem is

$$\mathscr{H}(q,x,u,p_0,p_1,p_2) = p_1^T\dot{q} + p_2^T\dot{x} - p_0\mathscr{L}(q,x,u) = p_1^T G(q)u - p_0\mathscr{L}(q,x,u), \tag{10.10}$$

where $p_0 \in R$, $p_1 \in R^n$, $p_2 \in R^p$ denote adjoint variables. The terminal manifold

$$M_T = \{(q,x) \in R^{n+p} | y_d - k(q,x) = 0\}.$$

We have the following control theoretic characterization of singular endogenous configurations

Theorem 10.1 *The set of singular endogenous configurations coincides with the set of singular optimal control-parameter pairs of the system (10.9) with Hamiltonian (10.10).*

Proof First, suppose that $(u(t), q(t), x)$ is a singular extremal of the parametric optimal control problem. This means that along this extremal $p_0 = 0$, $p_1^T(t)G(q(t)) = 0$, and that the adjoint vector $p(t) = (p_1(t), p_2(t)) \neq 0$ satisfies the Hamiltonian equations

$$\dot{p}_1^T = -\frac{\partial H}{\partial q} = -p_1^T(t)\frac{\partial (G(q(t))u(t))}{\partial q}, \quad \dot{p}_2^T = -\frac{\partial H}{\partial x} = 0. \tag{10.11}$$

The transversality conditions at T take the form

$$\left(p_1^T(T), p_2^T(T)\right) = \rho^T\left[\frac{\partial k(q(T),x)}{\partial q}, \frac{\partial k(q(T),x)}{\partial x}\right]$$

for a non-zero vector $\rho \in R^r$. Using (10.11), we deduce that $p_2(t) = const$ and, since there are no a priori initial or terminal conditions for x, we get $p_2 = 0$. Taking into account (10.6), the transversality conditions imply that

$$\rho^T C(T, x) = p_1^T(T) \text{ and } \rho^T D(T, x) = 0. \tag{10.12}$$

Furthermore, the singularity condition and the first equation in (10.11) yield

$$p_1^T(t)G(q(t)) = p_1^T(t)B(t) = 0 \quad \text{and} \quad \dot{p}_1^T = -p_1^T(t)A(t). \tag{10.13}$$

To proceed further, let us define a matrix

$$\mathscr{M}(t) = \int_0^t \Phi(t, s)B(s)R^{-1}(s)B^T(s)\Phi^T(t, s)ds, \tag{10.14}$$

such that the dexterity matrix

$$\mathscr{D}_{q_0,T}(u(\cdot), x) = D(T, x)W^{-1}D^T(T, x) + C(T, x)\mathscr{M}(T)C^T(T, x).$$

It is easily seen that the matrix (10.14) satisfies a Lyapunov matrix differential equation

$$\frac{d\mathscr{M}}{dt} = B(t)R^{-1}(t)B^T(t) + A(t)\mathscr{M}(t) + \mathscr{M}(t)A^T(t), \tag{10.15}$$

with initial condition $\mathscr{M}(0) = 0$. After the multiplication from the left of this equation by $p_1^T(t)$ and suitable substitutions from (10.13), we obtain

$$p_1^T(t)\frac{d\mathscr{M}}{dt} = -\dot{p}_1^T(t)\mathscr{M}(t) + p_1^T(t)\mathscr{M}(t)A^T(t),$$

so consequently,

$$\frac{d}{dt}(p_1^T\mathscr{M})(t) = (p_1^T\mathscr{M})(t)A^T(t). \tag{10.16}$$

Now, since $(p_1^T\mathscr{M})(0) = 0$, from the existence and uniqueness of the solution of (10.16) we deduce that $p_1^T(t)\mathscr{M}(t) = p_1^T(T)\mathscr{M}(T) = 0$. Together with (10.12) and

(10.14) this results in $\rho^T \mathscr{D}_{q_0,T}(u(\cdot), x)\rho = p_1^T(T)\mathscr{M}(T)p_1(T) = 0$. In conclusion, we get a rank deficiency of the dexterity matrix, what means that the endogenous configuration $(u(\cdot), x)$ is singular. This shows that every singular optimal pair control-parameter is a singular endogenous configuration.

Conversely, now let $(u(\cdot), x)$ be a singular endogenous configuration, and let the triple $(u(t), x, q(t))$ denote the corresponding control-parameter-trajectory of the control system (10.9). From the singularity of the configuration, it follows that there exists a non-zero vector $\rho \in R^r$ such that $\rho^T \mathscr{D}_{q_0,T}(u(\cdot), x)\rho = 0$, i.e.

$$\rho^T D(T, x) W^{-1} D^T(T, x)\rho + \rho^T C(T, x)\mathscr{M}(T)C^T(T, x)\rho = 0.$$

Since both above components are ≥ 0, we get

$$\rho^T D(T, x) = 0 \text{ and } \int_0^T ||R^{-1/2}(s)B^T(s)\Phi^T(T, s)C^T(T, x)\rho||^2 ds = 0,$$

$R^{-1/2}(t)$ denoting the square root of $R^{-1}(t)$. The latter equality implies that for $t \leq T$

$$\rho^T C(T, x)\Phi(T, t)B(t) = 0. \tag{10.17}$$

Now, by the definition of the Hamiltonian (10.10), we have at $p_0 = 0$

$$\dot{p}_1^T(t) = -\frac{\partial H}{\partial q} = -p_1^T(t)A(t),$$

so

$$p_1^T(t) = p_1^T(T)\Phi(T, t), \tag{10.18}$$

for $p_1^T(T) = \rho^T C(T, x)$. Finally, using (10.17) we conclude that $p_1^T(t)B(t) = p_1^T(t)G(q(t)) = 0$, i.e. $(u(t), x)$ is a singular optimal control-parameter pair for the system (10.9). □

The observation that singular endogenous configurations of the mobile manipulators coincide with singular optimal control-parameter pairs sometimes facilitates the computation of the singular endogenous configurations for mobile manipulators or mobile robots. In [9] an example of the kinematic car has been studied in this way. Furthermore, in the specific case when the system (10.1) is analytic, and so are the control functions $u(t)$ driving this system, the computation of singular endogenous configurations can be based on the necessary and sufficient rank condition for controllability of the variational system (10.7), see [20, Proposition 3.5.15] or [19]. An extension of this condition to the case of the output controllability results in the following.

Proposition 10.1 *In the analytic case the endogenous configuration* $(u(\cdot), x)$ *is singular, if and only if the infinite matrix*

$$\mathscr{T}_{q_0,T}(u(\cdot), x) = \big[D(T, x), C(T, x)[B_0(T), \ldots, B_k(T), \ldots]\big]$$

has rank $< r$*, where for* $k = 0, 1, \ldots$

$$B_0(t) = B(t), \quad B_{k+1}(t) = \dot{B}_k(t) - A(t)B_k(t).$$

This Proposition generalizes to mobile manipulators the condition used in [18] for the singularity assessment of the mobile platform kinematics.

10.2.4 Adjoint Jacobian

As has already been mentioned, the mobile manipulator's Jacobian

$$J_{q_0,T}(u(\cdot), x) : \mathscr{X} \longrightarrow R^r$$

is a linear transformation of the endogenous configuration space into the operational space. Its dual operator,

$$J^*_{q_0,T}(u(\cdot), x) : R^{r*} \longrightarrow \mathscr{X}^*,$$

acting between respective dual spaces is called the adjoint Jacobian of the mobile manipulator [26]. The canonical pairings between $\mathscr{X}^*$ and $\mathscr{X}$ as well as between R^{r*} and R^r are defined by the respective inner products (10.2) and (10.3).

Choosing a dual vector $\eta^* \in R^{r*}$, and an endogenous configuration $(v(\cdot), w) \in \mathscr{X}$, we define the adjoint Jacobian in the following way

$$\begin{aligned}\Big(J^*_{q_0,T}(u(\cdot), x)\eta^*\Big)(v(\cdot), w) &= \eta^* J_{q_0,T}(u(\cdot), x)(v(\cdot), w) = \langle \eta, J_{q_0,T}(u(\cdot), x)(v(\cdot), w)\rangle_Q \\ &= \eta^T QC(T, x)\int_0^T \Phi(T, t)B(t)v(t)dt + \eta^T QD(T, x)w \\ &= \langle (R^{-1}(\cdot)B^T(\cdot)\Phi^T(T, \cdot)C^T(T, x)Q\eta, W^{-1}D^T(T, x)Q\eta), (v(\cdot), w)\rangle_{RW}.\end{aligned}$$

It is straightforward to see that the last identity allows to define a function

$$(J^*_{q_0,T}(u(\cdot), x)\eta^*)(t) = \big[R^{-1}(t)B^T(t)\Phi^T(T, t)C^T(T, x),\ W^{-1}D^T(T, x)\big]Q\eta. \tag{10.19}$$

Being adjoint to the compact operator, the adjoint Jacobian is also compact for every fixed endogenous configuration. The operator norm of the adjoint Jacobian can be computed as

$$||J^*_{q_0,T}(u(\cdot),x)|| = \sup_{\eta\neq 0} \frac{||J^*_{q_0,T}(u(\cdot),x)\eta^*||_{RW}}{||\eta||_Q}.$$

We have the following proposition:

Proposition 10.2 *The norm of the adjoint Jacobian*

$$||J^*_{q_0,T}(u(\cdot),x)|| = \bar{\lambda}^{1/2}_{Q^{1/2}\mathscr{D}_{q_0,T}(u(\cdot),x)Q^{1/2}} = \bar{\lambda}^{1/2}_{\mathscr{D}_{q_0,T}(u(\cdot),x)Q}, \tag{10.20}$$

where $\bar{\lambda}_M$ denotes the maximal eigenvalue of a symmetric, positive matrix M.

Proof Substituting the identity (10.19) into the definition of the RW-norm, and using the formula (10.8) for the dexterity matrix, we obtain

$$||J^*_{q_0,T}(u(\cdot),x)\eta^*||^2_{RW} = \eta^T Q^T \mathscr{D}_{q_0,T}(u(\cdot),x)Q\eta.$$

An application of the Rayleigh–Ritz inequality yields

$$||J^*_{q_0,T}(u(\cdot),x)|| = \bar{\lambda}^{1/2}_{Q^{1/2}\mathscr{D}_{q_0,T}(u(\cdot),x)Q^{1/2}}. \tag{10.21}$$

Furthermore, since the matrix Q is positive definite, there exists a positive definite square root $Q^{1/2}$, so we have

$$\mathscr{D}_{q_0,T}(u(\cdot),x)Q = Q^{-1/2}(Q^{1/2}\mathscr{D}_{q_0,T}(u(\cdot),x)Q^{1/2})Q^{1/2}.$$

Now, the second identity of (10.20) follows from the matrix similarity. □

10.2.5 Jacobian Norm

Taking into account Proposition 10.2, from the identity of the operator norms of the Jacobian and of the adjoint Jacobian [14], we deduce a straightforward

Corollary 10.1 *The norm of the Jacobian is given by*

$$||J_{q_0,T}(u(\cdot),x)|| = \bar{\lambda}^{1/2}_{Q^{1/2}\mathscr{D}_{q_0,T}(u(\cdot),x)Q^{1/2}} = \bar{\lambda}^{1/2}_{\mathscr{D}_{q_0,T}(u(\cdot),x)Q}. \tag{10.22}$$

Notice that a computation of the maximal eigenvalue of a symmetric matrix means solving the following optimization problem [5, Corollary III.1.2]

$$\bar{\lambda}_{Q^{1/2}\mathscr{D}_{q_0,T}(u(\cdot),x)Q^{1/2}} = \max_{||w||=1} w^T Q^{1/2}\mathscr{D}_{q_0,T}(u(\cdot),x)Q^{1/2}w, \tag{10.23}$$

where $||\cdot||$ denotes the standard Euclidean norm in R^r. It is instructive to see this problem in the context of the parametric time optimal control version of the problem studied in the previous subsection, compare [9]. The following statement can be made:

Proposition 10.3 *The computation of the norm of the Jacobian involves the vector $\phi_{x,w,T,Q}(\cdot)$ of switching functions associated with time optimal control in (10.9), i.e.*

$$\bar{\lambda}_{Q^{1/2}\mathscr{D}_{q_0,T}(u(\cdot),x)Q^{1/2}} = \max_{||w||=1} ||(\phi_{x,w,T,Q}(\cdot), D^T(T,x)Q^{1/2}w)||^2_{RW}, \tag{10.24}$$

where $\phi_{x,w,T,Q}(\cdot)$ will be specified in the course of the proof.

Proof We shall use the Hamiltonian (10.10) with $\mathscr{L}(q,x,u) = 1$ and $p_0 = 1$. Let $(u(t), q(t), x)$ be an extremal of the problem. Along this extremal we get

$$\mathscr{H}(q(t), x, u(t), p_1(t), p_2) = p_1^T(t)B(t)u(t) - 1 \text{ and } \dot{p}_1^T = -p_1^T(t)A(t).$$

The components of the vector

$$p_1^T(t)B(t) = (p_1^T(t)B_1(t), \ldots, p_1^T(t)B_m(t)),$$

where $B_i(t)$ stands for the ith column of the matrix $B(t)$, are called switching functions. By (10.18) it follows that $p_1^T(t) = p_1^T(T)\Phi(T,t)$. We choose a vector $w \in R^r$, and let $p_1(T) = C^T(T,x)Q^{1/2}w$. Then, the following identity holds

$$w^TQ^{1/2}\mathscr{D}_{q_0,T}(u(\cdot),x)Q^{1/2}w = w^TQ^{1/2}D(T,x)W^{-1}D^T(T,x)Q^{1/2}w + \int_0^T p_1^T(t)B(t)R^{-1}(t)B^T(t)p_1(t)dt.$$

Denote by

$$\phi_{x,w,T,Q}(t) = \left(p_1^T(t)B(t)\right)^T \tag{10.25}$$

the vector of switching functions satisfying $p_1(T) = C^T(T,x)Q^{1/2}w$. A substitution of this vector into the above identity results in

$$w^TQ^{1/2}\mathscr{D}_{q_0,T}(u(\cdot),x)Q^{1/2}w = ||(\phi_{x,w,T,Q}(\cdot), D^T(T,x)Q^{1/2}w)||^2_{RW}. \tag{10.26}$$

A comparison with (10.25) concludes the proof. □

10.3 Inverse Kinematics

One of the fundamental problems of robotics is the inverse kinematics problem. Its formulation in the context of mobile manipulators is the following: given the kinematics (10.4) and a desired point y_d in the operational space, find an endogenous configuration $(u_d(\cdot), x_d)$ such that $K_{q_0,T}(u_d(\cdot), x_d) = y_d$. Equivalently, this means that the control $u_d(t)$ and the joint position x_d will drive the output of the control system (10.1) initialized at q_0 to $y(T) = y_d$. The inverse kinematics problem is usually solved numerically by means of a Jacobian inverse kinematics algorithm.

10.3.1 Dynamic Inverses

The Jacobian algorithms can be conveniently derived using a reasoning that originates in the continuation method [2, 22]. Given the inverse kinematics problem, we begin with any initial endogenous configuration $(u_0(\cdot), x_0) \in \mathscr{X}$. If the initial choice does not solve the problem, i.e. $K_{q_0,T}(u_0(\cdot)) \neq y_d$, we choose in $\mathscr{X}$ a differentiable curve $(u_\theta(\cdot), x(\theta))$, defined for $\theta \in R$ and passing at $\theta = 0$ through the initial configuration, and compute along this curve the operational space error

$$e(\theta) = K_{q_0,T}(u_\theta(\cdot), x(\theta)) - y_d. \tag{10.27}$$

Next, we request that the error dynamics obey a differential equation

$$\frac{d\,e(\theta)}{d\theta} = -\gamma S(\theta) e(\theta), \tag{10.28}$$

for a certain $r \times r$ matrix $S(\theta)$ whose properties will be specified later, and a positive number γ. After differentiating the formula (10.27) with respect to θ, we arrive at the Ważewski–Davidenko equation [10, 35]

$$J_{q_0,T}(u_\theta(\cdot), x(\theta)) \frac{d}{d\theta}(u_\theta(\cdot), x(\theta)) = -\gamma S(\theta) e(\theta), \tag{10.29}$$

containing the Jacobian (10.5). Now, let us choose an operator

$$J^{\#}_{q_0,T}(u(\cdot), x) : R^r \to \mathscr{X}$$

and define the dynamic system

$$\frac{d}{d\theta}(u_\theta(\cdot), x(\theta)) = -\gamma J^{\#}_{q_0,T}(u_\theta(\cdot), x(\theta)) e(\theta), \tag{10.30}$$

initialized at $(u_0(\cdot), x_0)$. The choice of $J^{\#}_{q_0,T}(u(\cdot), x)$ should guarantee at least a local existence of solutions of this system. Now, a substitution of (10.30) into the Ważewski–Davidenko equation results in the error Eq. (10.28) with

$$S(\theta) = J_{q_0,T}(u_\theta(\cdot), x(\theta)) J^{\#}_{q_0,T}(u_\theta(\cdot), x(\theta)). \tag{10.31}$$

wherever a trajectory $(u_\theta(t), x(\theta))$ of (10.30) exists, the solution of the error equation (10.28) satisfies the Ważewski inequality [36]

$$||e(0)||_Q \exp\left(\int_0^\theta \underline{\lambda}_{M(\theta)} ds\right) \le ||e(\theta)||_Q \le ||e(0)||_Q \exp\left(\int_0^\theta \overline{\lambda}_{M(\theta)} ds\right), \quad (10.32)$$

with $\underline{\lambda}_M$ and $\overline{\lambda}_M$ denoting, respectively, the minimum and the maximum eigenvalue of the matrix

$$M(\theta) = -\frac{1}{2}\gamma \left(Q^{-1/2}(QS(\theta) + S^T(\theta)Q)Q^{-1/2}\right). \quad (10.33)$$

The matrix $S(\theta)$ should make all eigenvalues of $M(\theta)$ to be ≤ 0. Moreover, if $(u_\theta(t), x(\theta))$ is defined for every $\theta \ge 0$ and the integral $\lim_{\theta\to+\infty} \int_0^\theta \overline{\lambda}_{M(\theta)} ds = -\infty$, then the error $e(\theta)$ vanishes asymptotically. When these conditions are satisfied, the operator $J^{\#}_{q_0,T}(u(\cdot), x)$ will be called a dynamic inverse of the mobile manipulator's Jacobian. The term "dynamic inverse" has been borrowed from [13] to denote an inversion operation performed by a dynamic system. After plugging the dynamic inverse into the dynamic system (10.30), we transform this system into an inverse kinematics algorithm producing in the limit a solution to the inverse kinematics problem, so that

$$\lim_{\theta\to+\infty} \left(u_\theta(\cdot), x(\theta)\right) = \left(u_d(\cdot), x_d\right).$$

By design, the dynamic inverse transforms an initial endogenous configuration $(u_0(\cdot), x_0) \in \mathscr{X}$ into a solution $(u_d(\cdot), x_d)$ of the inverse kinematics problem, exploiting, as a vehicle, the dynamics (10.30). In general, the dynamic inverse may be defined only locally. This being so, a challenging task in designing the inverse kinematics algorithms consists in guaranteeing the completeness of (10.30), i.e. the existence of its solutions for every initial configuration $(u_0(\cdot), x)$ and every $\theta \in R$ (obviously, $\theta \ge 0$ suffices). In this context the following result can be proved.

Theorem 10.2 *Let $J^{\#}_{q_0,T}(u(\cdot), x)$ be a local dynamic inverse of the Jacobian. Suppose additionally that*

$$||J^{\#}_{q_0,T}(u(\cdot), x)|| \le a||(u(\cdot), x)||_{RW} + b, \quad (10.34)$$

where $||\cdot||$ stands for the operator norm, and $a, b \ge 0$ are certain numbers dependent on q_0 and T. Then, $J^{\#}_{q_0,T}(u(\cdot), x)$ is a global dynamic inverse.

Proof Let $(u_\theta(\cdot), x(\theta))$ denote a solution of (10.30) defined for $0 \le \theta < \alpha$. Since along this solution $\lambda_{M(\theta)} \le 0$, the Ważewski inequality (10.32) results in $||e(\theta)||_Q \le ||e(0)||_Q$, so the error is bounded for every $\theta < \alpha$. Now, using (10.30) and invoking (10.34) we deduce

$$||(u_\theta(\cdot), x(\theta))||_{RW} \le ||(u_0(\cdot), x_0)||_{RW} + ||e(0)||_Q \int_0^\theta ||J^{\#}_{q_0,T}(u_\sigma(\cdot), x(\sigma))||d\sigma$$
$$\le ||(u_0(\cdot), x_0)||_{RW} + a||e(0)||_Q \int_0^\theta ||(u_\sigma(\cdot), x(\sigma))||_{RW} d\sigma + b||e(0)||_Q \theta.$$

Let $a > 0$. Then, an application of the Gronwall theorem [33, Theorem 1.3] yields for $0 \le \theta < \alpha$

$$||(u_\theta(\cdot), x(\theta))||_{RW} \le \left(\frac{b}{a} + ||(u_0(\cdot), x_0)||_{RW}\right) \exp(a\theta||e(0)||_Q) - \frac{b}{a}. \tag{10.35}$$

Using (10.34) and (10.35), we show the boundedness of the derivative,

$$||\frac{d}{d\theta}\big(u_\theta(\cdot), x(\theta)\big)||_{RW} \le \gamma||J^{\#}_{q_0,T}(u(_\theta\cdot), x(\theta))||||e(\theta)||_Q$$
$$\le \gamma(b + a||(u_0(\cdot), x_0)||_{RW}) \exp(a\theta||e(0)||_Q)||e(0)||_Q$$
$$\le \gamma(b + a||(u_0(\cdot), x_0)||_{RW}) \exp(a\alpha||e(0)||_Q)||e(0)||_Q = A.$$

To proceed, choose a sequence $\theta_n \to \alpha$, and let θ_m be another sequence such that $\theta_n < \theta_m$. As in [1, Proposition 4.1.22], we compute

$$||(u_{\theta_m}(\cdot), x(\theta_m)) - (u_{\theta_n}(\cdot), x(\theta_n))||_{RW} \le \int_{\theta_n}^{\theta_m} ||\frac{d}{d\sigma}\big(u_\sigma(\cdot), x(\sigma)\big)||_{RW} d\sigma \le A|\theta_m - \theta_n|,$$

and conclude that the sequence $(u_{\theta_n}(\cdot), x(\theta_n))$ is Cauchy, so it converges when $\theta_n \to \alpha$. This means that the solution $(u_\theta(\cdot), x(\theta))$ of (10.30) can be extended to $\theta = \alpha$, so it exists for every θ. If $a = 0$, we get the same conclusion for $A = \gamma b||e(0)||_Q$. □

With reference to the condition (10.34), in the following subsections we compute the norms of several most often encountered dynamic inverses. We shall start from the Jacobian pseudo-inverse.

10.3.2 Jacobian Pseudo-inverse

For a fixed configuration $(u(\cdot), x)$ and a given vector $\eta \in R^r$, let us consider a Jacobian equation

$$J_{q_0,T}(u(\cdot), x)(v(\cdot), w) = \eta. \tag{10.36}$$

By definition, at a regular $(u(\cdot), x)$ the Eq. (10.36) is solvable with respect to $(v(\cdot), w)$, for any η. Typically, a solution of this equation is found by the least squares method that is tantamount to minimizing the squared norm

$$\min_{(v(\cdot),w)} ||(v(\cdot), w)||^2_{RW}$$

under the equality constraint (10.36). A standard application of the optimization techniques results in the following formula

$$\begin{aligned}\big(v(t), w\big) &= \Big(J^{\#P}_{q_0,T}(u(\cdot), x)\eta\Big)(t)\\ &= \big[R^{-1}(t)B^T(t)\Phi^T(T,t)C^T(T,x),\ W^{-1}D^T(T,x)\big]\mathscr{D}^{-1}_{q_0,T}(u(\cdot), x)\eta, \quad (10.37)\end{aligned}$$

see [26], dependent on the inverse dexterity matrix (10.8). The operator

$$J^{\#P}_{q_0,T}(u(\cdot), x) : R^r \longrightarrow \mathscr{X} \qquad (10.38)$$

defined by (10.37) is called the Jacobian pseudo-inverse (the Moore–Penrose inverse of the Jacobian). By the smoothness of the Jacobian, outside singular configurations, we have the local existence of solutions of the dynamic system (10.30) containing the operator (10.38). Also, it is easily checked that along the solution $(u_\theta(\cdot), x(\theta))$

$$S(\theta) = J_{q_0,T}(u_\theta(\cdot), x(\theta))J^{\#P}_{q_0,T}(u_\theta(\cdot), x(\theta)) = I_r,$$

and $M(\theta) = -\gamma I_r$. Summarizing these observations we have

Corollary 10.2 *Suppose that $(u_\theta(t), x(\theta))$ exists for every $\theta \geq 0$ and stays away of singular configurations. Then, the Jacobian pseudo-inverse is a local dynamic inverse of the Jacobian. Its error (10.28) vanishes exponentially with decay rate γ.*

The inverse kinematics algorithm (10.30) that involves the operator (10.38) will be referred to as the Jacobian pseudo-inverse kinematics algorithm. The completeness of this algorithm for mobile manipulators has not been established, however specific results are known for mobile platforms like the unicycle and the kinematic car [9], [8], as well as for the rolling ball [7].

Our next result refers to the norm of the Jacobian pseudo-inverse.

Proposition 10.4 *The operator norm of the Jacobian pseudo-inverse*

$$||J^{\#P}_{q_0,T}(u(\cdot), x)|| = \underline{\lambda}^{-1/2}_{Q^{1/2}\mathscr{D}_{q_0,T}(u(\cdot),x)Q^{1/2}} = \underline{\lambda}^{-1/2}_{\mathscr{D}_{q_0,T}(u(\cdot),x)Q}, \qquad (10.39)$$

where $\underline{\lambda}_M$ denotes the minimal eigenvalue of a symmetric, positive matrix M.

Proof The equality of eigenvalues on the right hand side results from similarity of the corresponding matrices, therefore it suffices to prove the first equality. By definition

$$||J^{\#P}_{q_0,T}(u(\cdot), x)|| = \sup_{\eta\neq 0}\frac{||J^{\#P}_{q_0,T}(u(\cdot), x)\eta||_{RW}}{||\eta||_Q}.$$

After a substitution from the formula (10.37), we obtain

$$||J^{\#P}_{q_0,T}(u(\cdot), x)\eta||^2_{RW} = \eta^T\mathscr{D}^{-1}_{q_0,T}(u(\cdot), x)\eta,$$

so, in consequence,

$$||J^{\#P}_{q_0,T}(u(\cdot),x)|| = \overline{\lambda}^{-1/2}_{(Q^{1/2}\mathscr{D}_{q_0,T}(u(\cdot),x)Q^{1/2})^{-1}} = \underline{\lambda}^{-1/2}_{Q^{1/2}\mathscr{D}_{q_0,T}(u(\cdot),x)Q^{1/2}}.$$

□

It follows that the norm explodes at the singular endogenous configurations. The result (10.39) generalizes the norm estimate for mobile platforms exploited in [9]. We shall conclude this subsection with the following consequence of [5, Corollary III.1.2] and Proposition 10.5.

Corollary 10.3 *The computation of the minimal eigenvalue is equivalent to solving the minimization problem*

$$\begin{aligned}\underline{\lambda}_{Q^{1/2}\mathscr{D}_{q_0,T}(u(\cdot),x)Q^{1/2}} &= \min_{||w||=1} w^T Q^{1/2}\mathscr{D}_{q_0,T}(u(\cdot),x)Q^{1/2}w \\ &= \min_{||w||=1} ||(\phi_{x,w,T,Q}(\cdot), D^T(T,x)Q^{1/2}w)||^2_{RW}, \quad (10.40)\end{aligned}$$

involving the vector of switching functions (10.25).

10.3.3 Singularity Robust Jacobian Inverse

From the previous subsection, we deduce that the performance of the Jacobian pseudo-inverse algorithm deteriorates in a vicinity of singular configurations of the mobile manipulator. In order to prevent this deterioration, a modification of the formula (10.37) can be made, leading to the concept of a singularity robust inverse of the Jacobian [26]. This type of inverse results from the following optimization problem

$$\min_{(v(\cdot),w)} \left(\kappa||(v(\cdot),w)||^2_{RW} + ||J_{q_0,T}(u(\cdot),x)(v(\cdot),w) - \eta||^2_Q\right),$$

$\kappa > 0$ being a regularizing parameter, whose solution provides the singularity robust dynamic Jacobian inverse

$$J^{\#SRI}_{q_0,T}(u(\cdot),x) : R^r \longrightarrow \mathscr{X},$$

defined as

$$\begin{aligned}&\left(J^{\#SRI}_{q_0,T}(u(\cdot),x)\eta\right)(t) \\ &= \left[R^{-1}(t)B^T(t)\Phi^T(T,t)C^T(T,x),\ W^{-1}D^T(T,x)\right]Q\left(\kappa I_r + \mathscr{D}_{q_0,T}(u(\cdot),x)Q\right)^{-1}\eta. \quad (10.41)\end{aligned}$$

The operator (10.41) is well defined both at regular and at singular configurations, and ensures the local existence of solutions of (10.30). Let $(u_\theta(t), x(\theta))$ denote such a solution. Then, the matrix $S(\theta)$ appearing in the error equation (10.28) is equal to

$$S(\theta) = J_{q_0,T}(u_\theta(\cdot), x(\theta))J^{\#SRI}_{q_0,T}(u_\theta(\cdot), x(\theta))$$
$$= \mathscr{D}_{q_0,T}(u_\theta(\cdot), x(\theta))Q(\kappa I_r + \mathscr{D}_{q_0,T}(u_\theta(\cdot), x(\theta))Q)^{-1}. \quad (10.42)$$

In this context we have the following results, for proofs consult [29].

Proposition 10.5 *The operator norm of the singularity robust Jacobian inverse*

$$||J^{\#SRI}_{q_0,T}(u(\cdot), x)|| = \overline{\lambda}^{1/2}_{(\kappa I_r + \mathscr{D}_{q_0,T}(u(\cdot),x)Q)^{-2}\mathscr{D}_{q_0,T}(u(\cdot),x)Q} \quad (10.43)$$

is upper bounded by a constant

$$||J^{\#SRI}_{q_0,T}(u(\cdot), x)|| \leq \frac{1}{2}\kappa^{-1/2}. \quad (10.44)$$

Combined with Theorem 10.2, Proposition 10.5 implies the following

Theorem 10.3 *The singularity robust dynamic Jacobian inverse is global. The singularity robust Jacobian inverse kinematics algorithm converges provided that the following integral diverges,*

$$\lim_{\theta\to+\infty}\int_0^\theta \frac{\underline{\lambda}_{\mathscr{D}_{q_0,T}(u_\alpha(\cdot),x(\alpha))Q}d\alpha}{\kappa + \underline{\lambda}_{\mathscr{D}_{q_0,T}(u_\alpha(\cdot),x(\alpha))Q}} = +\infty. \quad (10.45)$$

10.3.4 Adjugate Dexterity Matrix Jacobian Inverse

A dynamic inverse Jacobian operator, alternative to the singularity robust inverse, that also remains well defined at singular configurations, comes from a generalization of the Newton method [4, Sect. 2.7.3]. The introduction of this operator proceeds in the following way. Given a desired operational space point $y_d \in R^r$, we compute the map $\tilde{K}_{q_0,T}(u(\cdot), x) = K_{q_0,T}(u(\cdot), x) - y_d$, and define a ray

$$\rho = \{\alpha\eta | \alpha \in R\}$$

passing through a point $\eta \in R^r$. Then, we assume that the kinematics (10.4) of the mobile manipulator are transverse to the ray, what means that if $\tilde{K}_{q_0,T}(u(\cdot), x) \in \rho$ then

$$dim\left(\{D\tilde{K}_{q_0,T}(u(\cdot), x))\mathscr{X}, \rho\} = \{J_{q_0,T}(u(\cdot), x)\mathscr{X}, \rho\}\right) = r,$$

or, equivalently,

$$rank\left[\mathscr{D}_{q_0,T}(u(\cdot), x), \eta\right] = r. \quad (10.46)$$

The transversality condition means that the columns of the dexterity matrix together with the ray span the operational space, so the transversality takes place not only at regular configurations ($rank\mathscr{D}_{q_0,T}(u(\cdot),x)=r$), but also at those singular configurations, where the rank of $\mathscr{D}_{q_0,T}(u(\cdot),x)$ drops by 1. In the transverse situation the inverse image $M_\rho=\tilde{K}^{-1}_{q_0,T}(\rho)$ of the ray becomes a codimension $r-1$ Hilbert submanifold of the endogenous configuration space, [1]. Suppose that $(u(\cdot),x)\in M_\rho$, and let $(v(\cdot),w)$ be tangent to M_ρ at this point. Then, the image of $(v(\cdot),w)$ by the Jacobian lies in the tangent space to the ray (identified with the ray), i.e.

$$J_{q_0,T}(u(\cdot),x)(v(\cdot),w)=\alpha\eta, \tag{10.47}$$

for a certain $\alpha=\alpha(u(\cdot),x)\in R$. The Eq. (10.47) is a sort of Jacobian equation whose least squares solution is equal to

$$v(t)=-R^{-1}(t)B^T(t)\Phi^T(T,t)C^T(T,x)\lambda,\;\; w=-W^{-1}D^T(T,x)\lambda, \tag{10.48}$$

with $\lambda\in R^r$ denoting a vector of Lagrange multipliers. A substitution of (10.48) into (10.47) results in

$$\mathscr{D}_{q_0,T}(u(\cdot),x)\lambda=\alpha\eta. \tag{10.49}$$

Observe that the existence of a solution (λ,α) of (10.49) is guaranteed by the transversality condition (10.12). A multiplication of (10.49) by the adjugate dexterity matrix $\mathrm{ad}\mathscr{D}_{q_0,T}(u(\cdot),x)$ (by definition, for an $n\times n$ matrix M, we have $\mathrm{ad}M\,M=M\mathrm{ad}M=I_n\det M$) yields

$$\lambda=\mathrm{ad}\mathscr{D}_{q_0,T}(u(\cdot),x)\eta,\;\; \alpha=\det\mathscr{D}_{q_0,T}(u(\cdot),x). \tag{10.50}$$

Finally, the solution (10.48) gives rise to the adjugate dexterity matrix Jacobian inverse operator

$$J^{\#ADM}_{q_0,T}(u(\cdot),x):R^r\to\mathscr{X}$$

such that

$$\begin{aligned}&\left(J^{\#ADM}_{q_0,T}(u(\cdot),x)\eta\right)(t)\\&=\left[R^{-1}(t)B^T(t)\Phi^T(T,t)C^T(T,x),\;W^{-1}D^T(T,x)\right]\mathrm{ad}\mathscr{D}_{q_0,T}(u(\cdot),x)\eta.\end{aligned} \tag{10.51}$$

Strictly speaking, this inverse operator is defined locally, in the domain

$$U_{(u(\cdot),x)}=\{\eta\in R^r|rank[\mathscr{D}_{q_0,T}(u(\cdot),x),\eta]=r\},$$

since $U_{(u(\cdot),x)}=R^r$ only at a regular configuration $(u(\cdot),x)$. Plugged into (10.30), the inverse (10.51) produces a local solution $(u_\theta(t),x(\theta))$, along which the matrix

$$S(\theta) = J_{q_0,T}(u_\theta(\cdot), x(\theta)) J^{\#ADM}_{q_0,T}(u_\theta(\cdot), x(\theta)) = \det \mathscr{D}_{q_0,T}(u_\theta(\cdot), x(\theta)) I_r,$$

therefore $M(\theta) = -\gamma S(\theta)$. We have come to the following

Corollary 10.4 *Suppose that $(u_\theta(t), x(\theta))$ exists for every $\theta \geq 0$. Then, the operator (10.51) becomes a local dynamic inverse of the Jacobian provided that*

$$\lim_{\theta \to +\infty} \int_0^\theta \det \mathscr{D}_{q_0,T}(u_\alpha(\cdot), x(\alpha)) d\alpha = +\infty. \tag{10.52}$$

Finally, let us compute the norm of the adjugate dexterity matrix inverse. Since for a symmetric matrix M there holds $(\mathrm{ad}M)^T = \mathrm{ad}M$, we get

Proposition 10.6 *The norm of the adjugate dexterity matrix dynamic Jacobian inverse*

$$\begin{aligned} ||J^{\#ADM}_{q_0,T}(u(\cdot), x)|| &= \sqrt{\det \mathscr{D}_{q_0,T}(u(\cdot), x)) \overline{\lambda}^{-1/2}_{Q^{-1/2}\mathrm{ad}\mathscr{D}_{q_0,T}(u(\cdot),x)Q^{-1/2}}} \\ &= \sqrt{\det \mathscr{D}_{q_0,T}(u(\cdot), x)) \overline{\lambda}^{-1/2}_{\mathrm{ad}\mathscr{D}_{q_0,T}(u(\cdot),x)Q^{-1}}}. \end{aligned} \tag{10.53}$$

As a consequence of this result, the adjugate dexterity matrix inverse kinematics algorithm has equilibrium points at singular configurations, what means that the algorithm is unable to leave a singularity. On the other hand, if a singular configuration is the solution of the inverse kinematics problem, we may expect that the algorithm will find such a solution. This type of behaviour has been confirmed by computer simulations presented in [26].

10.3.5 Adjoint Jacobian Inverse

In order to define this inverse, we shall exploit as a dynamic Jacobian inverse in (10.30) the adjoint Jacobian operator (10.19). Let us denote by $(u_\theta(t), x(\theta))$ a local solution of (10.30). Along this solution the matrix

$$S(\theta) = J_{q_0,T}(u_\theta(\cdot), x(\theta)) J^*_{q_0,T}(u_\theta(\cdot), x(\theta)) = \mathscr{D}_{q_0,T}(u_\theta(\cdot), x(\theta)) Q.$$

After invoking (10.33) and the inequality (10.32), we obtain the following

Corollary 10.5 *Suppose that $(u_\theta(t), x(\theta))$ exists for every $\theta \geq 0$. Then, the adjoint Jacobian operator defines a local dynamic inverse of the Jacobian on condition that*

$$\lim_{\theta \to +\infty} \int_0^\theta \underline{\lambda}_{\mathscr{D}_{q_0,T}(u_\alpha(\cdot),x(\alpha))Q} d\alpha = +\infty. \tag{10.54}$$

Obviously, the above condition will be fulfilled, if all eigenvalues of the dexterity matrix are uniformly bounded away from zero. The norm of the adjoint Jacobian operator has been provided by (10.21). Similarly to the adjugate dexterity matrix algorithm, the adjoint Jacobian inverse kinematics algorithm has also equilibrium points at singular endogenous configurations.

10.3.6 Extended Jacobian Inverse

A concept of the extended Jacobian inverse for mobile manipulators can be introduced in several steps [24, 27, 28]. The first step consists in extending the original mobile manipulator kinematics (10.4) to a map of the endogenous configuration space into itself. To this aim we need to decompose the endogenous configuration space into a pair of linear subspaces

$$\mathscr{X} \cong R^r \oplus \mathscr{X}/R^r, \tag{10.55}$$

of which the former corresponds to the operational space, and the latter forms the remaining quotient space. Next, we introduce an augmenting kinematics map

$$H_{q_0,T} : \mathscr{X} \longrightarrow \mathscr{X}/R^r \tag{10.56}$$

that takes values in the quotient space. The original kinematics together with the augmenting map (10.56) define the extended kinematics

$$L_{q_0,T} = (K_{q_0,T}, H_{q_0,T}) : \mathscr{X} \to \mathscr{X}$$

of the mobile manipulator. The derivative of the extended kinematics

$$D\,L_{q_0,T}(u(\cdot), x) = (J_{q_0,T}, D\,H_{q_0,T})(u(\cdot), x)) = \bar{J}_{q_0,T}(u(\cdot), x) \tag{10.57}$$

is called the extended Jacobian. It would be desirable that $\bar{J}_{q_0,T}(u(\cdot), x)$ be a linear isomorphism of the endogenous configuration space. However, usually this map not only suffers from the singularities of the original kinematics, but also exhibits some extra singularities, called algorithmic, that result from the extension procedure.

Given the extended kinematics, in the region of regular endogenous configurations, the extended Jacobian inverse operator

$$J^{\#E}_{q_0,T}(u(\cdot), x) : R^r \longrightarrow \mathscr{X}$$

is defined as

$$J^{\#E}_{q_0,T}(u(\cdot), x)\eta = \bar{J}^{-1}_{q_0,T}(u(\cdot), x)(\eta, 0(\cdot)), \tag{10.58}$$

where $\eta \in R^r$, and $0(\cdot) \in \mathscr{X}/R^r$ denotes the zero element of the quotient space. By definition, the operator (10.58) has two important properties: the identity property

$$J_{q_0,T}(u(\cdot),x)J^{\#E}_{q_0,T}(u(\cdot),x) = I_r, \tag{10.59}$$

and the annihilation property

$$D\,H_{q_0,T}(u(\cdot),x)J^{\#E}_{q_0,T}(u(\cdot),x) = 0(\cdot). \tag{10.60}$$

The former property means that the extended Jacobian inverse is a right inverse of the Jacobian, while the latter implies that the augmenting kinematics map remains constant on the trajectories produced by the extended Jacobian inverse kinematics algorithm. Equivalently, the distribution associated with the algorithm is involutive, ensuring its repeatability, see [23] for details.

Having plugged the inverse (10.58) into (10.30), we get a trajectory $(u_\theta(\cdot), x(\theta))$. The following result is a direct consequence of the identity (10.59).

Corollary 10.6 *Suppose that the trajectory $(u_\theta(t), x(\theta))$ exists for every $\theta \geq 0$, and stays away of singular configurations of the extended Jacobian (10.57). Then, the operator $J^{\#E}_{q_0,T}(u(\cdot),x)$ is a local dynamic inverse of the Jacobian. The error (10.28) decreases to* 0 *exponentially, with the rate γ. The extended Jacobian inverse kinematics algorithm is repeatable.*

10.3.7 Lagrangian Jacobian Inverse

This is an extension of the Jacobian pseudo-inverse, introduced recently in [31]. The Jacobian equation (10.36) is regarded as an equality constraint for the minimization of the Lagrange-type objective function

$$\min_{(v(\cdot),w)} \int_0^T \left(\xi^T(t)P(t)\xi(t) + v^T(t)Rv(t)\right)dt + w^T W w,$$

addressed in the variational system (10.7), where $P(t) = P^T(t) \geq 0$, and the remaining weight matrices come from the norm $||\cdot||_{RW}$. Using the variational calculus the following solution can be derived

$$\begin{aligned}(v(t), w) &= \left(J^{\#L}_{q_0,T}(u(\cdot),x)\eta\right)(t) \\ &= \Big[-R^{-1}(t)B^T(t)\psi_{22}(t)K(0),\, W^{-1}D^T(T,x)\mathscr{D}^{-1}_{q_0,T}(u(\cdot),x) \\ &\quad \times\,(\eta + C(T,x)\psi_{32}(T)K(0))\Big],\end{aligned} \tag{10.61}$$

where

$$K(0) = -\left(\psi_{22}(T) + C^T(T,x)\mathscr{D}^{-1}_{q_0,T}(u(\cdot),x)C(T,x)\psi_{32}(T)\right)^{-1} C^T(T,x)\mathscr{D}^{-1}_{q_0,T}(u(\cdot),x)\eta,$$

and the matrix functions $\psi_{22}(t)$, $\psi_{32}(t)$ satisfy the matrix differential equation

$$\dot{\Psi}(t) = \begin{bmatrix} \dot{\psi}_{11}(t) & \dot{\psi}_{12}(t) & \dot{\psi}_{13}(t) \\ \dot{\psi}_{21}(t) & \dot{\psi}_{22}(t) & \dot{\psi}_{23}(t) \\ \dot{\psi}_{31}(t) & \dot{\psi}_{32}(t) & \dot{\psi}_{33}(t) \end{bmatrix} = \begin{bmatrix} A(t) & -B(t)R^{-1}(t)B^T(t) & 0 \\ -P(t) & -A^T(t) & 0 \\ H(t)P(t) & 0 & A(t) \end{bmatrix} \Psi(t),$$

with initial condition $\psi_{ij}(0) = \delta_{ij} I_n$, δ_{ij} denoting the Kronecker's delta. The matrix $H(t)$ is a solution of the Lyapunov equation

$$\dot{H}(t) = B(t)R^{-1}(t)B^T(t) + A(t)H(t) + H(t)A^T(t),$$

with zero initial condition, so that the dexterity matrix (10.8)

$$\mathscr{D}_{q_0,T}(u(\cdot), x) = D(T, x)W^{-1}D^T(T, x) + C(T, x)H(T)C^T(T, x).$$

It is easily checked that $P(t) = 0$ implies $\psi_{32}(t) = 0$ and $\psi_{22}(t)\psi_{22}^{-1}(T) = \Phi^T(T, t)$, therefore the Lagrangian Jacobian inverse simplifies to the Jacobian pseudo-inverse. Plugged into a motion planning algorithm the Lagrangian Jacobian inverse allows to shape trajectories of the system (10.1), e.g. by repelling them from obstacles.

10.4 Performance

Kinematic performance of the mobile manipulator depends on the properties of the Jacobian operator. An assessment of the local performance can be made in terms of certain numerical indices associated with an endogenous configuration, called local performance measures. Given an endogenous configuration $(u(\cdot), x)$, these performance measures are related to the directions of motion in the operational space corresponding to the unit sphere

$$S_{q_0,T}(u(\cdot), x)) = \{(v(\cdot), w) \in \mathscr{X} \,|\, ||(v(\cdot), w)||_{RW} = 1\}$$

that describes possible variations of this configuration. Using the decomposition

$$\mathscr{X} = Ker\, J_{q_0,T}(u(\cdot), x) \oplus Im\, J^{\#P}_{q_0,T}(u(\cdot), x)$$

of the endogenous configuration space, we obtain

$$\begin{aligned} J_{q_0,T}(u(\cdot), x)S_{q_0,T} &= \{\eta \in R^r \mid \eta^T \mathscr{D}^{-1}_{q_0,T}(u(\cdot), x)\eta = 1\} \\ &= \{\eta \in R^r \mid (Q^{1/2}\eta)^T \left(Q^{1/2}\mathscr{D}_{q_0,T}(u(\cdot), x)Q^{1/2}\right)^{-1} Q^{1/2}\eta = 1\} \\ &= E_{q_0,T}(u(\cdot), x). \end{aligned} \tag{10.62}$$

The object appearing on the right hand side of (10.62) is called the dexterity ellipsoid of the mobile manipulator at the configuration $(u(\cdot), x)$. It is easily seen that the dexterity ellipsoid is inscribed into the sphere in R^r of radius $\overline{\lambda}^{1/2}_{Q^{1/2}\mathscr{D}_{q_0,T}(u(\cdot),x)Q^{1/2}} = \overline{\lambda}^{1/2}_{\mathscr{D}_{q_0,T}(u(\cdot),x)Q}$ (equal to $||J_{q_0,T}(u(\cdot), x)||$), and circumscribed on the sphere of radius $\underline{\lambda}^{1/2}_{Q^{1/2}\mathscr{D}_{q_0,T}(u(\cdot),x)Q^{1/2}} = \underline{\lambda}^{1/2}_{\mathscr{D}_{q_0,T}(u(\cdot),x)Q}$. Various functions of eigenvalues of the matrix $\mathscr{P}_{q_0,T}(u(\cdot), x) = \mathscr{D}_{q_0,T}(u(\cdot), x)Q$ may be used as local performance measures of the mobile manipulator [32]. Since, as we have shown, the performance of Jacobian algorithms deteriorates at singular configurations, the performance measures should be non-negative functions of eigenvalues of $\mathscr{P}_{q_0,T}(u(\cdot), x)$, vanishing at singular configurations. The best known example is the determinant,

$$d_{q_0,T}(u(\cdot), x) = \sqrt{\det \mathscr{P}_{q_0,T}(u(\cdot), x)} = \sqrt{\det Q}\sqrt{\det \mathscr{D}_{q_0,T}(u(\cdot), x)},$$

called the dexterity of $(u(\cdot), x)$. The dexterity establishes a volume measure of the dexterity ellipsoid. By definition, the dexterity vanishes at singular configurations. Another local performance measure is defined as the condition number of the matrix $\mathscr{P}_{q_0,T}(u(\cdot), x)$

$$\mathrm{cond}\mathscr{P}_{q_0,T}(u(\cdot), x) = ||J_{q_0,T}(u(\cdot), x)||||J^{\#P}_{q_0,T}(u(\cdot), x)|| = \overline{\lambda}^{1/2}_{\mathscr{D}_{q_0,T}(u(\cdot),x)Q}\underline{\lambda}^{-1/2}_{\mathscr{D}_{q_0,T}(u(\cdot),x)Q}$$

that characterizes anisotropy of the configuration $(u(\cdot), x)$. When the condition number equals 1, the dexterity ellipsoid becomes a sphere, and the configuration is called isotropic. At singular configurations the condition number grows up to infinity.

The local performance measures averaged over a prescribed (finite-dimensional) region of the endogenous configuration space provide global performance measures characterizing the mobile manipulator as a whole. The optimization of local performance measures leads to obtaining motion patterns of mobile manipulators. On the other hand, global performance measures provide objectives for the optimal design of the mobile manipulator's kinematics. A review of performance measures for mobile manipulators, including local and global, kinematic and dynamic measures can be found in [32, 37].

A specific performance feature of the mobile manipulator driven by an inverse kinematics algorithm is the repeatability [23]. To explain this concept, suppose that the inverse kinematics algorithm solves a sequence of inverse kinematics problems, with the same initial state of the platform q_0 and the control horizon T, defined by a number of successive desired points $y_{d1}, \ldots, y_{dk}$ in the operational space. It is assumed that in order to solve a given problem the algorithm starts from the solution of the preceding problem in the sequence. The algorithm, which for every repetition of a given problem in the sequence yields the same solution in the endogenous configuration space is called repeatable. Consider a dynamic inverse Jacobian operator $J^{\#}_{q_0,T} = J^{\#}_{q_0,T}(u(\cdot), x)$ appearing in (10.30). Geometrically, this operator generates over the endogenous configuration space an r-dimensional distribution

$$D^{\#}_{q_0,T} = span_{C^\infty(\mathscr{X})}\{J^{\#}_{q_0,T}e_1, \ldots, J^{\#}_{q_0,T}e_r\},$$

e_i denoting the ith unit vector in R^r. It has been proved that the inverse kinematics algorithm determined by $J^{\#}_{q_0,T}(u(\cdot), x)$ is repeatable whenever the distribution $D^{\#}_{q_0,T}$ is integrable. All solutions of the sequence of the inverse kinematics problems lie then on an integral manifold of this distribution, and there is only one endogenous configuration on this manifold corresponding to a prescribed point in the operational space.

10.5 Conclusion

We have presented an overview of the endogenous configuration space approach that provides a uniform conceptual basis for the theory of mobile manipulators, nonholonomic mobile platforms and holonomic manipulation robots. The presentation has been focused on control theoretic aspects of this approach. Specific results highlight the connection between singular endogenous configurations and the parametric optimal control, introduce a number of Jacobian inverses, provide explicit expressions for norms of inverse Jacobian operators, offer conditions for the existence of local and global dynamic inverses and establish foundations for performance measures of mobile manipulators.

Acknowledgments This research was supported by the Wrocław University of Technology under a statutory research project.

References

1. Abraham, R., Marsden, J.E., Ratiu, T.: Manifolds, Tensor Analysis, and Applications. Springer, New York (1998)
2. Allgower, E.L., Georg, K.: Numerical Continuation Methods. Springer, Berlin (1990)
3. Alouges, F., Chitour, Y., Long, R.: A motion-planning algorithm for the rolling-body problem. IEEE Trans. Robot. **26**, 827–836 (2010)
4. Aubin, J.P., Ekeland, I.: Applied Nonlinear Analysis. Wiley, New York (1984)
5. Bhatia, R.: Matrix Analysis. Springer, New York (1996)
6. Bismut, J.M.: Large Deviations and the Malliavin Calculus. Birkhäuser, Boston (1984)
7. Chelouah, A., Chitour, Y.: On the motion planning of rolling surfaces. Forum Math. **15**, 727–758 (2003)
8. Chitour, Y.: A homotopy continuation method for trajectories generation of nonholonomic systems. ESAIM: Control Optim. Calc. Var. **12**, 139–168 (2006)
9. Chitour, Y., Sussmann, H.J.: Motion planning using the continuation method. In: Baillieul, J., Sastry, S.S., Susmann, H.J. (eds.) Essays on Mathematical Robotics, pp. 91–125. Springer, New York (1998)
10. Davidenko, D.: On a new method of numerically integrating a system of nonlinear equations. Dokl. Akad. Nauk SSSR **88**, 601–604 (1953)
11. Divelbiss, A.W., Seereeram, S., Wen, J.T.: Kinematic path planning for robots with holonomic and nonholonomic constraints. In: Baillieul, J., Sastry, S.S., Susmann, H.J. (eds.) Essays on Mathematical Robotics, pp. 127–150. Springer, New York (1998)

12. Fliess, M., Levine, J., Martin, P., Rouchon, P.: Flatness and defect in non-linear systems: introductory theory and examples. Int. J. Control **61**, 1327–1361 (1995)
13. Getz, N.H.: Dynamic inversion of nonlinear maps with applications to nonlinear control and robotics. Ph.D. Thesis, UCB (1995)
14. Hutson, V.C.L., Pym, J.S.: Applications of Functional Analysis and Operator Theory. Academic Press, London (1980)
15. Janiak, M., Tchoń, K.: Constrained motion planning of nonholonomic systems. Syst. Control Lett. **60**, 625–631 (2011)
16. Li, Z., Ge, S.: Fundamentals in Modelling and Control of Mobile Manipulators. CRC Press, Boca Raton (2013)
17. Nijmeijer, H.: Zeros at infinity for nonlinear systems, what are they and what are they good for? In: Jakubczyk, J., Respondek, W., Tchoń, K. (eds.) Geometric Theory of Nonlinear Control Systems, pp. 105–129. Wrocław University of Technology Publishers, Wrocław (1985)
18. Popa, D.O., Wen, J.T.: Singularity computation for iterative control of nonlinear affine systems. Asian J. Control **2**, 57–75 (2000)
19. Rugh, W.J.: Linear System Theory. Prentice Hall, Englewood Cliffs (1993)
20. Sontag, E.D.: Mathematical Control Theory. Springer, New York (1990)
21. Sontag, E.D.: A general approach to path planning for systems without drift. In: Baillieul, J., Sastry, S.S., Susmann, H.J. (eds.) Essays on Mathematical Robotics, pp. 151–168. Springer, New York (1998)
22. Sussmann, H.J.: A continuation method for nonholonomic path finding problems. In: Proceedings of the 32nd IEEE CDC, pp. 2718–2723. San Antonio (1993)
23. Tchoń, K.: Repeatability of inverse kinematics algorithms for mobile manipulators. IEEE Trans. Autom. Control **47**, 1376–1380 (2002)
24. Tchoń, K.: Repeatable, extended Jacobian inverse kinematics algorithm for mobile manipulators. Syst. Control Lett. **55**, 87–93 (2006)
25. Tchoń, K., Nijmeijer, H.: On output linearization of observable dynamics. Control Theory Adv. Technol. **9**, 819–857 (1993)
26. Tchoń, K., Jakubiak, J.: Endogenous configuration space approach to mobile manipulators: a derivation and performance assessment of Jacobian inverse kinematics algorithms. Int. J. Control **76**, 1387–1419 (2003)
27. Tchoń, K., Jakubiak, J.: A repeatable inverse kinematics algorithm with linear invariant subspaces for mobile manipulators. IEEE Trans. Syst. Man Cybern. Part B Cybern. **35**, 1051–1057 (2005)
28. Tchoń, K., Jakubiak, J.: Extended Jacobian inverse kinematics algorithm for nonholonomic mobile robots. Int. J. Control **79**, 895–909 (2006)
29. Tchoń, K., Małek, Ł.: On dynamic properties of singularity robust Jacobian inverse kinematics. IEEE Trans. Autom. Control **54**, 1402–1406 (2009)
30. Tchoń, K., Muszyński, R.: Instantaneous kinematics and dexterity of mobile manipulators. In: Proceedings of the 2000 IEEE International Conference on Robotics and Automation, pp. 2493–2498. San Francisco (2000)
31. Tchoń, K., Ratajczak, A., Góral, I.: Lagrangian Jacobian inverse for nonholonomic robotic systems. Nonlinear Dyn. **82**, 1923–1932 (2015)
32. Tchoń, K., Zadarnowska, K.: Kinematic dexterity of mobile manipulators: an endogenous configuration space approach. Robotica **21**, 521–530 (2003)
33. Verhulst, F.: Nonlinear Differential Equations and Dynamical Systems. Springer, Berlin (1996)
34. Vincent, T.L., Grantham, W.J.: Nonlinear Optimal Control Systems. Wiley, New York (1997)
35. Ważewski, T.: Sur l'évaluation du domaine d'existence des fonctions implicites réelles ou complexes. Ann. Soc. Pol. Math. **20**, 81–120 (1947)
36. Ważewski, T.: Sur la limitation des intégrales des systemes d'équations différentielles linéaires ordinaires. Studia Math **10**, 48–59 (1948)
37. Zadarnowska, K., Tchoń, K.: A control theory framework for performance evaluation of mobile manipulators. Robotica **25**, 703–715 (2007)

Printed in the United States
By Bookmasters